AF781126

Springer Monographs in Mathematics

Springer-Verlag London Ltd.

E.G. Peter Rowe

# Geometrical Physics in Minkowski Spacetime

With 112 Figures

Springer

Springer Monographs in Mathematics ISSN 1439-7382

British Library Cataloguing in Publication Data
Rowe, E.G. Peter
Geometrical physics in Minkowski spacetime. - (Springer monographs in mathematics)
1. Generalized spaces 2. Space and time - Mathematics
I. Title
516.3'74

Library of Congress Cataloging-in-Publication Data
Rowe, E.G. Peter, 1938-1998
Geometrical physics in Minkowski spacetime / E.G. Peter Rowe.
p. cm. — (Springer monographs in mathematics)
Includes bibliographical references and index.
ISBN 978-1-84996-866-9 ISBN 978-1-4471-3893-8 (eBook)
DOI 10.1007/978-1-4471-3893-8
1. Special relativity (Physics) I. Title. II. Series.
QC173.65.R68 2000
530.11—dc21 00-061905

Mathematics Subject Classification (1991): 51B20, 83A05

Apart from any fair dealing for the purposes of research or private study, or criticism or review, as permitted under the Copyright, Designs and Patents Act 1988, this publication may only be reproduced, stored or transmitted, in any form or by any means, with the prior permission in writing of the publishers, or in the case of reprographic reproduction in accordance with the terms of licences issued by the Copyright Licensing Agency. Enquiries concerning reproduction outside those terms should be sent to the publishers.

© Springer-Verlag London 2001
Originally published by Springer-Verlag London Berlin Heidelberg in 2001
Softcover reprint of the hardcover 1st edition 2001

The use of registered names, trademarks, etc. in this publication does not imply, even in the absence of a specific statement, that such names are exempt from the relevant laws and regulations and therefore free for general use.

The publisher makes no representation, express or implied, with regard to the accuracy of the information contained in this book and cannot accept any legal responsibility or liability for any errors or omissions that may be made.

Typesetting: Camera-ready by Marcus Tindall

12/3830-543210 Printed on acid-free paper SPIN 10778825

# Foreword

Dr E.G. Peter Rowe had almost completed writing this book when his life was brutally terminated in the Yemen in December 1998.

Peter came to Durham from Canada via London in 1964 and quickly became a very popular and engagingly eccentric member of our Department. He was friendly and generous, informal, charitable, warm and full of diverse interests which, apart from mathematics and physics, ranged from anthropology and archaeology to foreign cultures and travel. This led him to spend his sabbatical leave in places such as northern Nigeria. His interest in other people's cultures took him to many parts of the world (from Ladakh in North West India, to many places in Africa, motorbiking in Saskatchewan, Canada and South Africa and finally to the fatal trip to Yemen).

Peter was witty and was not afraid to speak his mind; at our Departmental meetings we now miss his throw-away but very much to-the-point comments and suggestions.

In his teaching and research Peter was somewhat unconventional. He did not follow fashionable trends in research but worked on what interested him most, namely, the geometrisation of physics.

He considered refereeing of research papers to be an important task, writing long reports full of helpful suggestions to the authors. Peter also took his teaching seriously. His courses, often perceived as difficult by the students, were always quite advanced, as if designed to draw the very best from his audience. His view was that it was better to say something new and stimulating to the interested students than show routine steps to the uninterested ones.

The present book grew out of his course on special and general relativity given to our third year students. I first became aware of Peter's approach to relativity when, having taught a similar course before him, I was asked to check his examination questions. While some of them were routine, others demanded deeper thought and when I looked at his solutions I became aware of the merits of his more geometric approach. I was among those who encouraged him to write a book in order to make this approach available to a wider audience.

Peter's book puts an emphasis on geometry in the description of physical phenomena in Minkowski spacetime. In this it emphasises the covariance properties of the equations of motion, trying as much as possible to avoid working in any particular frame of reference. And the book achieves this aim, probably, more than any other book that I know.

I am very pleased that Springer-Verlag have published Peter's book. The book will not only help many people to understand physics in a more geometrical setting, but also it will be a lasting reminder of our colleague and friend, complementing our personal memories of him.

Wojtek J. Zakrzewski
University of Durham

# Preface

This book is **not** meant for the complete beginner in special relativity, **nor** for anyone wanting an account of the numerous and interesting experiments that support the theory. Instead, it is intended to be a description of the geometry of spacetime, and an aid in the creation and development of intuition in four-dimensional Minkowski space. The emphasis on the *geometry* means an emphasis on the ***absolutes*** which underlie relative descriptions. For example, the Poincaré transformation links different relative sets of coordinates, $x^\mu$, $x'^\mu$, but the underlying absolute is simply a point $P$ in spacetime (the coordinates are the relative descriptions). The deepest understanding, perhaps the only understanding, of relativity and spacetime is in terms of the geometrical absolutes, and this is what the book seeks to develop. Whereas the beginner in special relativity must have help in making the transition between his nonrelativistic view of physics as a time-development in space (his space) to a four-dimensional view of physics as a complete history in spacetime, it is hoped that the reader of this book is ready to study the subject in its final, unified (and beautiful) form.

The mathematical prerequisites for the early chapters of the book are very few, just linear algebra and elementary geometry (done using vectors and a scalar product). For the later chapters multivariable calculus and ordinary differential equations are often needed. No detailed knowledge of the experimental background to relativity is needed, nor any detailed knowledge of electromagnetism, but in both these areas, the more sophistication and sympathy is available for the subjects, the better.

The book aims to cover the most interesting topics requiring special relativity. It is an outgrowth of lectures on special and general relativity given to final year undergraduate students of theoretical physics in the Department of Mathematics. It could be presumed that the students had all had half a dozen or a dozen lectures in earlier years covering the experimental foundations of special relativity and the first, surprising consequences of Einstein's new kinematics. However, the book goes well beyond what was ever taught in practice. Although in a real sense special relativity is the culmination of classical physics, and worthy on that account of detailed study, in the lecture theatre time is limited and the attractions of gravity, with its curved spacetime, become overwhelming. In practice, a natural climax for special relativity is the definition of the energy tensor (which becomes the source of gravitational curvature) and its use in deriving equations of motion. Some of the more difficult aspects of the energy tensor, and most areas of electromagnetism, were left for self study (in the future). The material in the book, therefore, is partly at an undergraduate level and partly at a postgraduate level.

In the first chapter, Spacetime, the idea of a four-dimensional space having special coordinates (arising from the inertial frames of reference) is developed. An attempt is made to distinguish between the mathematical side of the exposition, where clarity and logic can be expected, and the real-world side, still partly unknown and mysterious, where our understanding advances in a series of temporary world views. The present model is described in natural language (not mathematical); it is a familiar world of clocks and spatial frameworks, but mysteriously without gravity. The mathematics we develop is put into correspondence with this model. The Lorentz transformation and the Poincaré transformation are discussed (as distinct from being postulated, or derived from an artificial starting point). The importance of the lightcone in the theory is exemplified by the way it creates a significant division into regions of the spacetime around any given event. In the whole of the chapter, the emphasis is on spacetime and how we can begin to picture events and processes (and inertial frames, which may be *relative* yet are also physical objects) in it.

In the second chapter, the most important one for building intuition in Minkowski space, vectors in spacetime are defined as transformations of points in spacetime (the geometrical or absolute concept), simply expressed in terms of the inertial frames, which both contribute to the definition and provide the relative expressions of the concept. The scalar product of vectors is constructed to provide the vector expression of the division of spacetime determined by the lightcone. All the famous kinematical effects can be given completely transparent discussions in terms of spacetime diagrams and simple vector geometry.

(The first time I taught the course on which this book is based, I attempted to begin with a discussion of vectors in Minkowski spacetime, without any discussion of spacetime as a manifold. Only a few students found this direct approach attractive and were able to build a useful intuition from it. Student discontent resulted in what is now Chapter 1 to fill in all the background material.)

The third chapter, Asymptotic Momentum Conservation, is devoted to the four-momentum of elementary particles and the relations that follow from the simple idea of equating momentum in the past with momentum in the future. All relations can be expressed in purely geometrical terms. The definition of the centre of momentum frame is particularly simple when it is expressed by its geometrically defining property (rather than in terms of its relation with other, irrelevant frames).

In Chapter 4, covectors and dyadics, which are generalisations of vectors, are defined and their properties developed. The gradient of scalar functions in spacetime is defined as a covector, then converted to a vector, then generalised to the gradient of vector fields and beyond. The concept of volume is discussed, as is the divergence theorem in spacetime.

In Chapter 5, the geometrical formulation of electromagnetism is given. The early sections deal with the decomposition of the field dyadic into relative electric and magnetic fields, and the relation of the different expressions of Maxwell's equations. The geometrical discussion of charge density and three-current is given in terms of a model of charged dust. Conservation of charge then has a visualisable form. The electomagnetism of point particles is begun. Because the consideration of point particles involves delta functions, the topic is technically more difficult and may be omitted at a first reading.

The energy tensor is the subject of Chapter 6. The meaning of its different components is developed with the example of flowing dust. Local conservation of four-momentum is expressed by the vanishing of the divergence of the energy tensor. The equation of motion for flowing, charged dust can be derived from this condition. The general definition of the energy tensor can be developed from a Lagrangian in those cases where the equations of motion can be derived from a variational principle.

A point particle with an accelerating timelike worldline creates some special, peculiarly relativistic problems. It is not self-evident how to define the time development of the rest frame. Two solutions, the Fermi-Walker transported frame and the frame which is boosted from the laboratory, both correspond, nonrelativistically, to the unique "nonrotating" frame of Newtonian mechanics. Yet there is a relative rotation between them, the Thomas precession. These problems and their solutions, and the equation of motion of the spin of a point particle with a magnetic moment, are discussed in Chapter 7*.

After every chapter, but especially the first four, are many exercises and problems which supply lots of opportunity to practise the skills and techniques appropriate to special relativistic geometry. And at the very end of each chapter are listed some references for supplementary reading on particular points. No attempt has been made to provide a complete bibliography.

E.G.P.R
University of Durham
England

* Chapter 7 was incomplete at the time of the author's death, and so is not included in the present volume.

# Contents

# 1. Spacetime

## *Inertial Frames and Poincaré Transformations*

Special relativity is not a subject that can be comprehended at a single sitting. When one is ready to embark on a study of it, one has a whole lifetime behind one of Newtonian intuitions, preconceptions (the most significant of the pre-relativistic prejudices is the belief in absolute time), concepts, habits and shortcuts in thinking, some appropriate only to pre-relativity which must be identified and abandoned (not always a quick process), but some adaptable and some immediately transferable to the new way of thinking. Even the still-useful concepts must be identified and deliberately cherished. Relativity must therefore be tackled in at least two stages. In the first stage, the old prejudices must be attacked and demolished (if possible), and a preliminary study made of simultaneity, the classical experiments of Michelson and Morley, the Lorentz transformation and its elementary consequences. The subconscious must then have time to work its magic and adjust to the new circumstances. It is assumed here that this stage has already been completed. In the second approach to the subject, beginning now, one considers the whole discipline *ab initio*, building up a structure in terms of those concepts that will last, using what has been learned before to try to avoid false starts or concentration on inappropriate aspects.

The domain of special relativity is the classical (non-quantum) description of the whole of nongravitational physics. Special relativity is a comparatively simple subject because in it spacetime is regarded as a flat manifold, whereas in general relativity gravity generates a curved spacetime, which must be treated by more advanced methods.

"The whole of nongravitational physics" can be apprehended in several different ways. Without attempting any philosophical depth, we have three levels:

1. The real, physical world, running as it will, including everything from everyday objects to laboratory experiments. We restrict attention to aspects of nature where gravity can be regarded as negligible or can be allowed for. We see this (non-quantum) world from the outside. All we can do is admire it and attempt to measure it. Mathematics does not appear at this stage and nor does explanatory physics; there is only the raw sensory data.

2. Physical concepts, a verbal description of categories that seem useful in simplifying the description of objects and processes of nature, and experiments to probe them: concepts of particles, position, mass, clocks, inertial frames, light rays, space, time. These are not yet mathematical objects but our mathematics makes models for them and then experimental physicists try to prove or disprove connections with the real world. We can play around with the category of physical concepts in a very free and inventive way, but we must remember: these concepts have no necessary correspondence with the real world, nor do they necessarily allow consistent mathematics.
3. Finally, we have mathematical models for the physical concepts. Here we must argue logically and mathematically. We have full control of the mathematics, but unfortunately no guarantee of being physically relevant. As an initial step we build a model of spacetime as a manifold, a continuum of point events.

The distinction between physical concepts and mathematical models has been made so that one can recognise where the limitations of each occur and what the character of the limitations is. The final aim is a mathematical model of reality so closely accurate that one is tempted to regard the model and the reality as indistinguishable. The great property of mathematics is its logical structure, but it has no necessary connection with physical reality; these links are made separately and we must always be suspicious of them. Arguments in terms of physical concepts are not always mathematical; they sometimes allow assumptions to be implicit or hidden, sometimes contradictory. To understand special relativity requires both mathematical and physical arguments; the new kinematical regions are unfamiliar and mean that, at the beginning, both mathematical and physical assumptions are regarded as tentative. In the end physical "proofs" need not be mathematical (in the mathematical literature[1] they are called "motivation"), and one must guard against expecting too much from them. (One cannot get blood out of a stone.) Equally well, mathematical discussions cannot lead to new truths if the links with physical reality are inappropriate. The book of David Bohm contains a very helpful discussion of the physical and philosophical problems involved in linking physical reality with the mathematical model.

In special relativity we first attempt to understand the structure of spacetime itself, after which mathematical methods can be developed which are appropriate to that structure (vectors, dyadics, tensors, differential operators, integrals), and which can be used to formulate physical theories. Inertial frames have a dual role: on the one hand they are part of the conceptual framework that underlies physical measurements, and on the other hand they

[1] An example is the book of Sachs and Wu. (See the short bibliography at the end of the chapter.)

generate the coordinates that help to define the mathematical notion of spacetime. Poincaré transformations relate the coordinates provided by different inertial frames.

## 1.1 Spacetime is a Four-dimensional Continuum

Spacetime $\mathcal{M}$ is a four-dimensional continuum whose points $P, Q, \ldots$ are called *events* (Figure 1.1). (A four-dimensional continuum is characterised by the fact that its neighbourhoods may be put in 1-1 correspondence with neighbourhoods of $\mathcal{R}^4$, which provide general coordinates.) The points of $\mathcal{M}$ are the mathematical counterparts of ***physical*** events, or rather, of their idealisations, the "here and nows" of the real world, of negligible duration and spatial extent. The picture can illustrate only some discrete points from the continuum. The dashed frame for the picture is purely aesthetic; it is not a mathematical boundary. The spacetime of special relativity has no boundary; it is open and non-compact, just like $\mathcal{R}^4$.

According to the present model, the points of $\mathcal{M}$ are in a fixed 1-1 relation to the physical events of the real world. Hence in the diagram points on the paper are imagined to be in fixed relation to physical events. (Such a diagram is called an ***absolute spacetime diagram.***) This is to be distinguished from the relation of physical events to coordinates in $\mathcal{R}^4$; this relation depends also on the coordinate system and the associated diagram is called a *relative spacetime diagram.* The distinction is important; it corresponds to two different ways of thinking about special relativity. With coordinates one gets a *relative* point of view, relative to the coordinate system, or from the physical side, relative to the observer. With the points of $\mathcal{M}$ one has an absolute, geometrical structure in unique relation to the physical world. Since observers apprehend the world in a relative way, the recognition of an underlying physical reality requires a process of abstraction. The mathematical

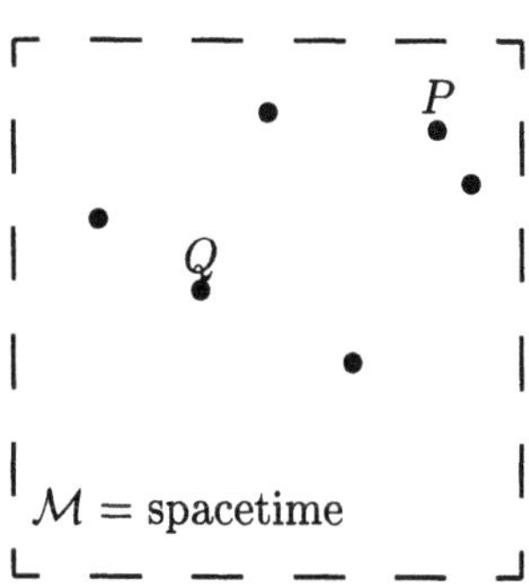


**Figure 1.1**

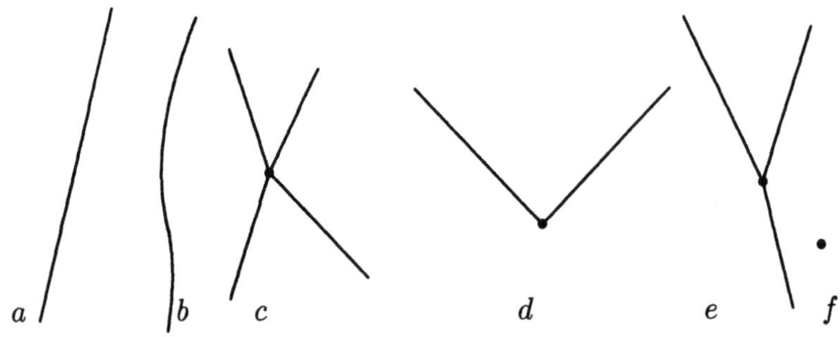


**Figure 1.2**

result is a single, concrete structure (like the unique physical world) which can be described from many viewpoints.

As examples of physically significant sets of events one may think of any point on the (straight) worldline $a$ of a free particle or the (curved) worldline $b$ of an accelerating particle (Figure 1.2). In both cases the worldline of a particle consists of all point events in the history of the particle. A single point is specified by the collision $c$ of infinitesimal billiard balls, or the lighting $d$ of a microscopic match or the decay $e$ of an elementary particle. But also, one may think of a physical event $f$ where nothing physical occurs, for, in special relativity, spacetime is regarded as an arena in which the universe plays out its destiny but which is unaffected by the physics which takes place within it. In cases $d$ and $e$ we see a distinction between the "future" (towards the top of the diagram) and the "past" (towards the bottom). Thus in $d$ light spreads in spacetime from its point source into the future on a characteristic lightcone (the projection on two dimensions of an expanding sphere in three dimensions). The mysterious difference between past and future is not due to the structure of spacetime (at least not at the level of special relativity) but is due to the processes occurring in it (life and death, increase of entropy).

Physicists have attempted to understand more of the properties of spacetime by assuming additional structure in the four-dimensional continuum. Two of the pre-Einsteinian models are worth noting, the so-called Aristotelian spacetime and the so-called Galilean spacetime. More extensive discussions of the pre-relativistic models may be found in a paper of Ehlers and in the first chapter of the book by Kopczynski and Trautman. (The full references may be found at the end of the chapter.)

## 1.2 Aristotelian Spacetime (Pre-relativistic)

The so-called Aristotelian view of spacetime is the simplest one, the one we first adopt in making sense of the world. It is reinforced by the early stages of school and is the view of spacetime held by everyone until, for each person, in a private revolution, a more sophisticated, less egocentric view is adopted.

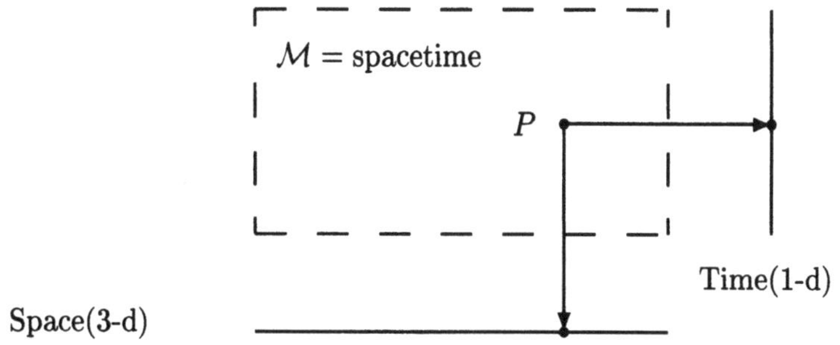


**Figure 1.3**

According to this view spacetime, as a set of points (events), is a Cartesian product:

$$\mathcal{M} = \text{Time} \times \text{Space}.$$

Time and Space are two separate (absolute) manifolds. For each $P \in \mathcal{M}$, specific points in absolute Time and absolute Space are determined (Figure 1.3).

The reason why the Aristotelian conception of spacetime is so common is that the solid earth provides such an immediate frame of reference that it takes imagination and an act of will to consider anything else. We (as egocentric observers) unthinkingly accept the status of our local reference frame (which is not even inertial) as special (and absolute). Even when we start to investigate kinematics and dynamics, there is often a tendency to posit "space" as $\mathcal{R}^3$, never to consider anything else, and thereby to accept the Aristotelian view by default.

The Aristotelian view is no longer accepted as appropriate. It implies a special "Space", which would then bring with it a special conception of "absolute rest" with respect to that Space. However, there appears to be no evidence for such a special Space, nor any need or role for it. The last candidate for "Space" was the putative rest frame of the aether, and the two concepts faded from significance together after Einstein removed the necessity for the aether.

## 1.3 Galilean Spacetime

In Aristotelian spacetime there are concepts of absolute time and absolute space. In Galiliean spacetime, there remains a concept of absolute time, but there is no absolute space and therefore no conception of absolute rest.

In Galilean spacetime, for each event $P \in \mathcal{M}$ there is determined an absolute time and the whole of spacetime is sliced up into three-dimensional subspaces of simultaneity, all the events that have the same absolute time

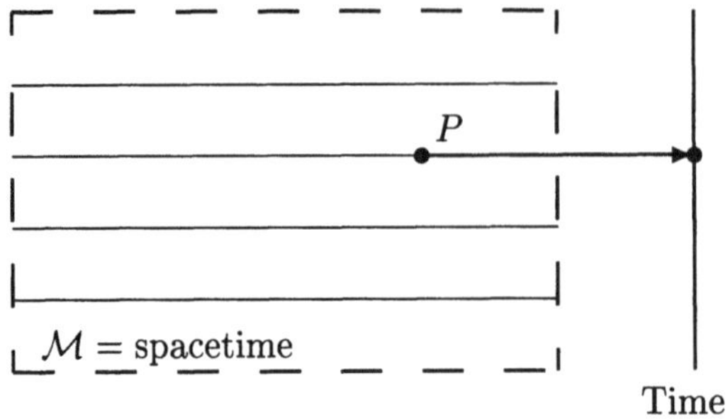


**Figure 1.4**

(Figure 1.4). In special relativity we will lose also the concept of absolute time. From the more advanced standpoint of Minkowskian spacetime the explanation of this structure in Galilean spacetime is that, because of the very high speed of light compared with terrestrial speeds, different relative time systems are all well approximated at low speeds by any one system, which then appears to be absolute time.

Galilean spacetime is an affine space, which provides a concept of straightlines and an associated class of special coordinate systems. Straightlines can be time simultaneous (for example, $tt$) or not, in which latter case they are possible tracks (for example, $ff$) of free particle motions (Figure 1.5). No particular motion among them is picked out, hence there is no special (absolute) state of rest.

Any of the free motion tracks can be made the basis for a Galilean inertial coordinate system. Each system has as its spatial origin at a given time the location of that track at that time. Two tracks which pass through a common event $O$ give rise to coordinate systems for the space at that time (the points (Figure 1.6) simultaneous with $O$) with a common grid and common coordinates. But this does not amount to a privileged, absolute space in spacetime, because the coincidence is momentary and accidental.

On a later hyperplane of simultaneity the spatial origins of the two systems have diverged (Figure 1.7). Each one has an equal right to describe itself

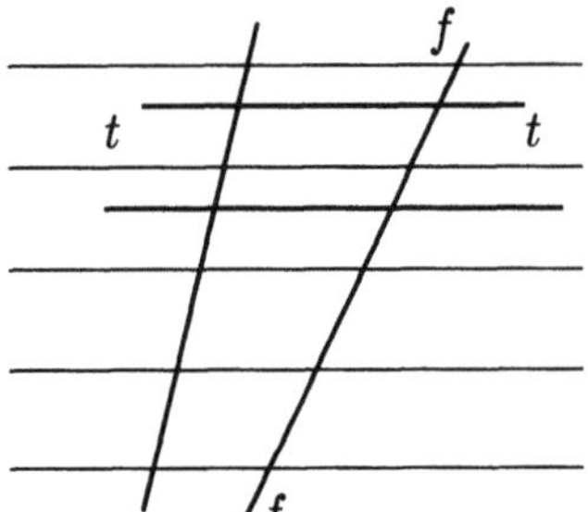


**Figure 1.5**

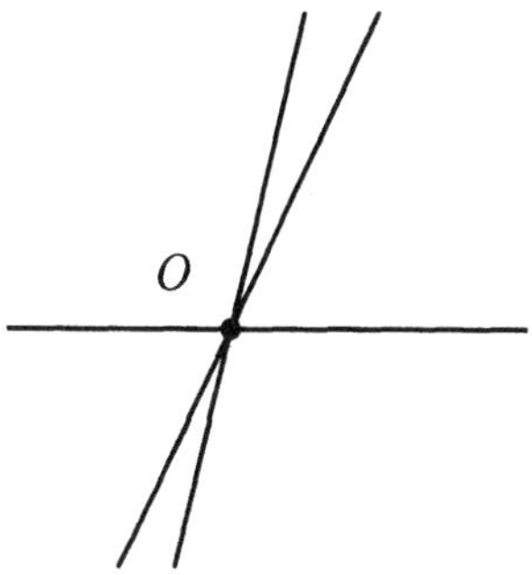


Figure 1.6

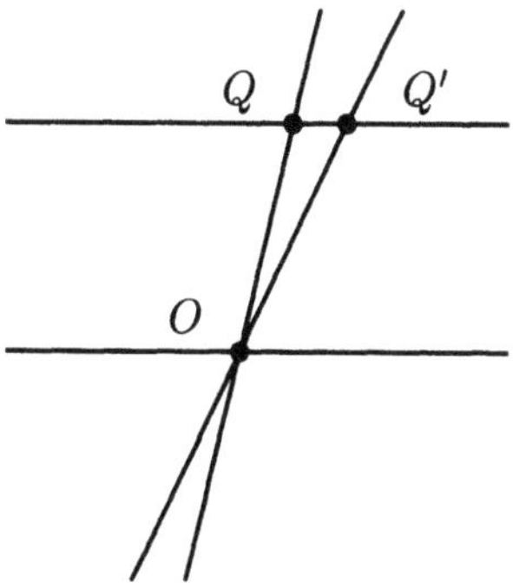


Figure 1.7

as at rest. This concept (of being at rest) is therefore an individual, subjective one, not an objective, absolute one. What we do get from the affine space structure is an idea of parallel spatial directions on the different hyperplanes of simultaneity. It therefore does make sense to say that the two frames are rotating or not rotating with respect to each other. The collection of special frames gives rise to a notion of Galilean relativity: Newton's equations take exactly the same form in all of them.

The pre-Einsteinian views of spacetime can be elaborated further but at the present stage it defeats our purpose (and may be poor psychology) to set up too clear a mental model in competition with Minkowskian spacetime, which will be our preoccupation for the remainder of this study.

## 1.4 Principles of Special Relativity

The two principles which inspired Einstein to create special relativity are:

1. the principle that no inertial frame (a frame in which free particles travel on straight lines) has a privileged status, so that there is a whole class of equivalent reference frames, called inertial, which must be treated equally;
2. the principle that the speed of light is constant, the same in all inertial frames using the same units, independent of the motion of the source of light, and independent of position and direction. (In these respects light behaves completely differently from particles with mass.)

These statements are not themselves mathematical principles; they are physical ideas at a verbal level. When they are developed and made more precise in the mathematical implementation of the physical theory they are to some extent absorbed and embodied by it so that they disappear from immediate view except in some kinematical discussions.

On the grounds of the first principle it is supposed that physical laws take exactly the same form, no matter what frame, when expressed in the inertial coordinates of the frame and components (of vectors and tensors) with respect to it. Tensor expressions for such local differential laws can be transformed from one system to another (hence expressing the equivalence of different *relative* frames of reference). Alternatively, the tensor expressions are recognised as the component forms with respect to different bases of *absolute*, geometrical physical laws.

According to the second principle, in each inertial frame the speed of light is constant, independent of direction and motion of the source. This principle reflects a complete acceptance of the result of the Michelson-Morley experiment, which failed to find a difference in the speed of light in perpendicular directions, nor any evidence of an aether. The fact that the speed of light is independent of the motion of the source distinguishes light from "ballistic" particles. The light from stars in a binary system has been observed from the earliest days of relativity to study this phenomenon.

It is convenient, given the special properties of the speed of light, to choose units of time and space in each inertial frame so that the speed of light is unity. For example, for space the centimetre, and for time the time interval required for light to travel one centimetre. The unit of time may then also be called a centimetre. (If the units for both time and space are now scaled by the same factor then the speed of light will remain unity.)

## 1.5 Minkowskian Inertial Frames of Reference

From the mathematical point of view, inertial frames of reference are special coordinate systems in spacetime: for each frame $K$, there is an invertible mapping $K : \mathcal{M} \to \mathcal{R}^4, \quad P \mapsto x^\mu(P)$, where it is understood here that $x^\mu$ stands for the *whole collection* $\{x^0, x^1, x^2, x^3\}$. It has become traditional in special relativity to use superscripts for the inertial coordinates in spacetime. This notation is then consistent with that for the components of vectors with respect to an inertial basis (vectors will be introduced in Chapter 2).

It is useful to regard the inertial frames as physical objects as well; not always with a genuine physical existence, but always with at least a potential existence; never a complete existence (we do not need an *infinite* three-dimensional grid), but at least a partial existence. The physical existence of the reference frames means that they appear integrated into the discussion of physical concepts. The most remarkable consequence of this integration was Einstein's discovery of the relative character of simultaneity. The inertial frames of reference function, from the physical point of view, literally as reference frameworks for the description of physical events, and also as special measuring apparatuses. The mathematical and physical concepts of special relativity are very close at this point. In general relativity the situation is

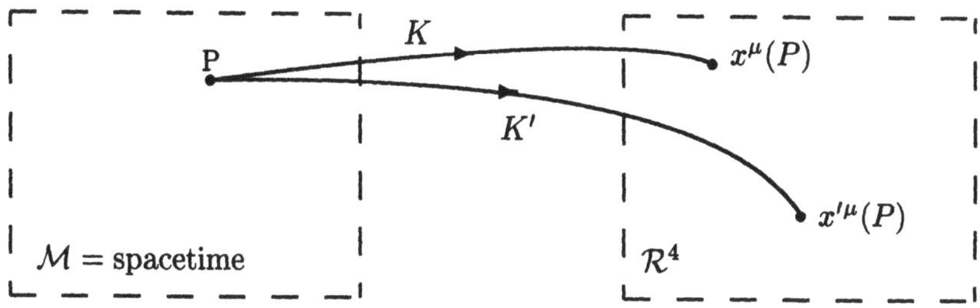


**Figure 1.8**

quite different; the coordinates are completely arbitrary, and no connection with the measurement procedure is imposed on them.

*The points $P \in \mathcal{M}$ provide the primary model for the real events in the physical world. Each coordinate system, $K, K', \ldots$, provides coordinates, four real numbers, $x^\mu(P), x'^\mu(P), \ldots$, for every P (Figure 1.8). Absolute spacetime diagrams are based on collections of events $P \in \mathcal{M}$. The various coordinate mappings provide the link with the corresponding relative spacetime diagrams.*

To give physical substance to the idea of an inertial frame as a measuring apparatus, and also to bring out the points that are significant in the mathematical model, we consider the elements that enter into its construction. It should be noted that such a discussion is in the arena of physical concepts, not mathematical ones. The consistency of the ideas is in large part a matter for experiment.

We attempt to establish, as an inertial frame, a system of clocks and a spatial grid which will provide what will look like the ordinary $(t, x, y, z)$ coordinates [2] of Galilean relativity, or indeed, pre-relativity. This will be only one system among many and to understand the relations between the different systems we will have to be clear what factors are relevant in setting up any one system. We expect these factors all to have mathematical counterparts when we discuss the mathematical model.

- The spatial origin for an inertial frame is taken to be the worldline of a freely moving particle, a particle with mass acted on by no forces (Figure 1.9). (This is an idealisation that is impossible to realise in practice. Gravitational forces are inevitable and must be negligible or corrected for. Other forces may be present and corrected for if they are significant.)
- We imagine that the free particle carries an orthogonal triad (Figure 1.10). Euclidean geometry is invoked here for the first time in a three-dimensional spatial subspace. The most important property of the triad is that it should not be rotating. In practice this idea of nonrotation has been implemented

[2] Two notations coexist: $t = x^0,\ x = x^1,\ y = x^2,\ z = x^3$. The scripted notation is mainly used in formal work where the summation convention may be useful; the unscripted notation is used more in problems and special situations.

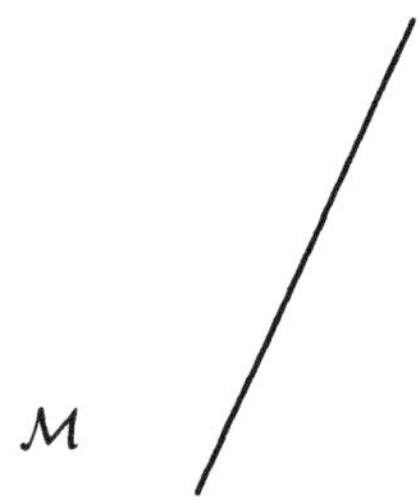


**Figure 1.9**

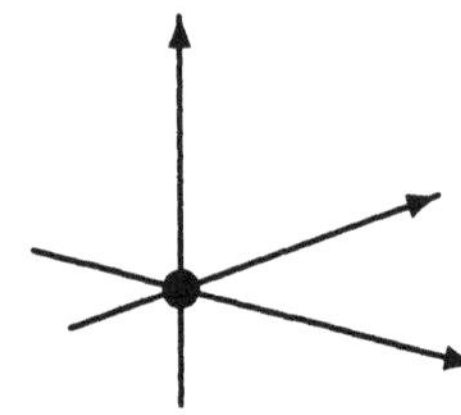


**Figure 1.10**

by comparison with the directions of the distant galaxies (the fixed stars). Alternatively, a gyroscope subject to no torques should provide a means of recognising a nonrotating triad.

- Extend the triad to a three-dimensional orthogonal Euclidean lattice with some fixed unit of distance providing the interval (Figure 1.11). We then have natural coordinates $(x, y, z) \equiv (x^1, x^2, x^3)$ for the vertices of the framework. There is available the possibility of refining the framework to provide more accurate coordinates for a general point (not a vertex) if it is needed.
- Construct a master clock for the inertial frame at the spatial origin, a quartz clock perhaps, based on some atomic frequency. We will be assuming that no inertial frame is in any way privileged or physically exceptional. We assume that the laws of physics are the same with respect to all inertial frames (in particular, the laws that enter into the construction of clocks). The *unit* of time should be chosen consistently with the previous unit for spatial measurements so that the speed of light in these units is one.
- Construct similar, secondary clocks at the vertices of the spatial framework. It should hardly need emphasis that the constructions sketched above are not meant to be taken literally; they are conceptual and metaphorical.
- Synchronise the secondary clocks with the master by using light signals: an apprentice at each secondary clock looks at the master clock, no matter how distant it is, and instantaneously sets the time on his secondary to

$$\text{time read at a distance from master clock} \\ + \text{distance to master clock.}$$

Every secondary clock is synchronised with the master by allowing in this way for the time taken by the light signal bringing the time. The procedure presupposes using units for time which are related to the unit of distance so that the speed of light is one. There is also a deep and important physical

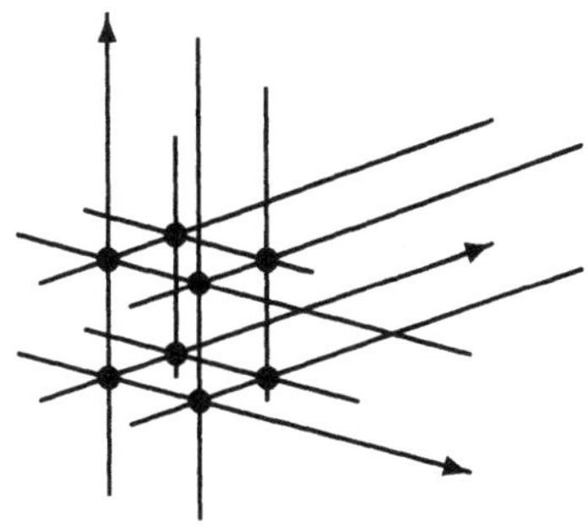

**Figure 1.11**

assumption about the nature of light, that its speed is independent of the motion of the source of the light and the direction the source is moving. We must assume that as a result of our synchronisation procedure the secondary clocks are synchronised with each other. (This is equivalent to assuming that, after synchronisation, every clock can serve as a master clock with all the others secondaries.) Whether this is true or not is a matter for experiment and if the secondary clocks are not synchronised the whole construction, at the heart of special relativity, fails.

The final three points refer not to a single inertial frame but specify a collection of all inertial frames with the same units and orientations.

- Rescale (if necessary) the units (spatial and temporal ones being linked as above to make the speed of light one) to a fixed standard for all inertial frames. This might be done by choosing as a unit the period of some particular atomic transition. The purpose of the rescaling is to achieve a homogeneous set of inertial frames. It is only after this rescaling that we have a set of equivalent inertial frames in the sense of Einstein's first principle.
- Time orientation. Clocks provide a run of ticks: tick, tick, tick,... We get a time coordinate (with a certain unit determined by the tick-interval) when we number them: $0,\ 1,\ 2,\ 3,\ldots$ or $0,-1,-2,-3,\ldots$ The two numbering systems give different time orientations. For simplicity we choose to use only one time orientation for the collection of inertial frames. In practice we choose the system in which the numbers increase as we experimenters get older. To remove the dependence on human physiology one can simply verify that the clocks of all inertial frames run the same way, and if any are running " backwards" they should be replaced.
- Space orientation. Verify that the spatial frameworks of all inertial frames have the same orientation, or handedness (Figure 1.12). If any are different, change them. This orientation will be called right-handed. (Mathematically, the choice is arbitrary; in practice, it depends on human anatomy.)

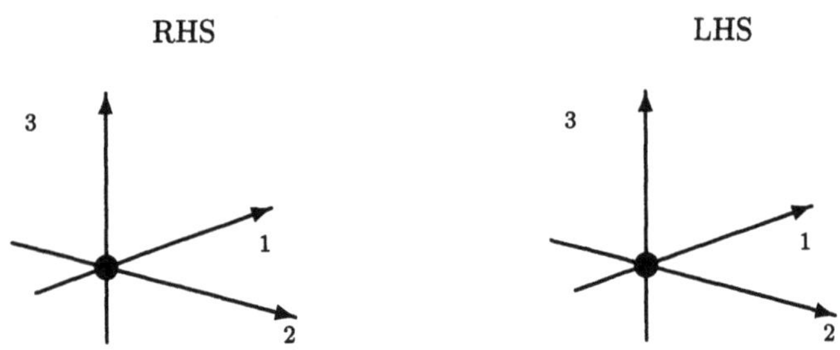


**Figure 1.12**

The choice of particular time and space orientations, like the choice of fixed units, is simply to get the smallest possible set of inertial frames.

For each inertial frame $K$ and for each event $P$ that occurs in spacetime we can now assign corresponding $K$-coordinates. The event $P$ will occur near (as near as we please, by refining the grid if necessary) to a vertex carrying a clock and a spatial label; we just record the readings: $(t \equiv x^0, x^1, x^2, x^3)$, or $x^\mu$ for short. If $K'$ is another inertial frame, a similar observation produces the $K'$-coordinates $x'^\mu$ for $P$. In the historical development of special relativity the practical procedure just described for assigning inertial coordinates to physical events was very important: it reduced the chance that notions like absolute time could slip in unnoticed.

We now turn to a discussion of the relation between the two sets of coordinates: it will result in the *Poincaré transformation.*

## 1.6 Poincaré Transformations

In this section the relation between the coordinates with respect to different inertial frames of reference is developed. The relation in the general case is called the Poincaré transformation, but if the two inertial frames in question share the same spacetime origin (where all four coordinates are zero), it is called a Lorentz transformation. We base the development of the form of the Poincaré transformation on three ideas:

1. A curve characterised as a straight line in one inertial frame $K$ will also be characterised as a straight line in any other inertial frame $K'$. (A curve in spacetime is characterised by $K$ as a straight line if its equations in $K$-coordinates are linear.) This will be shown to imply that for each pair of frames there are four numbers $a^\mu$and sixteen numbers $a^\mu{}_\nu$, these twenty numbers depending only on the frames $K$ and $K'$, such that for all events

$P$ the $K$-coordinates and $K'$-coordinates are related by[3]

$$x'^{\mu}(P) = a^{\mu} + a^{\mu}{}_{\nu} x^{\nu}(P). \tag{1.1}$$

This is a general affine transformation.

2. If two events $P$ and $Q$ can be connected by a straight line which, with respect to $K$, represents a motion with speed one (the speed of light), then with respect to $K'$ it again represents a motion with speed one. This condition implements Einstein's second principle of relativity, that the speed of light is one, independent of the frame, independent of the motion of the source. The condition gives rise to a restriction on the numbers appearing in (1.1)

$$a^{\mu}{}_{\rho} a^{\nu}{}_{\tau} \eta_{\mu\nu} = \Lambda^2 \eta_{\rho\tau}, \tag{1.2}$$

where $\Lambda^2$ is some real number and the nonzero values of $\eta_{\mu\nu}$ are $\eta_{00} = -1$, $\eta_{11} = \eta_{22} = \eta_{33} = 1$. (In the next chapter $\eta_{\mu\nu}$ is used to summarise the scalar products of basis vectors and in Chapter 4 it will emerge as a component form of the metric; for the present it is just a convenient shorthand for 16 numbers.)
3. The condition that the same units are used in $K$ and $K'$ leads to $\Lambda = 1$.

The general form of the eq. (1.1), with 4+6 parameters after the 10 independent restrictions of eq. (1.2) are taken into account, follows from the three physical ideas just described. We do not yet have a *physical* description of the relation between the different inertial frames. Our point of view is that we work out the physical relation of two inertial frames from the Poincaré transformation connecting them, rather than the other way round. This procedure avoids further, rather ad hoc assumptions about the physical relations of different frames. (It cannot avoid some apparently less dangerous presumptions about the relations between physical coordinates and mathematical coordinate systems.)

---

[3] For arguments involving the coordinates of unspecified inertial frames the summation convention is almost invariably used in place of a summation sign or an explicit sum. Summation convention: a Greek index repeated in a term, appearing once as a superscript and once as a subscript, is to be summed over $0, 1, 2, 3$. Any other Greek letter can be used in place of a repeated one without changing the meaning (the repeated indices are called dummy indices). Of course no letter can be used to generate a sum if it has already been used for a different sum or if it appears unrepeated. Changes of scripts during algebraic manipulations, to avoid violating the rules, are made without comment. Equations with unrepeated scripts hold for all possible values of the scripts. Each such scripts must appear in every term and in the same position, up or down. In particular, equation (1.1) holds for four different values of $\mu$; in each equation $\nu$ is understood to be summed over $0, 1, 2, 3$. Equation (1.2) holds for each of the 16 possible values of the pair $\rho, \tau$; in each equation it is understood that both $\mu$ and $\nu$ are summed over on the left-hand side. The applications of the convention will be extended as further theory is developed.

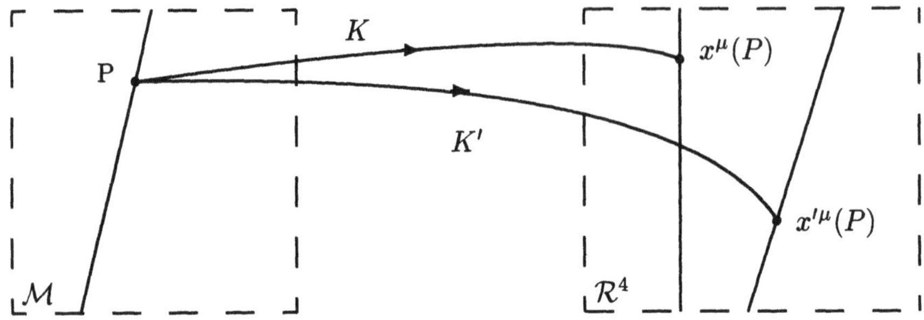


**Figure 1.13**

### 1.6.1 Straight Lines

We assume that a straight line as described by $K$ is also a straight line as described by $K'$. This assumption is based on the physical idea that the free motions of particles will be described as straight lines with respect to all inertial coordinate systems, but it is a little stronger because not all straight lines can represent free motions (in view of the fact, as it turns out, that particles move no faster than light).

We are not attempting in this section to deduce the Lorentz transformation from some minimal collection of postulates, but instead to show how reasonable it is, how sensible it would be to investigate the theory with this connection between inertial frames. (There are mathematical discussions in the papers of Zeeman and of Hegerfeldt concerning the *minimum* assumptions that imply the Lorentz transformation.)

A free motion in spacetime $\mathcal{M}$ is mapped by both $K$ and $K'$ into a straight line in coordinate space (Figure 1.13). A straight $K$-line is a curve, parameterised by $\lambda$, say, of the form

$$x^\mu(\lambda) = A^\mu + \lambda B^\mu, \tag{1.3}$$

with any constants $A^\mu$, $B^\mu$. If the general relation between the $K$-coordinates and $K'$-coordinates is [4]

$$x'^\mu = f^\mu(x^\nu),$$

then the expression for the curve, in $K'$-coordinates, is

$$x'^\mu(\lambda) = f^\mu(x^\nu(\lambda)) = f^\mu(A^\nu + \lambda B^\nu).$$

If this is to be a straight line with respect to the $K'$-coordinates, we must have

[4] To allay any doubt about the meaning of the symbols: there are four equations, in each one of which one of the four functions $f^\mu$ appears. $f^\mu$ is a function of four variables $x^0, x^1, x^2, x^3$, represented here by the notation $x^\nu$. One often writes $f^\mu(x)$, with the same meaning, $x$ being thought of as an element of $\mathcal{R}^4$.

$$x'^\mu(\lambda) = A'^\mu + \lambda B'^\mu = f^\mu(A^\nu + \lambda B^\nu) \tag{1.4}$$

for some new constants $A'^\mu$, $B'^\nu$.

Using the chain rule to differentiate Eq. (1.4) with respect to $\lambda$,

$$B'^\mu = B^\alpha \frac{\partial}{\partial x^\alpha} f^\mu(A^\nu + \lambda B^\nu),$$

and differentiating again (setting $\lambda = 0$ afterwards) we have[5]

$$0 = (B^\alpha \partial_\alpha)^2 f^\mu(A^\nu).$$

This must be true for all straight $K$-lines, for all positions and for all directions, that is, for all $A^\nu$ and all $B^\nu$. With $B^0 = 1$, $B^m = 0$ we have $(\partial_0)^2 f^\mu = 0$ for all $A^\nu$. Similarly, we get $(\partial_1)^2 f^\mu = 0$, etc. Then from $B^0 = 1$, $B^1 = 1$, $B^2 = B^3 = 0$ we get $\partial_0\partial_1 f^\mu = 0$. Hence, finally, we have

$$\partial_\alpha \partial_\beta f^\mu(A^\nu) = 0, \qquad \text{all } A^\nu. \tag{1.5}$$

This means that $\partial_\beta f^\mu$ is constant, equal to $a^\mu{}_\beta$(say), and that $f^\mu(x^\nu)$ must then have the simple form given by

$$x'^\mu = f^\mu(x^\nu) = a^\mu{}_\nu x^\nu + a^\mu \tag{1.6}$$

with four constants $a^\mu$ and sixteen constants $a^\mu{}_\nu$. The sixteen constants $a^\mu{}_\nu$ are not completely arbitrary. The matrix that they form must be invertible so that the coordinates of an event with respect to one frame can be determined from the coordinates with respect to any other frame. But, in addition to that general requirement, we will find that the $a^\mu{}_\nu$ must satisfy the *Lorentz condition* (1.12), which is a consequence of restricted units and the special requirement for the description of light.

The relation between the $K$-description and the $K'$-description of the straight lines can now be made completely explicit:

$$\begin{aligned} f^\mu(A^\nu + \lambda B^\nu) &= a^\mu{}_\nu(A^\nu + \lambda B^\nu) + a^\mu \\ &= (a^\mu + a^\mu{}_\nu A^\nu) + \lambda(a^\mu{}_\nu B^\nu) \\ &\equiv A'^\mu + \lambda B'^\mu. \end{aligned}$$

The spacetime $K$-origin (the point whose spatial and temporal $K$-coordinates are $x^\nu = 0$) has by Eq. (1.6) the $K'$-coordinates $x'^\mu = a^\mu$, which can have any values. So there is no restriction on the values of the constants $a^\mu$.

[5] In $(B^\alpha\partial_\alpha)^2$ the differential operator $B^\alpha\partial_\alpha$ is applied twice; another notation is $(B^\alpha\partial_\alpha)(B^\beta\partial_\beta)$, with independent sums. The abbreviation $\partial_\alpha \equiv \partial/\partial x^\alpha$ in the notation is not a matter just of laziness; it conforms to a wider notational scheme which will only gradually emerge. Eventually we will have two sorts of *vectors*, one whose components are labelled with superscripts, and one with subscripts.

If $K$ and $K'$ are related so $x'^{\mu} = x^{\mu} + a^{\mu}$, the spatial labels and clock readings at each event have constant differences, $a^m$ for space, $a^0$ for time. The spatial vertices and clocks of the two systems are relatively at rest. Hence the two inertial frames themselves are at rest with respect to each other, with parallel axes; each is simply translated with respect to the other, without rotation.

Since the simple translation has no further mysteries to study, we may regard this class of transformation as understood and turn our attention to the relation between inertial frames that share a common spacetime origin (which entails $a^{\mu} = 0$). Equation (1.6) reduces in this case to $x'^{\mu} = a^{\mu}{}_{\nu} x^{\nu}$.

The *spatial* origin of the $K$-system, being the worldline for a free motion, is an example of a straight line. Regarded as a parameterised curve its coordinate expressions in the two systems are

$$\begin{array}{lll} \text{K-coordinates :} & x^0 = \lambda, & x^n = 0 \\ \text{K}'\text{-coordinates:} & x'^0 = a^0{}_0 \lambda, & x'^m = a^m_{\ 0} \lambda. \end{array}$$

Note that $K$ and $K'$ will have the same time orientation only if $a^0{}_0 > 0$. Also to be noted is the speed of the $K$-spatial origin relative to the $K'$-frame. The three-geometry in the $K'$-space (or the "space" of any other inertial frame) is Euclidean so the speed in $K'$ is given[6] by $(\text{speed})^2 = (a^m_{\ 0} a^m_{\ 0})/(a^0{}_0)^2$. Since the speed of light has the built-in value $c = 1$, the statement that the speed of the $K$-spatial origin with respect to the $K'$-frame is less than that of light is equivalent to

$$a^m_{\ 0} a^m_{\ 0} - a^0{}_0 a^0{}_0 < 0. \tag{1.7}$$

This relation will prove to be a consequence of the restrictions due to the form of light rays.

### 1.6.2 Light Rays

Light and other electromagnetic phenomena can be approached on many different levels: from the geometrical concepts of light rays and plane waves to a complete field theory of the electromagnetic field tensor. Here we deal exclusively with *light rays.* (Electromagnetism as a field theory will be developed later, in Chapter 5.)

Suppose a light ray can travel from $O$, the common spacetime origin of frames $K$ and $K'$, to $P$, a point with $K$-coordinates $x^{\mu}$, and $K'$-coordinates $x'^{\mu}$. Since the spatial coordinates for an inertial frame are Cartesian, spatial distances are given by the Euclidean formula, and since the unit for the time coordinate has been chosen to make the speed of light $c = 1$ in each frame, we have, in terms of $K$-coordinates,

[6] The summation convention applies for repeated latin scripts (in any position, up or down), the sum being over the spatial 1,2,3.

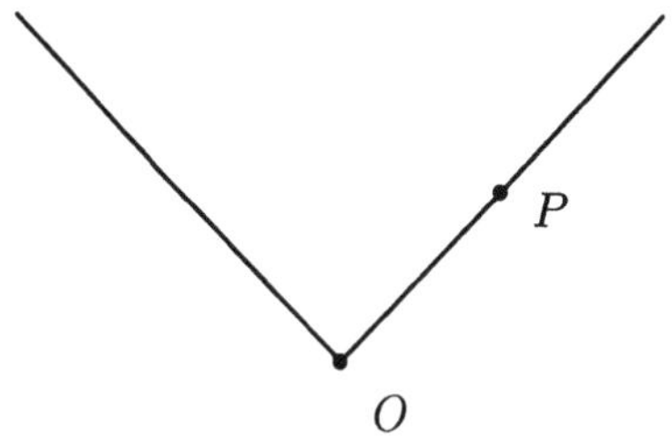


**Figure 1.14**

$$(a) \qquad x^m x^m = (x^0)^2,$$

and in terms of $K'$-coordinates,

$$(b) \qquad x'^m x'^m = (x'^0)^2.$$

The points which satisfy (a) may be described as lying on a uniformly expanding sphere with its centre at the spatial origin of $K$. Similarly, the points which satisfy (b) lie on a uniformly expanding sphere with respect to $K'$. The collection of points $P \in \mathcal{M}$ which can be reached by light rays from $O$ is called the *forward lightcone* from $O$ (Figure 1.14). It consists of all events at which the instantaneous flash at $O$ can be seen. Equations (a) and (b) show that the lightcone is expressed in exactly the same way in the two frames. Conventionally, even for absolute diagrams of $\mathcal{M}$, light rays are displayed at 45° to the vertical and horizontal. Using the notation[7]

$$\eta_{\mu\nu} \equiv \begin{cases} -1 & \text{if} \quad \mu = \nu = 0 \\ +1 & \text{if} \quad \mu = \nu = 1, 2, 3 \\ \ \ 0 & \text{if} \quad \mu \neq \nu \end{cases} \tag{1.8}$$

the condition (a) may be written with two simple implied sums as

$$(c) \qquad x^\mu x^\nu \eta_{\mu\nu} = 0,$$

and, in terms of the constants $a^\mu{}_\nu$, $a^\mu = 0$ which appear in (1.6), the condition (b) becomes

[7] The choice of numerical values with the opposite sign, but represented by the same symbol, is popular in the literature, especially that of high-energy physics. This leads to many superficial, but irritating, sign differences in the consequent theory. The advantage of the choice of Eq.(1.8) is that the "spatial sector" reproduces exactly the metric of three-dimensional Euclidean geometry, with the expected sign.

(*d*) $$x'^{\mu}x'^{\nu}\eta_{\mu\nu} = a^{\mu}{}_{\alpha}a^{\nu}{}_{\beta}\eta_{\mu\nu}x^{\alpha}x^{\beta} = 0.$$

The solutions of (c) must satisfy (d), and vice versa. This argument alone does not imply the Lorentz condition, $a^{\mu}{}_{\alpha}a^{\nu}{}_{\beta}\eta_{\mu\nu} = \eta_{\alpha\beta}$. That equation holds, it transpires, only when the same units are used in the two inertial frames.

In order to see how equations (c) and (d) produce restrictions on $a^{\mu}{}_{\nu}$, it is useful to look at some special cases, special points $P$ and so special coordinates $x^{\mu}$.

If $x^0 = 1,\ x^1 = \pm 1,\ x^2 = x^3 = 0$, then (c) holds and (d) becomes

$$a^{\mu}{}_{0}a^{\nu}{}_{0}\eta_{\mu\nu} + a^{\mu}{}_{1}a^{\nu}{}_{1}\eta_{\mu\nu} \pm 2a^{\mu}{}_{0}a^{\nu}{}_{1}\eta_{\mu\nu} = 0.$$

Subtracting one equation from the other gives

(*e*) $$a^{\mu}{}_{0}a^{\nu}{}_{1}\eta_{\mu\nu} = 0,$$

and then

(*f*) $$a^{\mu}{}_{1}a^{\nu}{}_{1}\eta_{\mu\nu} = -a^{\mu}{}_{0}a^{\nu}{}_{0}\eta_{\mu\nu} \equiv \Lambda^2;$$

similarly with $1 \to 2, 3$. (No assumption is really being made about the sign of the newly introduced $\Lambda^2$ ($\Lambda$ could be imaginary). Since the right-hand side of (f) does not involve the spatial index only one constant appears for 1,2,3.)

If $x^0 = \sqrt{2},\ x^1 = x^2 = 1,\ x^3 = 0$, then (c) holds, and after using (e), the condition (d) gives

$$2a^{\mu}{}_{0}a^{\nu}{}_{0}\eta_{\mu\nu} + a^{\mu}{}_{1}a^{\nu}{}_{1}\eta_{\mu\nu} + a^{\mu}{}_{2}a^{\nu}{}_{2}\eta_{\mu\nu} + 2a^{\mu}{}_{1}a^{\nu}{}_{2}\eta_{\mu\nu} = 0.$$

Using (f) twice we are left with

(*g*) $$a^{\mu}{}_{1}a^{\nu}{}_{2}\eta_{\mu\nu} = 0,$$

and similarly for other spatial pairs. Combining the results (e),(f),(g) with the definition Eq.(1.8) gives

$$a^{\mu}{}_{\alpha}a^{\nu}{}_{\beta}\eta_{\mu\nu} = \Lambda^2\eta_{\alpha\beta}. \qquad (1.9)$$

The number $\Lambda^2$ was introduced in (f). As the notation was chosen to suggest, it will prove to be positive. It certainly cannot be zero; zero would be inconsistent with the fact that the determinant of the left-hand side of Eq.(1.9) is nonzero.

$$(\det(a^{\mu}{}_{\nu}))^2\det(\eta_{\rho\tau}) = (\Lambda^2)^4\det(\eta_{\rho\tau}) \qquad (1.10)$$

But $\det(\eta_{\rho\tau}) = -1$ and $\det(a^{\mu}{}_{\nu})$ must be nonzero in order that (1.6) can be inverted. Hence $\Lambda^2 \neq 0$.

The constant $\Lambda^2$ cannot be negative either. Since the numbers $a^{\mu}{}_{\alpha}$ are real, the left-hand side of (1.9) can be regarded as giving new components for a quadratic form with original components $\eta_{\mu\nu}$. By Sylvester's law of inertia[8] the two sets of components must have the same signature, +2. This is only possible if $\Lambda^2 > 0$.

An alternative argument to show that $\Lambda^2 > 0$ uses a lemma which will be very helpful later in understanding the classification of vectors. (It turns out that in spacetime, unlike in Euclidean space, vectors come in several different varieties.) This lemma and several other similar results that are collected in Problem 1 (at the end of the chapter) come from Wigner's study of 1939.

**Lemma 1.1**

Consider nonzero numerical vectors $a = \{a^\mu\}$, $b = \{b^\mu\}$ in $\mathcal{R}^4$ with an indefinite scalar product

$$a \cdot b \equiv a^\mu b^\nu \eta_{\mu\nu} = -a^0 b^0 + a^n b^n.$$

Suppose $a \cdot a < 0$, that is, $(a^0)^2 > a^n a^n$, and $a \cdot b = 0$, that is, $a^0 b^0 = a^n b^n$. We say that $a$ is *timelike* and $a$ and $b$ are *orthogonal.* The assertion of the lemma is that $b \cdot b > 0$, or $b^n b^n > (b^0)^2$; the vector $b$ is necessarily *spacelike.*

*Proof* If $b^0 = 0$, then $b \cdot b = b^n b^n > 0$ since $b \neq 0$, and $b$ is spacelike. We need only consider further the case $b^0 \neq 0$. We must have $(a^0)^2 > 0$ for otherwise $a$ could not be timelike. Squaring the condition of orthogonality and using the Schwartz inequality gives

$$(a^0)^2 (b^0)^2 = (a^n b^n)^2 \leq (a^n a^n)(b^m b^m) < (a^0)^2 (b^m b^m).$$

Dividing by the positive $(a^0)^2$ gives $(b^0)^2 < b^m b^m$ and so $b$ is spacelike in this case too. □

*If a vector is orthogonal to a timelike vector it must be spacelike.*

Now look at equation (1.9) in the terms of the lemma. If we regard the sixteen numbers $a^\mu{}_\nu$ as supplying 4 numerical vectors $a_{(\alpha)} = \{a^\mu{}_\alpha\}$ then the left-hand side of (1.9) is the scalar product of $a_{(\alpha)}$ and $a_{(\beta)}$ :

$$a_{(\alpha)} \cdot a_{(\beta)} = \Lambda^2 \eta_{\alpha\beta}.$$

If $\alpha \neq \beta$, then the right-hand side vanishes, so the four vectors are orthogonal to each other. Recall that $\eta_{00} = -1$, $\eta_{11} = \eta_{22} = \eta_{33} = 1$.

- If $\Lambda^2 > 0$ we have $a_{(0)}$ timelike, and each $a_{(n)}$ spacelike; one timelike vector and three orthogonal spacelike vectors.
- If $\Lambda^2 < 0$ we would have $a_{(0)}$ spacelike and three timelike vectors $a_{(n)}$. This case is impossible by the lemma, which allows only one timelike vector in a collection of orthogonal vectors.

---

[8] Most algebra texts, for example, P.M.Cohn, Algebra, Volume 1 (Wiley, 1974), p.189.

Hence $\Lambda^2 > 0$.

From the $\alpha = \beta = 0$ case of (1.9) we have

$$a^m{}_0 a^m{}_0 - a^0{}_0 a^0{}_0 = a^\mu{}_0 a^\nu{}_0 \eta_{\mu\nu} < 0,$$

which is precisely the statement (1.7) that the speed of the $K$-spatial origin with respect to the $K'$-frame is less than one, less than the speed of light.

### 1.6.3 Units

In the discussion of the relation between coordinates with respect to $K$ and those with respect to $K'$ no account has yet been taken of the overall scale of the units. In each system the time units and the space units are related so as to achieve $c = 1$, but the overall scale remains unfixed. In specifying a minimum set of inertial frames we want to use a universal spacetime unit, for example, a multiple of a given atomic period.

If we choose a unit of time (for example, seconds), then to obtain $c = 1$ for the speed of light we must use an appropriate unit of distance, the distance travelled by light in a unit of time; in the example this unit of distance would also be called a second (it is approximately $3 \times 10^{10}$ cm). Correspondingly, a unit of distance (for example, a centimetre) gives rise to a unit of time (the time taken for light to travel the unit of distance) with the same name so that in this system $c = 1$. All the equations of the previous subsection, relating coordinates with respect to two such systems of units, remain true, no matter how different the scales, even if, for example, $K$ uses seconds and $K'$ uses light years.

The relation between the coordinates before we harmonise the units is

$$x'^\mu = a^\mu{}_\nu x^\nu \quad \text{subject to} \quad a^\mu{}_\alpha a^\nu{}_\beta \eta_{\mu\nu} = \Lambda^2 \eta_{\alpha\beta}.$$

This may be rearranged as

$$\frac{x'^\mu}{\Lambda} = \left(\frac{a^\mu{}_\nu}{\Lambda}\right) x^\nu \quad \text{subject to} \quad \frac{a^\mu{}_\alpha}{\Lambda}\frac{a^\mu{}_\beta}{\Lambda}\eta_{\mu\nu} = \eta_{\alpha\beta},$$

with $\Lambda > 0$ so that $a^0{}_0/\Lambda > 0$. It is clear that rescaling the temporal and spatial units by the same factor in the $K'$-frame gives the same results as changing $\Lambda$.

If we keep the $K$-units fixed we can choose all other $K'$-units to eliminate the $\Lambda$s in the relations connecting $K$- and $K'$-coordinates. To do this the $K'$-coordinates for each point $P$ may be rescaled by replacing $\frac{x'^\mu}{\Lambda} \to x'^\mu$, or $x'^\mu \to \Lambda x'^\mu$ (this corresponds to replacing the original unit by a new unit equal to $\frac{1}{\Lambda}$(original unit)). Then, using the symbols $a^\mu{}_\nu$ for the previous $\frac{a^\mu{}_\nu}{\Lambda}$, the relation between the coordinates after rescaling is

$$x'^\mu = a^\mu{}_\nu x^\nu \tag{1.11}$$

subject to

$$a^\mu{}_\alpha a^\nu{}_\beta \eta_{\mu\nu} = \eta_{\alpha\beta}. \tag{1.12}$$

Once we have verified that the inverse of a Lorentz transformation is also a Lorentz transformation it will be easy to see that, when all frames $K', K'', \ldots$ have been related to $K$ in the above way, the relation between *any* two frames, $K'$ and $K''$ say, will have the same form.

The procedure just described serves to fix the units in all inertial frames in terms of the units in one. We make the fundamental physical (not mathematical) postulate that the inertial frames so related have the same units. The reasons for this are very strong, for it is true for rotations and for the orthogonal directions in boosts (special Lorentz transformations corresponding to relative velocities). From the *physical* point of view, choosing the same units means, for example, relating the unit of time to some *specific* atomic wavelength and then choosing the unit of distance to make the speed of light one. By the two principles of relativity, it is expected (i.e. we expect nature to support this) that this can be done consistently in all inertial frames. *The postulate is that the inertial frames described in Section 1.5, with a chosen unit, are exactly those whose coordinates are connected by (1.6) subject to (1.12).*

If $a^\mu{}_\nu$ represents a *spatial rotation*, for which $a^0{}_0 = 1$, $a^m_0 = a^0{}_m = 0$, the condition (1.12) gives $a^m_a a^m_b = \delta_{ab}$, which means that $a^m_n$ is a standard 3-d rotation matrix connecting the Cartesian coordinates in Euclidean space with respect to relatively rotated frames, *using the same unit of distance.* The similar transformation (1.9) with $\Lambda \neq 1$ would relate relatively rotated Cartesian frames using different units of distance (and the unit of time would change by the same factor). Therefore, for rotations between frames using the same units we require $\Lambda = 1$.

The second basic Lorentz transformation is a *boost*, a transformation between relatively moving frames, without rotation. In its simplest form, satisfying (1.12) in the case $\Lambda = 1$ , it is

$$x'^0 = \gamma(x^0 - Vx^1), \quad x'^1 = \gamma(x^1 - Vx^0), \quad x'^2 = x^2, \quad x'^3 = x^3,$$

with $\gamma \equiv (1 - V^2)^{-1/2}$. We call this transformation the standard boost; it gives rise to (or explains) all the new phenomena of special relativity, as the reader is presumed already to know. In the directions orthogonal to the relative velocity, here the 2- and 3-directions, the coordinates are unchanged, hence give an unchanged measure of distance. The choice $\Lambda = 1$ for the boosts is equivalent to choosing equal distances measured by two frames in the spatial plane orthogonal to the relative motion.

Since all Lorentz transformations may be regarded as compositions of rotations and boosts (detailed relations are in the Problems), this argument connects the units of spatial measurements in all inertial frames.

We now accept that equations (1.11) and (1.12) constitute the most general Lorentz transformation, the general relation between the coordinates of two inertial frames with the same spacetime origin, *using the same units.* If the $K$-coordinates and $K'$-coordinates are rescaled by the same factor the equations remain the same (rescaled coordinates, unchanged $a^{\mu}{}_{\nu}$). The final, universe-wide choice of unit is now a matter of convention.

In their present form, however, the Lorentz transformations, satisfying only (1.11) and (1.12), connect inertial frames with possibly different orientations, different senses for the direction of the time coordinate and different handednessin space. (There is no occasion to be overexcited by these facts, they are just a matter of convention for the systems of measurement.)

### 1.6.4 Orientations and Definition of Lorentz Transformations

Time orientation of the system of inertial frames is achieved by including only those frames whose coordinates are connected by a Lorentz transformation with

$$a^{0}{}_{0} > 0. \tag{1.13}$$

This condition is just what is required to prove that all events on or within a forward lightcone have a positive time coordinate difference with the vertex of the cone, no matter which inertial frame is used. Similarly, all events on or within a backward lightcone have negative time coordinate differences. Equation (1.21) below may be used for a proof (in the Problems).

Regarding (1.12) as a matrix equation, the corresponding determinantal equation gives $(\det(a^{\mu}{}_{\nu}))^2 = 1$, hence $\det(a^{\mu}{}_{\nu}) = \pm 1$. With a time orientation fixed by choosing $a^{0}{}_{0} > 0$, a space orientation is achieved by using only inertial frames connected by Lorentz transformations with

$$\det(a^{\mu}{}_{\nu}) = +1. \tag{1.14}$$

Such transformations are continuously connected to the identity in $\mathcal{R}^4$ and so, with $a^{0}{}_{0} > 0$ ensuring that the time orientation is maintained, the spatial transformation in $\mathcal{R}^3$ is itself continuously connected to the identity. Hence the spatial orientation is preserved. These transformations are called the ***proper, orthochronous Lorentz transformations.*** They connect the coordinates assigned by the oriented inertial frames with a fixed, universal choice of unit that were constructed in Section 1.5.

Our conventional choice of spatial orientation is the Right-Handed System (RHS) illustrated in Figure 1.15, (as opposed to a LHS).

### 1.6.5 Inverse Lorentz Transformations

Every inertial coordinate map $\mathcal{M} \to \mathcal{R}^4$ is invertible, so by composing one such map with the inverse of another we see that the connection between different inertial coordinates is also invertible. For two frames sharing a common spacetime origin the coordinates are related by an equation of the form $x'^{\mu} = a^{\mu}{}_{\nu}x^{\nu}$. An explicit expression for the inverse relation,

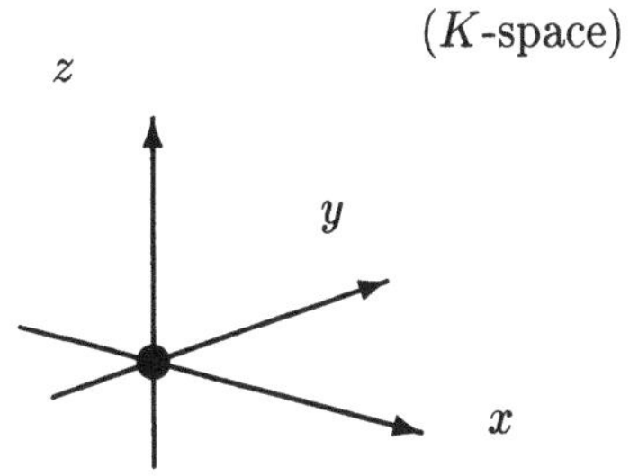


**Figure 1.15**

$$x^\mu = a^{-1\mu}{}_\nu x'^\nu, \tag{1.15}$$

may be obtained from the Lorentz condition.

Let $\eta$ be the matrix with entry $\eta_{\alpha\beta}$ in the $\alpha^{th}$row and $\beta^{th}$column, and let $a$ be the matrix with entry $a^\mu{}_\nu$ in the $\mu^{th}$row and $\nu^{th}$column. Then the Lorentz condition (1.12) may be written as the matrix equation

$$a^T \eta a = \eta, \tag{1.16}$$

where $a^T$ is the matrix transpose of $a$. Denote the inverse for $\eta$ by $\eta^{-1}$, with entries $\eta^{\alpha\beta}$ in the $\alpha^{th}$row and $\beta^{th}$column, such that

$$\eta^{\alpha\beta}\eta_{\beta\gamma} = \delta^\alpha_\gamma = \eta_{\gamma\beta}\eta^{\beta\alpha}. \tag{1.17}$$

(Superscripts are chosen for the elements of $\eta^{-1}$ just so that this component equation obeys the rules for the use of the summation convention. Of course, $\eta_{\alpha\beta}$ equals $\eta^{\alpha\beta}$ but we do not want to write this fact as an equation and break the convention that scripts must balance on the two sides.)

Multiplying equation (1.16) on the left by $\eta^{-1}$ and on the right by $a^{-1}$ gives

$$a^{-1} = \eta^{-1} a^T \eta. \tag{1.18}$$

In components the matrix inverse is

$$a^{-1\mu}{}_\nu = \eta^{\mu\alpha} a^\beta{}_\alpha \eta_{\beta\nu}. \tag{1.19}$$

Matrix inverses work on both right and left, so multiplying (1.18) on the left by $a$, one has $a\eta^{-1}a^T\eta = 1$, hence

$$a\eta^{-1}a^T = \eta^{-1}. \tag{1.20}$$

The component form of the last equation is

$$a^\mu{}_\alpha \eta^{\alpha\beta} a^\nu{}_\beta = \eta^{\mu\nu}, \tag{1.21}$$

which is equivalent to the condition (1.12), and is an alternative form of the Lorentz condition. Once an explicit formula for the inverse Lorentz transformation is available, it is easy to prove that the speed with respect to $K'$ of the spatial origin for $K$ equals the speed with respect to $K$ of the spatial origin for $K'$.

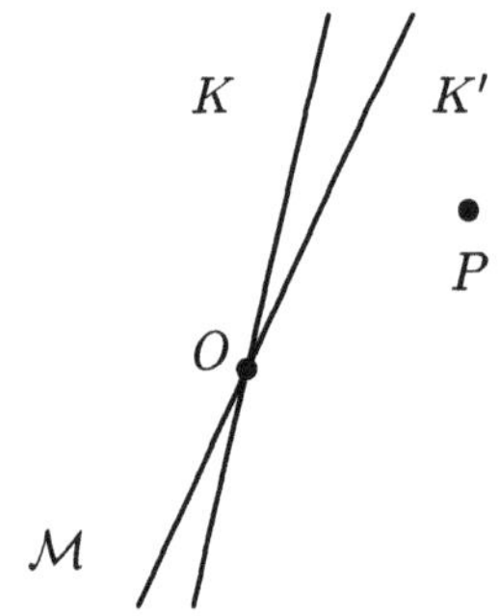


**Figure 1.16**

# 1.7 Inertial Coordinates in Spacetime

## 1.7.1 Absolute vs Relative Diagrams

As has already been mentioned, there are two types of spacetime diagram, *absolute* diagrams (of $\mathcal{M}$) and *relative* diagrams (of $\mathcal{R}^4$). For a given physical situation there is one absolute diagram, but as many relative diagrams as there are inertial coordinate systems.

An absolute spacetime diagram aims to give a pictorial representation of at least some aspects of the (absolute) manifold $\mathcal{M}$ and the physical processes unfolding in it. In addition to the history of experiments, the worldline of the master clock for an inertial frame, or part of its spatial grid, might very well be of just as much interest in this context as any bits of the "observed physics". Points on the paper stand for points in the manifold: $O$, $P$ (Figure 1.16). They are "absolute" in the sense that they are intended to be coordinated directly with physical events; they do not depend on any relative description. The straight lines $K$ and $K'$ similarly are an absolute representation of the worldlines of the free motions generating the frames (the worldlines of their master clocks). Where it is helpful to be specific we suppose that the inertial frames $K$ and $K'$ are related by the simple standard boost.

*Relative* spacetime diagrams give us a picture not directly of $\mathcal{M}$ but of a coordinate description of $\mathcal{M}$ in the coordinate space $\mathcal{R}^4$ in which we have made a choice of special inertial frame, $K$ say. The features of $\mathcal{M}$ are mapped into $\mathcal{R}^4$ by the coordinate map $K$ (Figure 1.17).

The points in the picture of $\mathcal{R}^4$ stand for $K$-coordinates because it arises from $\mathcal{M}$ by the $K$-map. The $K$-line stands for the points $\{(t, 0, 0, 0),\ \text{all } t\}$; the "line" $x^0 = 0$ is actually the projection on the two-dimensional paper of the three-dimensional hyperplane $\{(0, x, y, z),\ \text{all } x, y, z\}$. The $K'$-line satisfies $x = Vt$.

The $\mathcal{R}^4$–picture of spacetime relative to $K'$ has a slightly different appearance; it is a view of the same universe but from the perspective of $K'$ (Figure 1.18). Here it is the $K'$-line that is the set $\{(t', 0, 0, 0),\ \text{all } t'\}$, (where

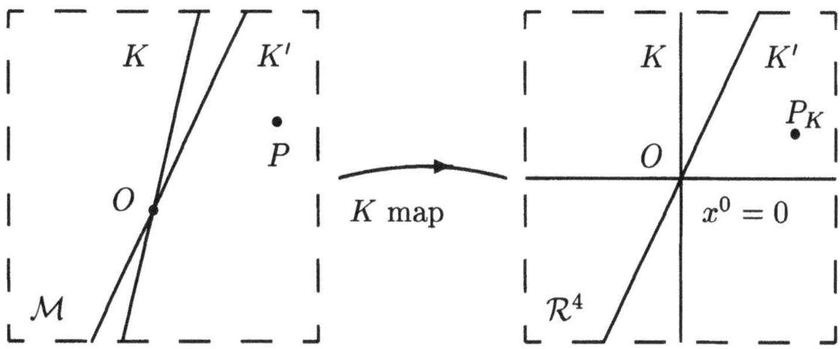


**Figure 1.17**

the dash on $t$ is merely cosmetic), and the $K$-line has the equation $x' = -Vt'$.

We can get the same information from any of these different types of diagram but the absolute diagram is preferred because it helps us to focus on the essentials, the *coordinate-independent geometrical relations.*

Generally speaking, ***the use of absolute diagrams for $\mathcal{M}$ helps us to concentrate on concepts with well-defined special relativistic meanings (events, worldlines, hyperplanes of simultaneity) and to avoid careless verbalisations with built-in nonrelativistic misconceptions.***

The geometrical features of an inertial frame may be illustrated in an absolute diagram of $\mathcal{M}$. The diagram might show the worldlines of master clocks and secondary clocks, hyperplanes of simultaneity and vertex points of the coordinate grid (Figure 1.19). The worldline of the master clock (also the spatial origin) for the inertial frame is $KK$. The arrow points to the future and indicates the time orientation.

The point $O$ is the spacetime origin for $K$, the event at the spatial origin when the master clock reads zero.

The worldline $SS$ is for a secondary clock, at rest with respect to the master clock (so their worldlines are parallel).

Two hyperplanes of simultaneity consist of all events for which $x^0 = 0$ or $x^0 = a^0$, respectively.

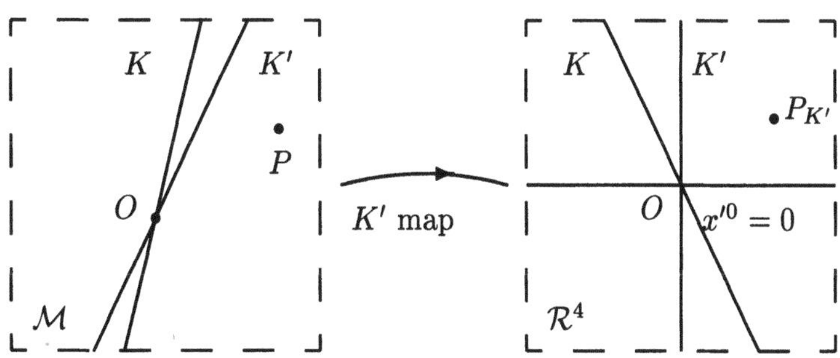


**Figure 1.18**

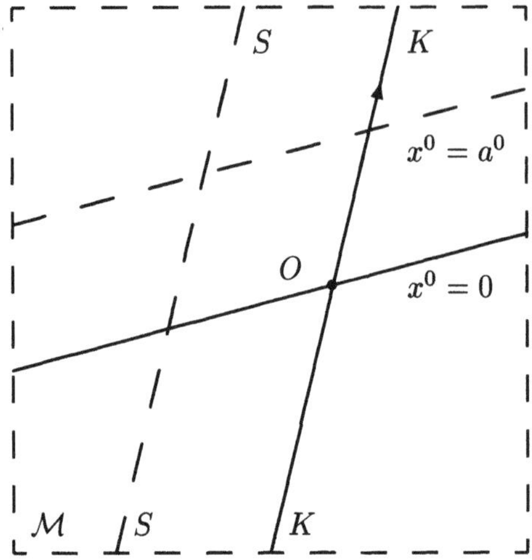


**Figure 1.19**

### 1.7.2 The Use of Inertial Coordinates

If we use inertial coordinates to analyse problems in special relativity we must always be aware that coordinates are (i) *for a specific point event* and (ii) *with respect to a specific frame.* No matter how obvious these remarks may seem, remembering them may help avoid the errors that one is sometimes led into by Newtonian habits. For example, in a Newtonian context one refers to a "length", or a "time interval"; one may have lost the more elementary and more important idea that these are *coordinate differences*, for which one must know "which events?" and "which frame?". The use of absolute diagrams keeps the relevant events in view and makes one aware how the use of different frames arises. Coordinates and coordinate differences are represented in absolute diagrams as in any affine space by using parallels (Figure 1.20). Euclidean orthogonality is meaningless in such diagrams. The time interval between $P$ and $Q$ is $\Delta x^0 = d - c$, while the spatial interval (in these coordinates) is $\Delta x^1 = b - a$.

### 1.7.3 Relation of Coordinates for Boosted Frames

Two distinct inertial frames will in general have different spacetime origins and nonparallel coordinate axes. Since the relation between frames that differ only in their spacetime origins or their spatial axes (but have the same time axis) is exactly as in Euclidean geometry, our interest is in frames $K$ and $K'$ related by a boost. If necessary after spatial rotations, the Lorentz transformation between coordinates with respect to the two frames is

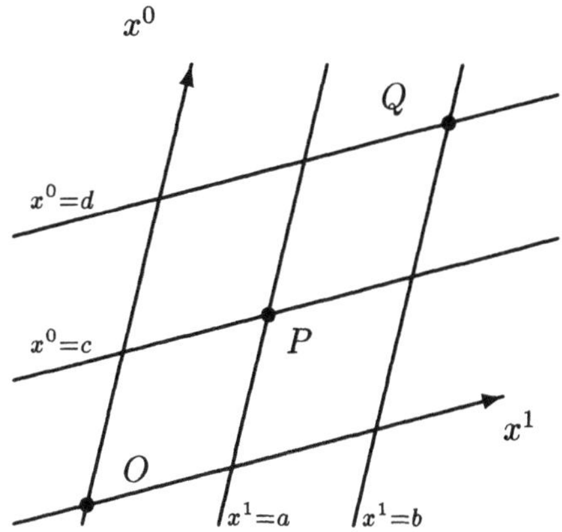


**Figure 1.20**

$$\left.\begin{aligned} x'^0 &= \gamma(x^0 - Vx^1), \qquad \gamma \equiv (1-V^2)^{-1/2} \\ x'^1 &= \gamma(x^1 - Vx^0) \\ x'^2 &= x^2, \quad x'^3 = x^3. \end{aligned}\right\} \tag{1.22}$$

The coordinate diagrams in $\mathcal{R}^4$ (really $\mathcal{R}^2$ of course, in the plane $x^2 = x'^2 = 0$, $x^3 = x'^3 = 0$) give rise to the diagram for $\mathcal{M}$ (Figure 1.21), shown here in the case $V > 0$, in the plane containing the worldlines for the master clocks of $K$ and $K'$. The $K$-worldline corresponds to $x^m = 0$ and the $K'$-worldline corresponds to $x'^m = 0$. The individual 3-d hyperplanes of simultaneity correspond to $x^0 = 0$ and $x'^0 = 0$. The great surprise in such diagrams, for anyone with Newtonian intuition, is that the hyperplanes $x^0 = 0$ and $x'^0 = 0$ do not coincide even though $O$ is their common spacetime origin. The larger is $\gamma$ (which can have any value greater than 1 as the speed $V$ ranges between

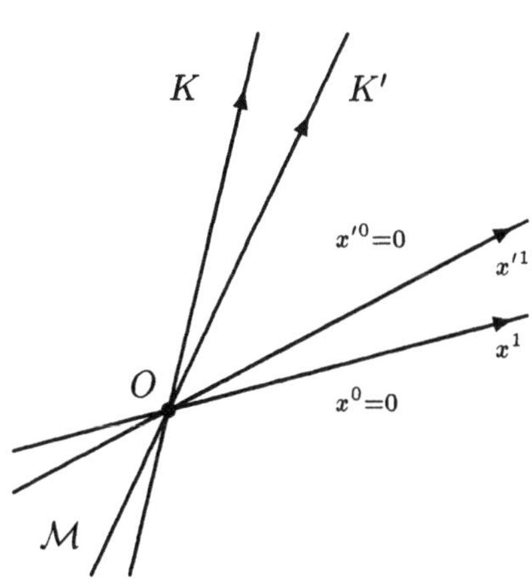


**Figure 1.21**

0 and 1) the more marked is this effect. Simultaneity is relative. (In Galilean spacetime simultaneity is absolute and there is only one hyperplane, as we saw on page 6.) If we accept the ideas in the description of inertial frames as physically appropriate, and the assumptions built into them, and the derivation of the Lorentz transformation, we must accept the relative character of simultaneity. This is a very physical statement. The $K$-clocks, synchronised with their master, differ (when we can compare them at a single event) from the $K'$-clocks, synchronised with *their* master, even though the two master clocks agree as they pass each other at $O$.

## 1.8 Geometrical Relations Between Events

### 1.8.1 Spacetime Interval

Consider any two events $P$ and $Q$ in spacetime, and any two inertial frames $K$ and $K'$. For each event the $K$- and $K'$-coordinates are related by a Poincaré transformation of the form $x'^{\mu} = a^{\mu} + a^{\mu}{}_{\nu} x^{\nu}$. Therefore, for the *differences* of coordinates the relation is a Lorentz transformation,

$$x'^{\mu}_{Q} - x'^{\mu}_{P} = a^{\mu}{}_{\nu}(x^{\nu}_{Q} - x^{\nu}_{P}),$$

or, for short,

$$\Delta'^{\mu} = a^{\mu}{}_{\nu} \Delta^{\nu}.$$

The (squared) *spacetime interval* between $P$ and $Q$ is defined by

$$\Delta s^2 \equiv \Delta^{\mu} \Delta^{\nu} \eta_{\mu\nu} = -(\Delta^0)^2 + (\Delta^1)^2 + (\Delta^2)^2 + (\Delta^3)^2.$$

It is the same no matter in which inertial frame it is evaluated. This is a consequence of the Lorentz condition (1.12):

$$\Delta'^{\mu} \Delta'^{\nu} \eta_{\mu\nu} = (a^{\mu}{}_{\alpha} \Delta^{\alpha})(a^{\nu}{}_{\beta} \Delta^{\beta}) \eta_{\mu\nu} = \Delta^{\alpha} \Delta^{\beta} \eta_{\alpha\beta}. \tag{1.23}$$

(Note that if the frames did not share a common unit, a scaling factor $\Lambda^2$ would appear in this formula.) It is easy to see that if (1.23) is satisfied for all $P$ and $Q$ (that is, for all $\Delta^{\alpha}$) then the Lorentz condition is satisfied; the two characterisations of the Lorentz transformations are equivalent.

As a consequence of (1.23) the spacetime interval expresses a geometrical property in a coordinate-independent way. The geometrical property depends on the relation between $P$ and $Q$.

For example, if $\Delta s^2$ is positive, then a frame exists in which $P$ and $Q$ are simultaneous, and the value of $\Delta s^2$ is the square of the distance between $P$ and $Q$ in that frame. So the expression $\Delta^{\mu} \Delta^{\nu} \eta_{\mu\nu}$ shows how to calculate, using components in a general frame, the squared distance in the special frame. If $\Delta s^2$ is negative, its value is minus the square of the time interval between $P$ and $Q$ in a frame in which they have the same position. If $\Delta s^2 = 0$, then $P$ and $Q$ can be connected by a light ray. Thus the sign of $\Delta s^2$ is extremely important, signalling the difference between a timelike or spacelike or lightlike relation between the points $P$ and $Q$.

In this context $\Delta s$ is never used and $\Delta s^2$ itself is often referred to as the spacetime interval.

If the spacetime interval $\Delta^\mu \Delta^\nu \eta_{\mu\nu} \leq 0$, then the sign of $\Delta^0$ is invariant too. The sign of $\Delta^0$ is the same no matter which inertial frame is used (here it is essential that the time orientation of the inertial frames has been fixed). (See Problem 12.)

### 1.8.2 Invariant Relations

The sign of the invariant $\eta_{\mu\nu}\Delta^\mu \Delta^\nu$ and the invariant sign of $\Delta^0$ (when it is invariant) are used to classify the geometrical relations between the points $P$ and $Q$ ($\Delta^\mu = x^\mu_Q - x^\mu_P$), the geometrical statements on which all inertial observers agree. It is simplest to express the relation of $Q$ to $P$ in terms of the lightcone with vertex $P$.

If $\eta_{\mu\nu}\Delta^\mu \Delta^\nu < 0$, we say that $Q$ is timelike displaced with respect to $P$.

If $\text{sign}(\Delta^0) > 0$ (Figure 1.22), then $Q$ is FPTL (future-pointing timelike) with respect to $P$.

If $\text{sign}(\Delta^0) < 0$ (Figure 1.23), then $Q$ is PPTL with respect to $P$.

In such cases, a frame exists (and so infinitely many exist) in which $P$ and $Q$ have the same spatial position.This fact is easily proven. The condition $\Delta^\mu \Delta^\nu \eta_{\mu\nu} = -(\Delta^0)^2 + (\Delta^1)^2 + (\Delta^2)^2 + (\Delta^3)^2 < 0$ may be expressed more simply in new coordinates $x'^\mu$ reached by a rotation $x'^0 = x^0$, $x'^m = a^m_{\ n} x^n$ chosen so that $(\Delta^1)^2 + (\Delta^2)^2 + (\Delta^3)^2 = (\Delta'^1)^2$ (this is always possible by Euclidean geometry). The condition is now $-(\Delta'^0)^2 + (\Delta'^1)^2 < 0$, or $|\Delta'^1/\Delta'^0| < 1$. If we transform to new coordinates $x''^\mu$, reached by a boost of the form (1.22), then we can choose $V$, $|V| < 1$ so that

$$\Delta''^1 = \gamma(\Delta'^1 - V\Delta'^0) = 0.$$

Since $\Delta''^2 = \Delta''^3 = 0$ also, in this coordinate system the events $P$ and $Q$ are at the same place and differ only in their time (hence the description

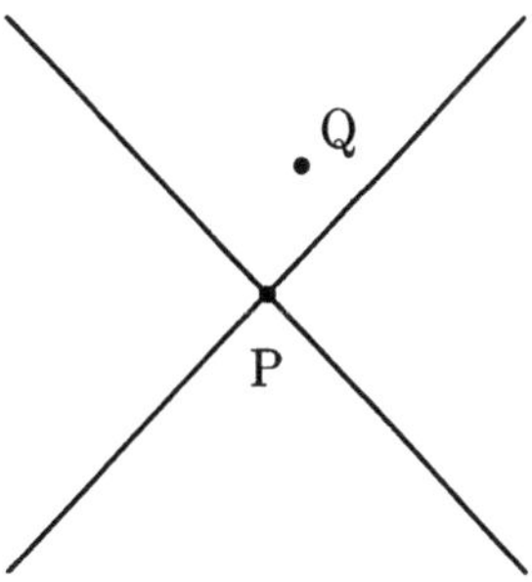


**Figure 1.22**

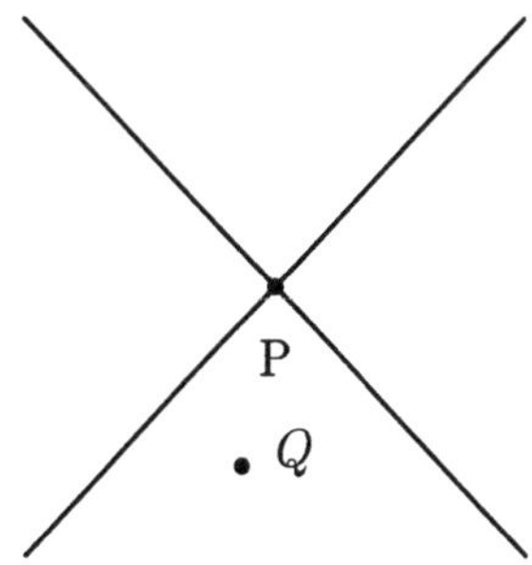


**Figure 1.23**

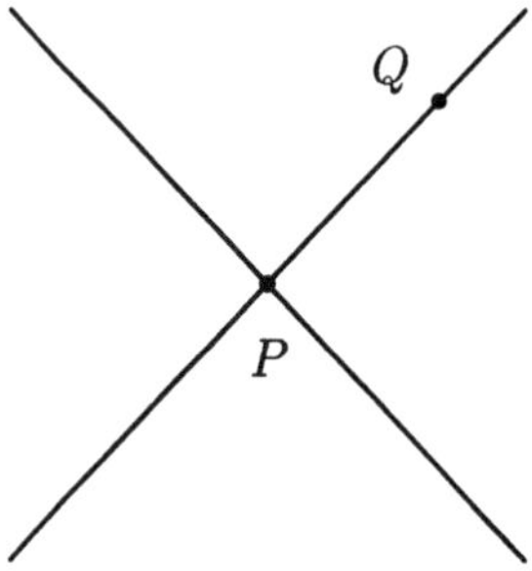


Figure 1.24

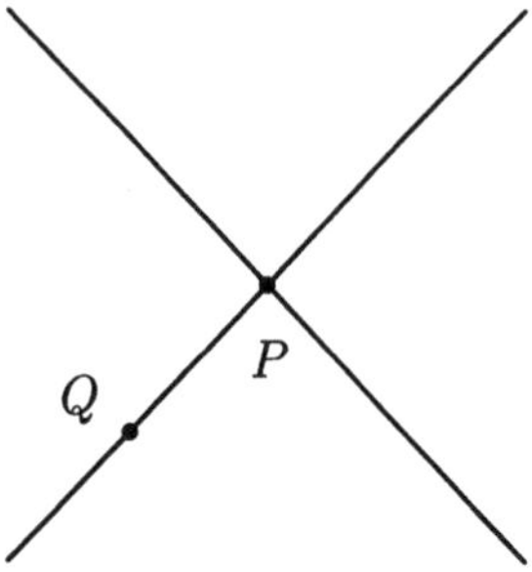


Figure 1.25

"timelike displacement"). The points $P$ and $Q$ will be at the same place in any inertial frame which is reached from the last one by a rotation; there is a 3-d continuum of such frames.

If $\eta_{\mu\nu}\Delta^\mu\Delta^\nu = 0$, we say that $Q$ is lightlike displaced with respect to $P$.

If $\text{sign}(\Delta^0) > 0$ (Figure 1.24), then $Q$ is FPLL (future-pointing lightlike) with respect to $P$.

If $\text{sign}(\Delta^0) < 0$ (Figure 1.25), then $Q$ is PPLL with respect to $P$.

No frames exist in which $P$ and $Q$ are either at the same place or at the same time. It is possible, however, to send a light ray either from $P$ to $Q$ or vice versa (only one of these alternatives is physically possible, the one in which the end point of the light ray is future-pointing with respect to the source).

If $\eta_{\mu\nu}\Delta^\mu\Delta^\nu > 0$, then $Q$ is spacelike displaced with respect to $P$. $Q$ is SL, spacelike, with respect to $P$ (Figure 1.26). A frame exists (and so many exist) in which events $P$ and $Q$ are simultaneous, and consequently have the same time coordinates. In some other frames $Q$ will be later than $P$ and in some it will be earlier.

Finally, if $\Delta^\mu = 0$, (in the previous cases we have tacitly assumed that not all $\Delta^\mu$ vanish) we have $P = Q$.

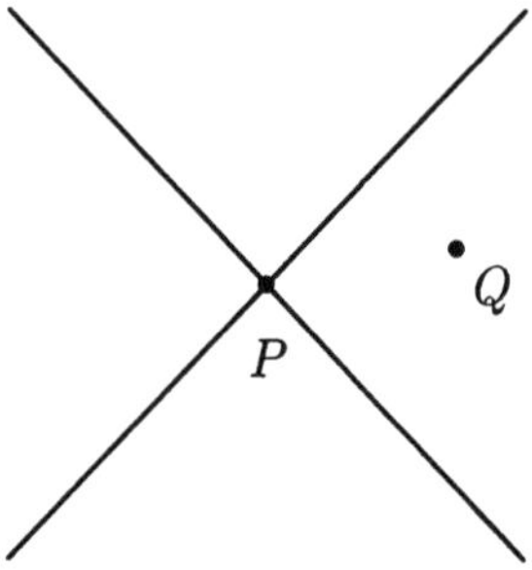


Figure 1.26

## 1.9 Poincaré Group

Having fixed a time orientation, a space orientation (RHS) and units, we have a class of inertial frames the members of which we can imagine continuously deforming from one to another, rather like the different positions of a rigid body in Newtonian mechanics. But, for many reasons, this cannot be a *physical* process of deformation. What is true mathematically is that the transformations connecting the coordinates of the different frames may be continuously changed by continuously changing the independent parameters. The collection of transformations $\mathcal{R}^4 \to \mathcal{R}^4$ appropriate to convert the coordinates of any one frame to those of an arbitrary frame,

$$(1) \qquad x'^{\mu} = a^{\mu}{}_{\nu} x^{\nu} + a^{\mu},$$

$$(2) \qquad a^{\mu}{}_{\alpha} a^{\nu}{}_{\beta} \eta_{\mu\nu} = \eta_{\alpha\beta},$$

$$(3) \qquad a^{0}{}_{0} > 0, \quad \det(a^{\mu}{}_{\nu}) = 1,$$

form a group under composition, the proper orthochronous Poincaré group, whose elements are continuously connected to the identity $x'^{\mu} = x^{\mu}$. (The continuous connection would no longer be true if we included among the transformations any reflections, which would change the time orientation or the handedness of the spatial triad.) The verification that the composition of two Poincaré transformations is again a transformation of the same type is an excellent exercise in index manipulation, and is left as a problem.

Equation (1) contains 4 + 16 parameters $a^{\mu}$ and $a^{\mu}{}_{\nu}$, but because of (2) they are not all independent. There are 16 equations in (2), of which only 10 are independent in view of the symmetry $\eta_{\mu\nu} = \eta_{\nu\mu}$. The conditions (3) do not reduce further the number of independent parameters (recall that (2) implies $\det(a^{\mu}{}_{\nu}) = \pm 1$). Equation (1), therefore, has $4 + (16 - 10) = 4 + 6 = 10$ independent parameters. The 4 refers to the four $a^{\mu}$, the translations in four coordinate directions. The 6 is most easily understood as 3 + 3, three parameters for a rotation (for example, the three Euler angles), and three components of a relative velocity.

### 1.9.1 Subgroup of Translations

When we set $a^{\mu}{}_{\nu} = \delta^{\mu}_{\nu}$ we get

$$x'^{\mu} = x^{\mu} + a^{\mu}, \tag{1.24}$$

representing the subgroup of translations. The frames $K$ and $K'$ are at rest with respect to each other and their axes are parallel. They differ in their choices of spacetime origin and consequently their assignments of coordinates differ by the four constant numbers $a^{\mu}$ (Figure 1.27).

Four of the ten parameters required to specify a particular Poincaré transformation are accounted for by the possibility of having the spacetime origins of different frames at different points.

There remain six parameters and to explain them it is sufficient to think of frames that share a common spacetime origin, so $x'^\mu = a^\mu{}_\nu x^\nu$.

### 1.9.2 Rotation Subgroup

If $K$ and $K'$ are at rest with respect to one another, and share a common spacetime origin, they can differ only in the direction of their spatial axes. In this case the general transformation

$$x'^\mu = a^\mu{}_\nu x^\nu \quad \text{with} \quad a^0{}_m = a^m_0 = 0, \;\; a^0{}_0 = 1$$

becomes

$$x'^0 = x^0, \qquad x'^m = a^m_n x^n.$$

The Lorentz condition, eq. (2), is satisfied if the $3 \times 3$ spatial submatrix of $a^\mu{}_\nu$ is a rotation matrix, the nontrivial condition on the third line:

$$\begin{aligned}
\alpha = \beta = 0: \qquad -1 &= a^\mu{}_0 a^\nu{}_0 \eta_{\mu\nu} = a^0{}_0 a^0{}_0(-1) = -1 \\
\alpha = 0,\; \beta = b \neq 0: \qquad 0 &= a^\mu{}_0 a^\nu{}_b \eta_{\mu\nu} = a^0{}_0 a^0{}_b(-1) = 0 \\
\alpha = a \neq 0,\; \beta = b \neq 0: \qquad \delta_{ab} &= a^\mu{}_a a^\nu{}_b \eta_{\mu\nu} = a^m_a a^n_b \delta_{mn}
\end{aligned}$$

The condition that the submatrix is a rotation matrix is $\delta_{ab} = a^m_a a^m_b$ or, bearing in mind the form of the 0-column and 0-row, $1 = a^T a$.

We know how rotations in three dimensions work and we know that three parameters are needed to describe them (for example, the Euler angles).

Hence three of the remaining six parameters represent the possibility of different frames having relatively rotated (*not rotating!*) axes. Only three parameters remain to be explained.

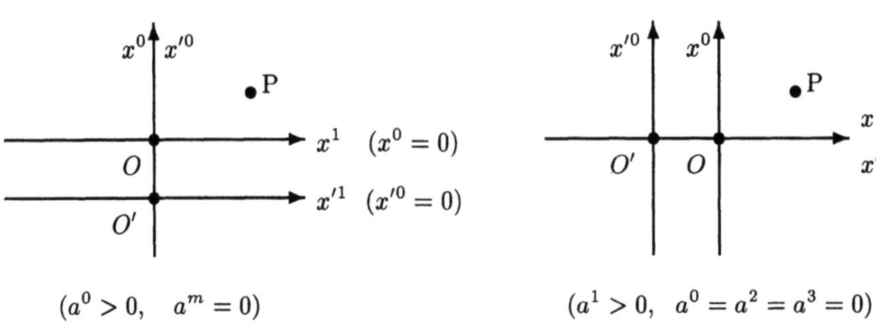


**Figure 1.27**

### 1.9.3 Boosts Do Not Form a Subgroup

The boosts are transformations of a special type which relate frames in relative motion. The simplest example has the form of (1.22). The characteristic features are the specific linear relation between time with one spatial variable, and the fact that two spatial directions are left invariant. It is worth verifying the Lorentz condition by writing down the matrix $a = (a^{\mu}{}_{\nu})$ for (1.22) and checking that $a^T \eta a = \eta$.

Any proper orthochronous Lorentz transformation may be expressed as the product of a rotation and a boost. Thus for many purposes a knowledge of rotations and boosts individually is sufficient.

Since $x'^{\mu} = 0$ when $x^{\mu} = 0$ the boosts relate frames which share a spacetime origin. The relation between the frames is almost completely determined by the motion of the spatial origin of each frame with respect to the other.

The spatial origin for $K'$ is $x'^m = 0$, which in $K$ is given by $x^1 = Vx^0$, $x^2 = x^3 = 0$, so it moves along the $x$-axis in $K$ at speed $|V|$ (Figure 1.28) ($V$ may be positive or negative but pictures are usually drawn for the $V > 0$ case).

Because $x'^2 = x^2$, $x'^3 = x^3$ there is a temptation to say that the frames $K$ and $K'$ are "parallel". From the spacetime point of view this would be misleading. The 2- and 3- axes for the two frames are parallel but the 1-axes are not. (Recall the diagram on page 27.) Yet this situation is very important because the standard boost, (1.22), appears in the decomposition of almost every nontrivial Lorentz transformation. When we need to refer to the geometrical configuration of the standard boost we may call the axes "pseudo-parallel".

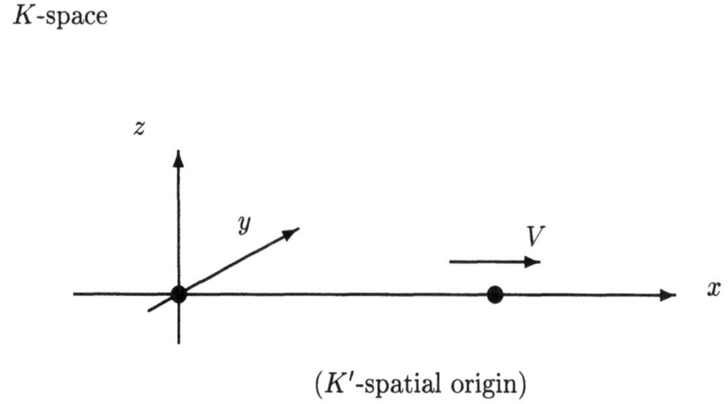


**Figure 1.28**

If we follow a boost in the $x^1$-direction with one in the $x^2$-direction (say), we do not get a boost as the composite result, but a boost together with a rotation, the *Thomas rotation*. This fact was only discovered in the 1920s and was said to have surprised Einstein. One consequence is that the collection of boosts does not constitute a subgroup of the Poincaré group. (Of course, there is a simple, one-parameter subgroup of boosts, those in which the subjective direction of the boosts does not change.)

Although the boosts do not form a subgroup they do account for the remaining three parameters. These can be used to fix the relative velocity in any direction of the two relatively moving frames.

## 1.10 Physical Spacetime Diagrams

In this section we return to the physics which takes place in spacetime. The sketches illustrate some of the physical effects that can be described and explained in later chapters.

*Freely Moving Particles With Mass*

Any particle moving freely (subject to no forces, so having a straight world-line) through $O$ lies within the lightcone with vertex at $O$ (Figure 1.29). This is a representation of the fact that with respect to any frame its speed must be less than one.

*Scattering of Particles*

Scattering experiments of elementary particles provide almost all the information we have about atomic and nuclear structure. The proper theory of such experiments is quantum mechanical, yet classical models are an essential

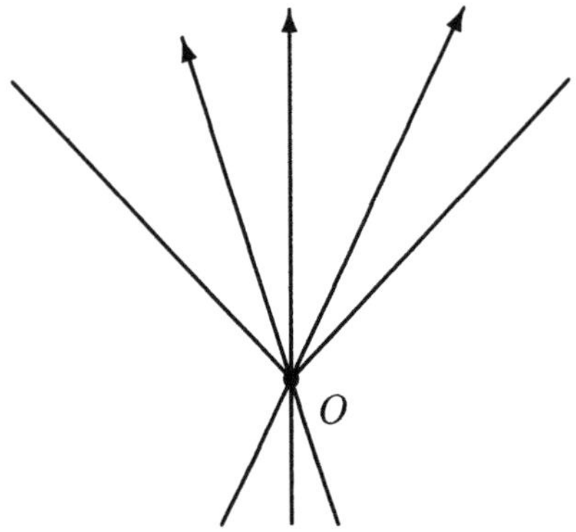


**Figure 1.29**

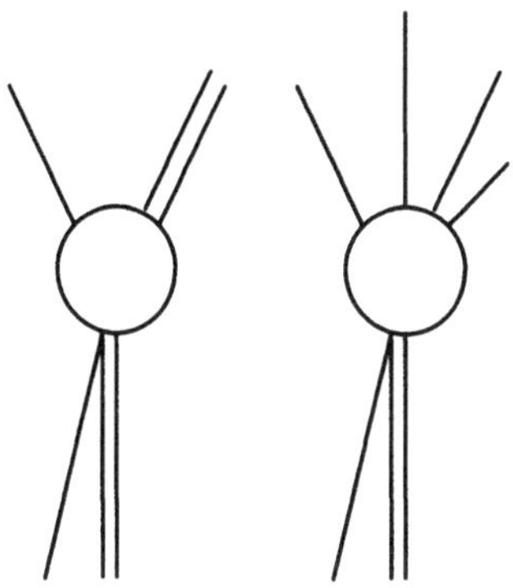

Figure 1.30

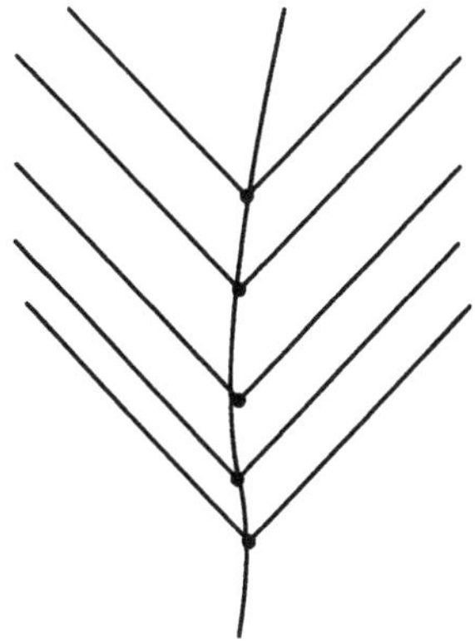

Figure 1.31

aid. In Figure 1.30, the elastic scattering on the left, with the same particles after scattering as before, might be electrons striking protons. In an inelastic scattering, as on the right, new particles may be created. No attempt is made in the classical theory to model the interaction region, which is therefore veiled from view with an opaque circle. Outside this region it is supposed that the particles travel freely.

*An Accelerating Charged Particle Radiates*
The worldline of an accelerating point particle lies within the lightcone emanating from any point on the worldline. If the particle is charged it emits radiation which itself may be illustrated by the succession of lightcones (Figure 1.31). The diagram is meant to suggest the smooth and continuous radiation which accompanies acceleration.

*A Rod At Rest in an Inertial Frame*
We have a sheet of worldlines for the particles constituting the rod, all par-

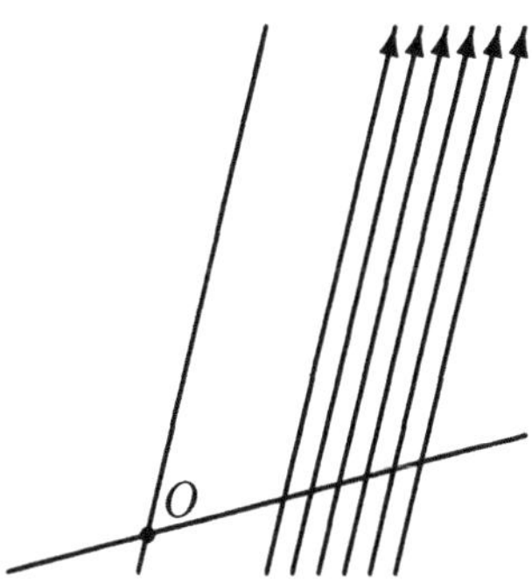


Figure 1.32

allel to the worldline of the master clock for the frame in which the rod is at rest (Figure 1.32).

This example illustrates how rigid bodies may be described in special relativity. In fact the concept of a rigid body is very limited. It makes sense only for those bodies which have a rest frame. The worldlines for the individual particles are then parallel, and with respect to any inertial frame the body moves uniformly. Rigidity is impossible for an accelerating body; the two concepts are incompatible.

## 1.11 Problems

1. Let $l = \{l^\mu\}$ and $m = \{m^\mu\}$ be two sets of four numbers, not all zero, and write $l.m = l^\mu m^\nu \eta_{\mu\nu}$. We call $l$ spacelike, timelike or lightlike according to whether $l.l$ is positive, negative or zero. When $l.l$ is not positive, we say $l$ is future-pointing if $l^0 > 0$, past-pointing if $l^0 < 0$.
   a) If $l$ is timelike and $l.m = 0$, then $m$ is spacelike.
   b) If $l$ and $m$ are timelike, and $l.m > 0$, then either both are future-pointing, or both are past-pointing.
   c) If $l$ and $m$ are both lightlike, and $l.m = 0$, then $l = \lambda m$ (means $l^\mu = \lambda m^\mu$) for some number $\lambda$.
   d) If $l$ is lightlike and $l.m = 0$, then either $m$ is spacelike or $l = \lambda m$.
   e) If $l$ and $m$ are both future-pointing lightlike then $l + m$ (means $\{l^\mu + m^\mu\}$) is future-pointing timelike unless $l = \lambda m$ for some $\lambda$. Conversely, any timelike $n$ can be written $n = l + m$ for lightlike $l$ and $m$. The $l$ and $m$ are not unique. What is the nature of the freedom in their choice?
   f) If $l, m, n, p$ are mutually orthogonal (meaning $l.m = 0$, etc) and linearly independent (meaning $\alpha l + \beta m + \gamma n + \delta p = 0$ implies $\alpha = \beta = \gamma = \delta = 0$), then one of them is timelike, three are spacelike. (Note that 4 mutually orthogonal $l, m, n, p$ are not necessarily linearly independent since a lightlike $l$ is orthogonal to itself, but if none is lightlike linear independence follows.)

2. Two problems which embody significant results but which are mainly useful for exercising and consolidating the summation convention:
   a) If $x'^\mu = a^\mu{}_\nu x^\nu$ and $x''^\mu = b^\mu{}_\nu x'^\nu$ are both Lorentz transformations, verify that $x''^\mu = c^\mu{}_\nu x^\nu$ is too. Besides checking that the composite transformation satisfies the Lorentz condition, verify that if the original transformations are proper and orthochronous, the final one is also.

b) Show that the *inverse* of a Lorentz transformation is also a Lorentz transformation. Give the argument in matrix form, and also (for practice) in superscript-subscript form.

3. All the Lorentz transformations form a group. There are two sorts of simple one-parameter subgroups, each involving only 2 coordinates in the simplest cases. These are rotations

$$\begin{pmatrix} x' \\ y' \end{pmatrix} = R(\theta) \begin{pmatrix} x \\ y \end{pmatrix} = \begin{pmatrix} \cos\theta & \sin\theta \\ -\sin\theta & \cos\theta \end{pmatrix} \begin{pmatrix} x \\ y \end{pmatrix}$$

and boosts

$$\begin{pmatrix} t' \\ x' \end{pmatrix} = B(V) \begin{pmatrix} t \\ x \end{pmatrix} = \begin{pmatrix} \gamma & -V\gamma \\ -V\gamma & \gamma \end{pmatrix} \begin{pmatrix} t \\ x \end{pmatrix}.$$

The rotation connects the coordinates of a fixed point with respect to two frames, $K$ and $K'$, the axes of $K'$ rotated by $\theta$ anticlockwise about the $z(=z')$-axis with respect to those of $K$. For the boost, $K'$ is moving at V with respect to $K$ along the x-axis.

Show that the rotations satisfy

$$R(\theta')R(\theta) = R(\theta' + \theta),$$

whereas for the boosts

$$B(V')B(V) = B\left(\frac{V' + V}{1 + VV'}\right).$$

The latter is a version of the simplest addition of velocities formula (note that $V'$ is a velocity with respect to $K'$, not $K$). We see that the angle $\theta$ is a nice parameter for a 1-d subgroup but that V is not ("nice" meaning that the subgroup is isomorphic to the real numbers under addition).

Note that $R'(0) = \begin{pmatrix} 0 & 1 \\ -1 & 0 \end{pmatrix}$, $B'(0) = \begin{pmatrix} 0 & -1 \\ -1 & 0 \end{pmatrix}$ and show, using a power series for the exponentials, that

$$\exp\left[\theta \begin{pmatrix} 0 & 1 \\ 1 & 0 \end{pmatrix}\right] = R(\theta) = \begin{pmatrix} \cos\theta & \sin\theta \\ -\sin\theta & \cos\theta \end{pmatrix}$$

and

$$\exp\left[\psi \begin{pmatrix} 0 & -1 \\ -1 & 0 \end{pmatrix}\right] = \begin{pmatrix} \cosh\psi & -\sinh\psi \\ -\sinh\psi & \cosh\psi \end{pmatrix} = B(V)$$

if $V = \tanh\psi$. We see that $\psi$ ($\simeq$V for small V) is the appropriate nice parameter for the boosts since

$$B(\psi')B(\psi) = B(\psi' + \psi),$$

a formula which is almost obvious from the exponential form.

4. Note that if $a, b, c = ba, \ldots$ are Lorentz matrices, the corresponding matrices $(a^{-1})^T, (b^{-1})^T, (c^{-1})^T, \ldots$ multiply to the same pattern. Later it will be seen that the two sets of matrices determine the transformation laws for components of vectors and covectors respectively. (The rotations are a special case in which the two sets are exactly the same.)

5. Two billiard balls approach one another, each at speed $V$ with respect to some inertial frame $K$, along its $x$-axis. They collide and bounce back, again at speed $V$. Draw a spacetime diagram to show what happens. Include coordinate axes for $K$, and also for $K'$ moving at $V$ with respect to $K$ in its $+x$-direction.

6. $K$ and $K'$ are the standard boost-related inertial frames. In $K'$ there is a spherical surface, of radius $R$, at rest, centred on the spatial origin. Describe the surface with respect to $K$. (You might say what the intersection is for $t = 0$, and what happens as $t$ increases.)

7. Verify that the simplest boost transformation $x'^0 = \gamma(x^0 - Vx^1)$, $x'^1 = \gamma(x^1 - Vx^0)$, $x'^2 = x^2$, $x'^3 = x^3$ satisfies $a^\mu{}_\alpha a^\nu{}_\beta \eta_{\mu\nu} = \eta_{\alpha\beta}$ where $x'^\mu = a^\mu{}_\nu x^\nu$. Write down the inverse transformation and confirm that it satisfies the condition too.

8. Show that for a Lorentz transformation it is not possible that $a^\mu{}_\nu = f^\mu g_\nu$ (for some numbers $f^0, f^1, \ldots g_3$). (Hint: consider whether the 00, 11, 01, say, sectors of the Lorentz condition can be satisfied.)

9. Frames $K$ and $K'$ are related by a standard boost.
   a) At $t' = 0$ the lattice clocks of the $K'$-frame all emit a flash. Show that in $K$ the flashes are emitted on a plane orthogonal to the $x$-axis and travelling in the $+x$-direction at speed $\frac{1}{V}$.

   b) Show that there is a plane, travelling with speed $(\gamma - 1)/(\gamma V) = \sqrt{(\gamma - 1)/(\gamma + 1)}$ with respect to $K$, on which the time in $K'$ agrees with that in $K$.

10. a) An event $E$ is spacelike with respect to an event $O$. Show that in some frames of reference $E$ is earlier than $O$, while in some others it is later. (In some frames, of course, the events are simultaneous.)

    b) Suppose that two events occur at the same place in a certain inertial frame $K$, with a $K$-time separation $T$. Show that in any other frame their time separation is $\geq T$.

11. The coordinates of two frames $K$ and $K'$ are connected by the general Poincaré transformation $x'^\mu = a^\mu + a^\mu{}_\nu x^\nu$. Show that the speed with

respect to $K$ of the spatial origin for $K'$ equals the speed with respect to $K'$ of the spatial origin for $K$.

12. The coordinate differences, in two different inertial frames, between events $P$ and $Q(\neq P)$ are $\Delta^\mu$, and $\Delta'^\mu = a^\mu{}_\nu \Delta^\nu$. Show that $(a^0{}_0)^2 = 1 + a^0{}_n a^0{}_n$. For the cases $\Delta^\mu \Delta^\nu \eta_{\mu\nu} \leq 0$, show that $\Delta^0$ and $\Delta'^0$ have the same sign if $a^0{}_0 > 0$.

13. For each set of (real) coordinates $x^\mu$, define a $2 \times 2$ hermitian matrix
$$h(x^\mu) = \begin{pmatrix} x^0 + x^3 & x^1 - ix^2 \\ x^1 + ix^2 & x^0 - x^3 \end{pmatrix}.$$
Let $B$ be a complex $2 \times 2$ matrix with $\det B = 1$, and define a new collection of coordinates $x'^\mu$ by $Bh(x^\mu)B^+ = h(x'^\mu)$, where $B^+$ stands for the hermitian conjugate (transpose and complex conjugate) of $B$. Show that the relation between $x^\mu$ and $x'^\mu$ is of the form $x'^\mu = b^\mu{}_\nu x^\nu$, and that it is a Lorentz transformation. Verify that the Lorentz transformation is proper ($\det = 1$) and orthochronous ($b^0{}_0 > 0$). What extra condition on the matrix $B$ reduces the transformation to a spatial rotation? Note that each $B$ produces a unique Lorentz transformation $\Lambda$, but that the same $\Lambda$ is produced by both $B$ and $-B$. If $B_1$ produces $\Lambda_1$, and $B_2$ produces $\Lambda_2$, then $B_1 B_2$ produces $\Lambda_1 \Lambda_2$. The matrices $B$ provide a double-valued representation of the restricted Lorentz group.

14. A very useful characterisation of Lorentz transformations is given by the following: If $x'^\mu = a^\mu{}_\nu x^\nu$, and $\eta_{\mu\nu} x'^\mu x'^\nu = \eta_{\mu\nu} x^\mu x^\nu$ for all $x^\mu$, then $a^\mu{}_\nu$ satisfies the Lorentz relation $a^\mu{}_\alpha a^\nu{}_\beta \eta_{\mu\nu} = \eta_{\alpha\beta}$.

15. The coordinates of events in spacetime with respect to two inertial frames of reference $K$ and $K'$ are related by
$$\begin{aligned} x' &= x \\ y' &= y + a(t - z) \\ z' &= z + ay + (a^2/2)(t - z) \\ t' &= t + ay + (a^2/2)(t - z), \end{aligned}$$
where $a$ is a real number. Verify that we really do have a Lorentz transformation, and that $K$ and $K'$ have the same time orientation, same space orientation same spacetime origin. Find the direction and speed, both with respect to $K$, of the $K'$-spatial origin. (Those with more time available may wish to relate $K$ and $K'$ with a boost and rotation(s).) Check that for the inverse transformation we need only replace $a$ with $-a$. Verify that the set of transformations with all values of $a$ constitutes a one-parameter subgroup of Lorentz transformations, with (nice) parameter $a$.

16. For what values of $u$ is

$$(a^{\mu}{}_{\nu}) = \begin{pmatrix} A & B & 0 & 0 \\ B & A & 0 & 0 \\ 0 & 0 & 1 & 0 \\ 0 & 0 & 0 & 1 \end{pmatrix},$$

with $A \equiv (u^2+1)/2u$ and $B = (u^2-1)/2u$, a Lorentz matrix? Of the allowed values, for which is the matrix proper? For which is it orthochronous?

17. For what values of the parameter $w$ is the transformation

$$t' + x' = w(t+x), \quad t' - x' = \frac{1}{w}(t-x), \quad y' = y, \quad z' = z$$

an orthochronous, proper Lorentz transformation? For an acceptable value of $w$ find an expression for the speed of the frame $K'$ with respect to $K$.

18. A completely general Poincaré transformation ($\mathcal{R}^4 \to \mathcal{R}^4$) has the form $x^{\mu} \mapsto a^{\mu}{}_{\nu}x^{\nu} + a^{\mu}$. It may be used passively, i.e. $x^{\mu} \mapsto x'^{\mu}$ gives the $K'$-coordinates for a fixed point $P$ in spacetime in terms of the $K$-coordinates, or it may be used actively, i.e. $x^{\mu} \mapsto y^{\mu}$ gives the $K$-coordinates of a new point $Q$ in terms of the $K$-coordinates of a point $P$. For each use one can describe the transformation physically when it is near the identity, that is, $a^{\mu} = \alpha^{\mu}$, $a^{\mu}{}_{\nu} = \delta^{\mu}_{\nu} + \alpha^{\mu}{}_{\nu}$ where $\alpha^{\mu}$ and $\alpha^{\mu}{}_{\nu}$ are small. Show that the Lorentz condition becomes, to lowest order in small quantities,

$$\alpha^{\mu\nu} + \alpha^{\nu\mu} = 0, \quad \text{where} \quad \alpha^{\mu\nu} \equiv \eta^{\nu\lambda}\alpha^{\mu}{}_{\lambda}.$$

Define cases (a) $\alpha^0 \neq 0$, (b) $\alpha^1 \neq 0$, (c) $\alpha^{01}(= -\alpha^{10}) \neq 0$, (d) $\alpha^{12}(= -\alpha^{21}) \neq 0$ (each time the unmentioned $\alpha$s vanish). For the passive use describe the relation between the frames $K$ and $K'$ so clearly that no one could be in any doubt about it (say exactly how $K'$ is reached from $K$; draw pictures when the nonzero parameter is positive). For the active use describe clearly the relation between the points $P$ ($K$-coordinates $x^{\mu}$) and $Q$ ($K$-coordinates $y^{\mu}$).

19. a) If $V > 0$, then $\gamma V > \gamma - 1$.
    b) If $0 < U, V < 1$, then $\frac{U+V}{1+UV} > U$ (and also $> V$ of course).
    c) If the moduli of the speeds $V$ and $V'$ are both less than one, show that the modulus of $(V+V')/(1+VV')$ is less than one too.

20. Suppose that $x^{\mu}$ and $x'^{\mu}$ are two sets of coordinates in Minkowski space, but *do not* assume that they are inertial coordinates. Show that if they satisfy

$$\frac{\partial x'^{\mu}}{\partial x^{\rho}}\frac{\partial x'^{\nu}}{\partial x^{\tau}}\eta_{\mu\nu} = \eta_{\rho\tau},$$

then $\frac{\partial^2 x'^{\mu}}{\partial x^{\rho}\partial x^{\omega}} = 0$, so $x'^{\mu} = a^{\mu} + a^{\mu}{}_{\nu}x^{\nu}$ (with constant $a^{\mu}$ and $a^{\mu}{}_{\nu}$). Hint: to start off, differentiate the given relation with respect to $x^{\omega}$, then combine with two permutations of $\omega, \rho, \tau$.

21. Suppose that $x'^{\mu} = L^{\mu}{}_{\nu}x^{\nu}$ is an orthochronous Lorentz transformation. Show that if

$$B^{\mu}{}_{0} \equiv L^{0}{}_{\mu}, \quad B^{0}{}_{m} \equiv L^{0}{}_{m}, \quad B^{m}{}_{n} \equiv \delta^{m}_{n} + \frac{L^{0}{}_{m}L^{0}{}_{n}}{L^{0}{}_{0}+1},$$

then $x''^{\mu} = B^{\mu}{}_{\nu}x^{\nu}$ is also an orthochronous Lorentz transformation; in fact, it is just such a transformation so $x'^{\mu} = R^{\mu}{}_{\nu}x''^{\nu}$ is a purely spatial rotation. Then the original transformation is decomposed into a boost and a rotation: $L^{\mu}{}_{\nu} = R^{\mu}{}_{\lambda}B^{\lambda}{}_{\nu}$.

22. If the real numbers $\Omega > 0,\ M > 0,\ K_n$ satisfy $\Omega^2 = M^2 + K_nK_n$, show that $x'^{\mu} = L^{\mu}{}_{\nu}x^{\nu}$ is an orthochronous Lorentz transformation, where $L^{0}{}_{0} = \Omega/M,\ \ L^{0}{}_{n} = L^{n}{}_{0} = -K_n/M$, and

$$L^{m}{}_{n} = \delta_{mn} + \frac{K_mK_n}{M(M+\Omega)}.$$

What is the inverse transformation?

23. Express the Poincaré transformation $x'^{\mu} = a^{\mu}{}_{\nu}x^{\nu} + a^{\mu}$ as a matrix transformation in a 5-dimensional space with column vectors of the form $x = \mathrm{col}\{x^0, x^1, x^2, x^3, 1\}$. Check that the $5\times 5$ matrix has unit determinant and find its inverse. Check that this gives the inverse Poincaré transformation. Suppose we regard a function $\phi(x)$ on the hyperplane $x^4 = 1$ of the 5-space as a function $\phi(x^{\mu})$ in Minkowski space. For each Poincaré transformation $\{a^{\mu},\ a^{\mu}{}_{\nu}\}$ and function $\phi(x)$, define $\phi_a(x) = \phi(a^{-1}x)$, where $a$ is the $5\times 5$ matrix. Show that this gives a representation of the Poincaré group.

## References

David Bohm, The Special Theory of Relativity (Routledge, 1965,1996)

Jürgen Ehlers, The Nature and Structure of Spacetime, in J. Mehra (ed.), The Physicist's Conception of Nature (D. Reidel, Dordrecht-Holland, 1973)

G.C. Hegerfeldt, The Lorentz Transformations: Derivation of Linearity and Scale Factor, Nuovo Cimento **10A** (1972) 257-267

W. Kopczynski and A. Trautman, Spacetime and Gravitation (Wiley, 1992)

H.A. Lorentz, A. Einstein, H. Minkowski, H. Weyl, The Principle of

Relativity (Dover Publications reprint, 1952, of Methuen and Co., 1923)
R.K. Sachs and H. Wu, General Relativity for Mathematicians (Springer, 1977)
E.P. Wigner, Unitary Representations of Lorentz Group, Annals of Mathematics **40** (1939) 149-204
E.C. Zeeman, Causality Implies the Lorentz Group, J. Math. Physics **5** (1964) 490-3

# 2. Vectors in Spacetime

*Geometry and Kinematics*

The simplest, yet most striking, effects in special relativity are consequences of elementary four-dimensional geometry. Kinematics is geometry with a temporal dimension. Vectors in four-dimensional spacetime allow us to express the geometry in its simplest form. As part of the algebraic structure there is also a coordinate-independent scalar product which reproduces the spacetime interval.

Vectors are the simplest geometrical objects (apart from numerical-valued functions) which arise in spacetime. They are at the heart of all further applications of special relativity. Many physical concepts, for example, particle momentum, are directly expressed mathematically as vectors. But also, as is developed in later chapters, more complicated geometrical objects, dyadics and tensors, are defined in terms of vectors. They too are essential in the construction of physical theories. (For example, the electromagnetic field is most usefully regarded as a dyadic.)

In the present chapter, with the help of vectors, we begin to study the description of physical phenomena in spacetime by looking at the very simplest cases. Many more elaborate applications may be constructed from these "building blocks".

- Straight timelike worldlines represent the tracks of free particles. The simplest instance of a *rest* frame occurs here: any inertial frame with respect to which the particle is stationary. Time in such a frame serves as *proper time* for the particle. We must discuss the spacetime velocity for a particle and its *relative* velocity with respect to an inertial frame.

  A free particle by definition is noninteracting. If the particle does interact, for example, if it is charged and in the presence of an electromagnetic field, it accelerates and its worldline is curved. The concept of a rest frame is now an instantaneous one because it changes from moment to moment as the particle accelerates. For the same reason the concept of proper time becomes differential. These generalisations are simple because the worldline is a one-dimensional curve, which is easy to describe and work with mathematically.

- Straight spacelike lines represent the instantaneous position of rods. To be considered rigid, such rods must have a rest frame. Otherwise, if they accelerate, they cannot be considered to be rigid[1], and some model of continuum mechanics must be constructed. (This will not be contemplated here.)
- Straight lightlike lines represent the tracks of light rays. Such rays give an adequate approximation for some electromagnetic phenomena. The full theory of electromagnetism, a field theory, will be developed later.

  Light rays are geometrical structures in spacetime. We must learn to transfer a description with respect to one frame to a description with respect to another, including associated frequencies. A scalar model of light allows us to discuss plane waves without dealing with the complications of polarization.

The study of vectors in special relativity begins with translation vectors in spacetime, defined here as a class of special mappings in $\mathcal{M}$. Applications demand a very much wider concept of vectors, a general vector space $\mathcal{V}$ which is naturally (independently of coordinates) and physically (physically meaningfully) related to the individual spaces of translation vectors, momentum vectors, and so on.

## 2.1 Translation Vectors in Spacetime

Our model for spacetime is a four-dimensional continuum $\mathcal{M}$ consisting of points $P, Q, \ldots$ We have inertial frames of reference $K,\ K', \ldots$ as discussed in the previous chapter. These give labels $x^\mu_P,\ x'^\mu_P, \ldots$ in $R^4$ to each event $P$. The labels are related by the Poincaré transformations

$$x'^\mu_P = a^\mu + a^\mu{}_\nu x^\nu_P$$

in terms of constants $a^\mu$ and $a^\mu{}_\nu$ (which depend on the relation of $K$ and $K'$, but not on the point $P$).

We now want to define *translation vectors.* These are specific mappings of $\mathcal{M}$ onto itself

$$\boldsymbol{A} : \mathcal{M} \to \mathcal{M}, \quad \boldsymbol{A} : P \mapsto Q. \tag{2.1}$$

[1] A much studied concept is Born-rigidity, an attempt at a relativistic replacement for Newtonian rigidity. According to Born-rigidity, nearby particles in a body must maintain a constant separation in their rest frame. However, the number of degrees of freedom for a Born-rigid motion is not six but only three, and even for the solutions that do exist, Born-rigid bodies may have different shapes with respect to different inertial frames. Thus little of the nonrelativistic conception remains. There is more discussion and some nontrivial analysis in the books of Rindler and of Synge.

By no means every mapping of $\mathcal{M}$ onto itself defines a vector. They are highly restricted by a very close relation to the inertial frames of reference. The relation is with the whole class of inertial frames, which provides a concept of parallel lines in spacetime.

To explain exactly what a vector is we choose some inertial frame $K$ (it does not matter which one). The effect of a particular vector $\boldsymbol{A}$ on some point $P$, with $K$-coordinates $x_P^\mu$, is to map it to a point $Q$, with $K$-coordinates $x_Q^\mu = A^\mu + x_P^\mu$. The numbers $A^\mu$, the components of $\boldsymbol{A}$ with respect to $K$, depend on $K$ but not on $P$. We have, for each point $P$,

$$\begin{aligned} \boldsymbol{A} \quad : \quad & P(\text{with K-coordinates } x_P^\mu) \\ & \mapsto Q(\text{with K-coordinates } x_Q^\mu = x_P^\mu + A^\mu). \end{aligned} \tag{2.2}$$

The $K'$-coordinates, or any others, could equally well be used to express the same mapping between the points. Transcribing the specification (2.2) into $K'$-coordinates by using the Poincaré transformation gives

$$\begin{aligned} \boldsymbol{A} \quad : \quad & P(\text{with K}'\text{-coordinates } x_P'^\mu = a^\mu + a^\mu{}_\nu x_P^\nu) \\ & \mapsto Q(\text{ with K}'\text{-coordinates } x_Q'^\mu = x_P'^\mu + a^\mu{}_\nu A^\nu). \end{aligned}$$

This shows that the components of $\boldsymbol{A}$ with respect to $K'$ are given in terms of the $K$-components by

$$A'^\mu = a^\mu{}_\nu A^\nu. \tag{2.3}$$

If we had started our specification of the vector $\boldsymbol{A}$ by choosing the frame $K'$ we would have used the numbers $A'^\mu$ as its $K'$-components. The transformation law for vector components, (2.3), is derived by looking at the coordinate-independent definition of the vector from two, coordinate-related points of view.

***Every translation vector has the form just described.***

Every $P$ is mapped into some $Q$ (Figure 2.1). For each $Q$ there is precisely one $P$ which is mapped to it. All the displacements $\overrightarrow{PQ}$ are parallel because inertial frames were used in the definition. It is important to appreciate that because vectors $\boldsymbol{A}, \ldots$ are defined as mappings $\mathcal{M} \to \mathcal{M}$ they are *not* tied to any particular frame of reference. The vectors are *geometrical*, or absolute. The components $A^\mu$, $A'^\mu, \ldots$ provide a *relative* description of them.

### 2.1.1 Vector Space

All possible vectors form a collection $\mathcal{V}$. We make $\mathcal{V}$ into a vector space by suitably defining $\boldsymbol{A} + \boldsymbol{B}$ and $\lambda \boldsymbol{A}$.

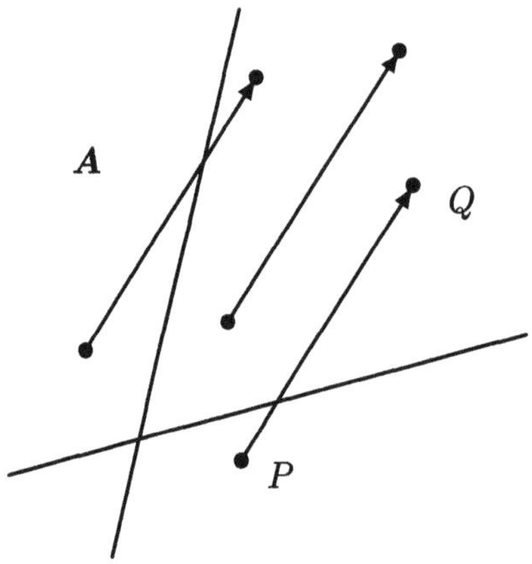


**Figure 2.1**

### 2.1.2 Addition

$$\begin{aligned} &\text{If} \quad \boldsymbol{A} : P \mapsto Q \\ &\text{and} \quad \boldsymbol{B} : Q \mapsto R, \end{aligned}$$

then the vector $\boldsymbol{A}+\boldsymbol{B}$ is defined by

$$\boldsymbol{A}+\boldsymbol{B} : P \mapsto R \qquad \text{(Figure 2.2)}.$$

Strictly, the vectors in the sum are ordered but it is easy to prove, using the components of the vectors with respect to some frame, that

$$\boldsymbol{A}+\boldsymbol{B} = \boldsymbol{B}+\boldsymbol{A},$$

and

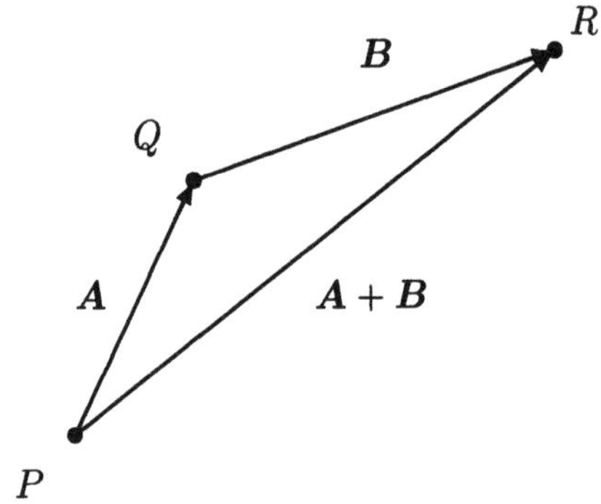


**Figure 2.2**

$$(\boldsymbol{A}+\boldsymbol{B})+\boldsymbol{C}=\boldsymbol{A}+(\boldsymbol{B}+\boldsymbol{C}).$$

With respect to a frame $K$, the components of the sum $\boldsymbol{A}+\boldsymbol{B}$ are $A^\mu + B^\mu$ in terms of the $K$-components of $\boldsymbol{A}$ and $\boldsymbol{B}$ individually.

### 2.1.3 Multiplication by a Scalar

To define $\lambda\boldsymbol{A}$ we again use components with respect to a frame $K$ (and again, it does not matter which one).

$$\begin{array}{rrl} \text{If} & \boldsymbol{A} & : \; P(x^\mu_P) \mapsto Q(x^\mu_Q = x^\mu_P + A^\mu) \\ \text{then} & \lambda\boldsymbol{A} & : \; P(x^\mu_P) \mapsto R(x^\mu_R = x^\mu_P + \lambda A^\mu). \end{array}$$

The $K$-components of $\lambda\boldsymbol{A}$ are $\lambda A^\mu$ (Figure 2.3). To see that it does not matter which frame is used, the $K'$-coordinates may be used to rewrite the previous definition.

$$\begin{array}{rl} \boldsymbol{A} & : P(x'^\mu_P = a^\mu + a^\mu{}_\nu x^\nu_P) \mapsto Q(x'^\mu_Q = a^\mu + a^\mu{}_\nu(x^\nu_P + A^\nu)) \\ \lambda\boldsymbol{A} & : P(x'^\mu_P) \mapsto R(x'^\mu_P + a^\mu{}_\nu(\lambda A^\nu)). \end{array}$$

We see that $R(x'^\mu_P + \lambda A'^\mu)$ may be written in terms of $A'^\mu$, the $K'$-components of $\boldsymbol{A}$, in the same way as $R(x^\mu_P + \lambda A^\mu)$ is written in terms of the $K$-components. The rule defining $\lambda\boldsymbol{A}$, although it uses a frame of reference, gives the same answer no matter which frame is used.

With the definitions of $\boldsymbol{A}+\boldsymbol{B}$ and $\lambda\boldsymbol{A}$ it is easy to check that the collection $\mathcal{V}$ now obeys all the axioms for a vector space. We can picture the whole collection by showing on a diagram for $\mathcal{M}$ the new points that one particular event would be mapped to by a selection of different vectors (Figure 2.4).

The picture is, of course, exactly the same for the effect of a collection of vectors on any other point. Spacetime is homogeneous. Yet we will come to regard such a picture as special in associating particular mappings in $\mathcal{M}$

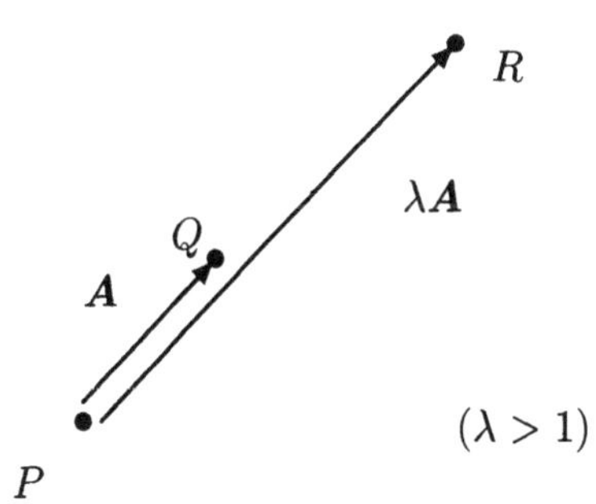


**Figure 2.3**

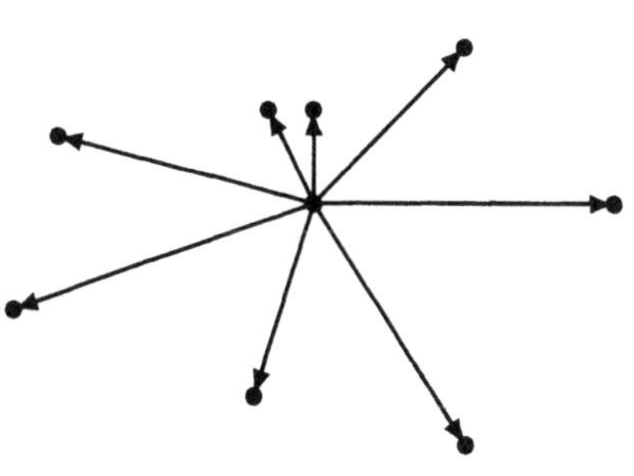

**Figure 2.4**

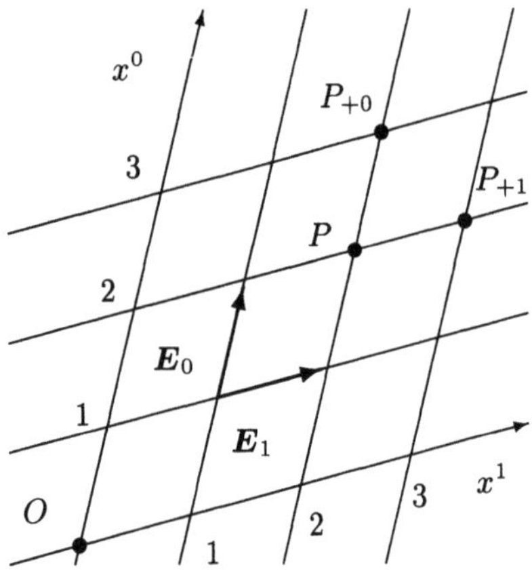


**Figure 2.5**

with vectors. We will want to consider the vectors in $\mathcal{V}$ as quite separate, free-standing mathematical objects, which can, however, be put in a 1-1 relation (an isomorphism) with the translation mappings. This generalisation will then give us a space of objects which can include other vectors, such as velocities, accelerations, momenta, electric fields, and so on.

### 2.1.4 Inertial Basis Vectors

Once we have learnt about addition of vectors and multiplication by scalars it is useful to introduce basis vectors. All other vectors can be written as linear combinations of them.

For each inertial frame $K$ we define four vectors $\boldsymbol{E}_0$, $\boldsymbol{E}_1$, $\boldsymbol{E}_2$, $\boldsymbol{E}_3$ (or, for short, $\boldsymbol{E}_\mu$ for the whole collection) by

$$\boldsymbol{E}_\nu : P(x^\mu_P) \mapsto P_{+\nu}(x^\mu_P + \delta^\mu_\nu), \tag{2.4}$$

where the points are recognised by their $K$-coordinates. The point $P_{+\nu}$ is reached from $P$ by a unit displacement in the $+x^\nu$-direction (Figure 2.5). For example,

$$\boldsymbol{E}_0 : P(x^0, x^1, x^2, x^3) \mapsto P_{+0}(x^0 + 1, x^1, x^2, x^3).$$

For any other inertial frame $K'$, the corresponding vectors $\boldsymbol{E}'_\nu$ are defined as unit $K'$-coordinate displacements in the corresponding directions along the $x'^\nu$-axes.

If $K'$ is another inertial frame such that $x'^\mu = x^\mu + a^\mu$, so $a^\mu{}_\nu = \delta^\mu_\nu$, and $K'$ differs from $K$ only in its choice of spacetime origin, not in its network of spatial and temporal coordinate lines, then $\boldsymbol{E}_\mu = \boldsymbol{E}'_\mu$. On the other hand, a completely general frame $K'$ from the full collection of frames is specified by a choice of spacetime origin $O'$ and an inertial basis $\boldsymbol{E}'_\mu$ for $\mathcal{V}$.

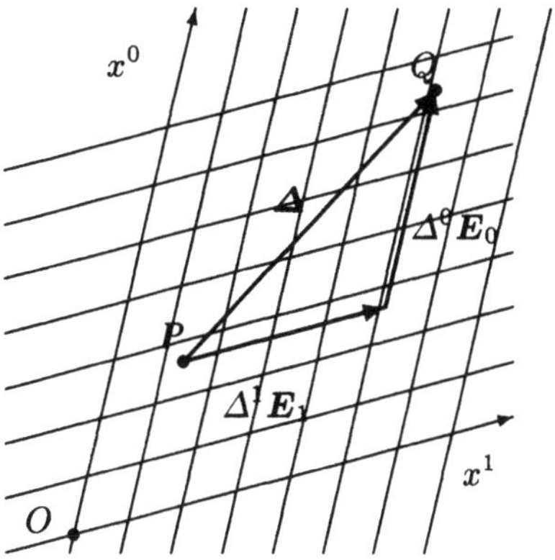


**Figure 2.6**

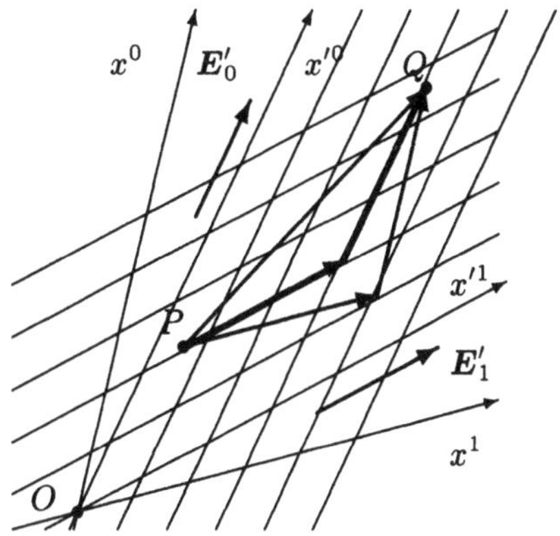


**Figure 2.7**

### 2.1.5 Decomposition

Once the $K$-basis vectors $\boldsymbol{E}_\mu$ are to hand, a general translation vector, say, $\boldsymbol{\Delta} = \overrightarrow{PQ}$, when it is written with $K$-coordinates and $K$-components $\Delta^\mu = x^\mu_Q - x^\mu_P$,

$$\boldsymbol{\Delta} : P(x^\mu_P) \mapsto Q(x^\mu_Q = x^\mu_P + \Delta^\mu)$$

may be decomposed as

$$\boldsymbol{\Delta} = \Delta^0\boldsymbol{E}_0 + \Delta^1\boldsymbol{E}_1 + \Delta^2\boldsymbol{E}_2 + \Delta^3\boldsymbol{E}_3 \equiv \Delta^\mu\boldsymbol{E}_\mu$$

by using vector addition and multiplication by scalars. The use of the summation convention is here extended to vectors (Figure 2.6). If the similarly defined basis vectors $\boldsymbol{E}'_\mu$ for a second general frame $K'$ are used to decompose $\boldsymbol{\Delta}$, then

$$\boldsymbol{\Delta} = \Delta'^\mu\boldsymbol{E}'_\mu = \Delta^\mu\boldsymbol{E}_\mu, \tag{2.5}$$

where the transformation of vector components,

$$\Delta'^\mu = x'^\mu_Q - x'^\mu_P = a^\mu{}_\nu(x^\nu_Q - x^\nu_P) = a^\mu{}_\nu\Delta^\nu, \tag{2.6}$$

an example of (2.3), comes from the transformation connecting the $K'$- and $K$-coordinates (Figure 2.7).

### 2.1.6 Transformation of Basis Vectors

For any vector $\boldsymbol{A}$, the components and basis vectors with respect to the frames $K$ and $K'$ are connected by the simple formula

$$\boldsymbol{A} = A'^\mu\boldsymbol{E}'_\mu = A^\mu\boldsymbol{E}_\mu, \tag{2.7}$$

which expresses the fact that a single vector may be decomposed with respect to any basis. The relation between the components (coordinate differences subject to Lorentz transformations, (2.6), for translation vectors) is

$$A'^{\mu} = a^{\mu}{}_{\nu} A^{\nu}. \tag{2.8}$$

Hence, since these formulae are true for all vectors, thus true for arbitrary components $A^{\mu}$, we may choose $A^{\mu} = \delta^{\mu}_{\nu}$, and deduce the transformation law for inertial basis vectors,

$$\boldsymbol{E}_{\nu} = a^{\mu}{}_{\nu} \boldsymbol{E}'_{\mu}. \tag{2.9}$$

Equation (2.9) may also be derived by expressing the definition (2.4) in $K'$-coordinates, and using vector algebra.

To get the relation in the opposite direction, $\boldsymbol{E}'_{\mu}$ in terms of $\boldsymbol{E}_{\nu}$, we need $a^{-1\mu}{}_{\nu}$, for which an expression may be derived from the Lorentz condition. It was given in (1.19), and results in the explicit formula

$$\boldsymbol{E}'_{\tau} = a^{-1\nu}{}_{\tau} \boldsymbol{E}_{\nu} = \eta^{\nu\rho} a^{\lambda}{}_{\rho} \eta_{\lambda\tau} \boldsymbol{E}_{\nu}. \tag{2.10}$$

The transformation law (2.10) is only superficially more complicated than (2.8), the law for vector components. If the transformation (2.8) is written as a matrix equation, with matrix $a$, then (2.10) is written with matrix $(a^{-1})^T$. When two Lorentz transformations, with matrices $b$ and $a$, are composed, then $c = b \cdot a$ and also $(b^{-1})^T \cdot (a^{-1})^T = (a^{-1} \cdot b^{-1})^T = ((ba)^{-1})^T$. We will find later that the transformation relation expressed in (2.10) is also that for the components of *covectors*, a different sort of geometrical object.

For the simple boost, (1.22), for which $x'^0 = \gamma(x^0 - Vx^1)$, $x'^1 = \gamma(x^1 - Vx^0)$, the transformation of any vector components is the same,

$$\left.\begin{array}{rcl} A'^0 &=& \gamma(A^0 - VA^1) \\ A'^1 &=& \gamma(A^1 - VA^0) \\ A'^2 &=& A^2, \quad A'^3 \;=\; A^3. \end{array}\right\} \tag{2.11}$$

while for the basis vectors the relations are

$$\left.\begin{array}{rcl} \boldsymbol{E}'_0 &=& \gamma(\boldsymbol{E}_0 + V\boldsymbol{E}_1) \\ \boldsymbol{E}'_1 &=& \gamma(\boldsymbol{E}_1 + V\boldsymbol{E}_0) \\ \boldsymbol{E}'_2 &=& \boldsymbol{E}_2, \quad \boldsymbol{E}'_3 \;=\; \boldsymbol{E}_3. \end{array}\right\} \tag{2.12}$$

Note the different signs in front of $V$ in the transformation formulae for components and basis vectors. The two transformations must compensate for each other to leave the two decompositions $\boldsymbol{A} = A'^{\mu}\boldsymbol{E}'_{\mu} = A^{\mu}\boldsymbol{E}_{\mu}$ valid with the same form. To verify (2.12) it is simplest just to check $A'^{\mu}\boldsymbol{E}'_{\mu} = A^{\mu}\boldsymbol{E}_{\mu}$ directly, rather than to use (2.10), a relation that is useful for general transformations but can be needlessly awkward for specific ones.

## 2.2 Scalar Product of Spacetime Vectors

The scalar product of Euclidean vectors in three-dimensional space is real, symmetric, bilinear and positive definite. Now we want to define a useful scalar product for the spacetime vectors in $\mathcal{V}$. It turns out that the positive definite property is inappropriate, but that there is a useful scalar product between $\boldsymbol{A}$ and $\boldsymbol{B}$ which is real, symmetric and bilinear:

$$\boldsymbol{A} \cdot \boldsymbol{B} = \boldsymbol{B} \cdot \boldsymbol{A} \in \mathcal{R} \text{ (the real numbers)} \tag{2.13}$$

such that

$$(\alpha \boldsymbol{A} + \beta \boldsymbol{B}) \cdot \boldsymbol{C} = \alpha(\boldsymbol{A} \cdot \boldsymbol{C}) + \beta(\boldsymbol{B} \cdot \boldsymbol{C}) = \boldsymbol{C} \cdot (\alpha \boldsymbol{A} + \beta \boldsymbol{B}). \tag{2.14}$$

In place of the positive-definite property (which means in Euclidean geometry that only the zero vector has a vanishing scalar product with itself) the new scalar product is *nondegenerate*: if $\boldsymbol{A} \cdot \boldsymbol{X} = 0$ for all vectors $\boldsymbol{A}$, then $\boldsymbol{X} = \boldsymbol{0}$. The new scalar product in spacetime is *not* positive definite: it transpires that all *lightlike* vectors (translations in lightlike directions) have zero scalar product with themselves.

Given a bilinear scalar product, when we decompose vectors with respect to the bases of frames $K$ and $K'$

$$\boldsymbol{A} = A^\mu \boldsymbol{E}_\mu = A'^\mu \boldsymbol{E}'_\mu, \qquad \boldsymbol{B} = B^\mu \boldsymbol{E}_\mu = B'^\mu \boldsymbol{E}'_\mu,$$

we have

$$\boldsymbol{A} \cdot \boldsymbol{B} = A^\mu B^\nu (\boldsymbol{E}_\mu \cdot \boldsymbol{E}_\nu) = A'^\mu B'^\nu (\boldsymbol{E}'_\mu \cdot \boldsymbol{E}'_\nu). \tag{2.15}$$

If the scalar product is to have a physical significance which reflects the requirement that the inertial frames are of equal status, that none is privileged, then (2.15) leads us to expect that it should satisfy $\boldsymbol{E}_\mu \cdot \boldsymbol{E}_\nu = \boldsymbol{E}'_\mu \cdot \boldsymbol{E}'_\nu$ (so the form with respect to both systems is the same, and the relative significance is the same).

In the search for a useful scalar product, an important clue is provided by the invariant interval. For two events $P$ and $Q$, and any two frames $K$ and $K'$, we have the invariant function (1.23) of coordinate differences in the two systems, $\Delta^\mu \Delta^\nu \eta_{\mu\nu} = \Delta'^\mu \Delta'^\nu \eta_{\mu\nu}$. Such an invariant signals something with absolute, or geometrical, significance. We can build the invariant interval into the geometry with the help of the scalar product. With $\boldsymbol{\Delta} \equiv \overrightarrow{PQ}$, we obtain the invariant interval from the geometrical scalar product, $\boldsymbol{\Delta} \cdot \boldsymbol{\Delta} = \Delta^\mu \Delta^\nu \eta_{\mu\nu}$, provided we define $\boldsymbol{E}_\mu \cdot \boldsymbol{E}_\nu \equiv \eta_{\mu\nu}$.

Therefore we are led to define, at first for the $K$-basis,

$$\boldsymbol{E}_\mu \cdot \boldsymbol{E}_\nu = \eta_{\mu\nu}. \tag{2.16}$$

Note that this means that for the spatial orthonormal basis vectors,

$$\boldsymbol{E}_m \cdot \boldsymbol{E}_n = \delta_{mn}, \tag{2.17}$$

exactly as in a three-dimensional Euclidean space. For $\boldsymbol{E}_0$, the basis vector in the time direction, we have $\boldsymbol{E}_0 \cdot \boldsymbol{E}_m = 0$, which seems uncontroversial, an expression of the orthogonality of time and space directions in a fixed frame. But (2.16) also gives

$$\boldsymbol{E}_0 \cdot \boldsymbol{E}_0 = -1. \tag{2.18}$$

The negative value for $\boldsymbol{E}_0 \cdot \boldsymbol{E}_0$ may seem at first unsettling, but this is only because of its unfamiliarity. It certainly makes a striking difference for the timelike basis vectors compared with the spacelike ones. Even more importantly, the negative value is essential if the scalar product is to give us a geometrical formulation of the invariant interval.

It is easy to verify that the scalar product of basis vectors has the same form in all inertial frames. Recalling the form (2.9), $\boldsymbol{E}_\mu = a^\nu{}_\mu \boldsymbol{E}'_\nu$, and assuming $\boldsymbol{E}'_\mu \cdot \boldsymbol{E}'_\nu = \eta_{\mu\nu}$ as the first choice (it does not matter which system is chosen first because any frame can be reached from any other by some Poincaré transformation),

$$\boldsymbol{E}_\mu \cdot \boldsymbol{E}_\nu = a^\alpha{}_\mu a^\beta{}_\nu \eta_{\alpha\beta} = \eta_{\mu\nu},$$

by the the Lorentz condition (1.12). The form of the scalar product, assumed for one inertial basis, therefore holds for all. As a consequence, the invariant interval for any displacement $\overrightarrow{PQ} = \boldsymbol{\Delta}$ is given in any coordinates by the expression $\Delta^\mu \Delta^\nu \eta_{\mu\nu} = \overrightarrow{PQ} \cdot \overrightarrow{PQ}$.

Having based the scalar product on the invariant form in terms of inertial frame basis vectors, we have by linearity the component form of the scalar product between any two vectors,

$$\boldsymbol{A} \cdot \boldsymbol{B} = A^\mu B^\nu \eta_{\mu\nu} = A'^\mu B'^\nu \eta_{\mu\nu}. \tag{2.19}$$

Because the scalar product in $\mathcal{V}$ is not positive definite, there is no general formula corresponding to the relation $\mathbf{a} \cdot \mathbf{b} = ab\cos\theta$ in Euclidean space. The indefinite scalar product effectively divides the space into different sectors and the geometrical meaning of the scalar product depends on which sectors the factors belong to. For vectors $\boldsymbol{C}$ which are spatial with respect to some inertial frame, so $\boldsymbol{C} \cdot \boldsymbol{E}_0 = 0$, this scalar product reproduces Euclidean geometry in the "space" of the frame, the hyperplanes of simultaneity for the frame. For *these* vectors one has $\boldsymbol{A} \cdot \boldsymbol{B} = AB\cos\theta$ with the usual Euclidean meanings in that frame.

If we have two vectors, $\boldsymbol{C}$ and $\boldsymbol{D}$, say, spatial with respect to a given frame, we can define a third spatial vector, $\boldsymbol{F}$, say, as the *vector product* of the first two exactly as it is done in three-dimensional vector algebra:

$$\boldsymbol{F} \equiv \boldsymbol{C} \times \boldsymbol{D} = \boldsymbol{E}_i \epsilon^{ijk} C^j D^k, \tag{2.20}$$

where $\boldsymbol{C} = C^j \boldsymbol{E}_j,\ \ \boldsymbol{D} = D^k \boldsymbol{E}_k$. This is an occasion on which we have spatial sums over $1, 2, 3$ indicated with repeated latin scripts either up or down. This vector product has all the properties of the three-dimensional vector product but it is essential to remember *which* frame defines the spatial vectors that appear. The concept is inherently *frame-dependent.*

The scalar product can be used to provide expressions for the components of vectors. The procedure is the same as in elementary Euclidean geometry but the expressions are slightly more complicated because of the sign difference for timelike directions. For any vector $\boldsymbol{A} = A^\mu \boldsymbol{E}_\mu$, when we take the scalar product with $\boldsymbol{E}_\nu$, we get

$$\boldsymbol{A} \cdot \boldsymbol{E}_\nu = A^\mu \boldsymbol{E}_\mu \cdot \boldsymbol{E}_\nu = A^\mu \eta_{\mu\nu}, \tag{2.21}$$

so

$$A^\mu = \eta^{\mu\nu} \boldsymbol{A} \cdot \boldsymbol{E}_\nu, \quad \text{and} \quad \boldsymbol{A} = \eta^{\mu\nu}(\boldsymbol{A} \cdot \boldsymbol{E}_\nu)\boldsymbol{E}_\mu, \tag{2.22}$$

using the inverse components $\eta^{\mu\nu}$ that appeared first in (1.17).

The scalar products of basis vectors have the same values, (2.16), for all inertial frames. For the basis vectors of two frames related by a boost we have the decomposition (2.12). Taking the scalar product with $\boldsymbol{E}_0$ and $\boldsymbol{E}_1$ gives the mixed products

$$\left.\begin{array}{lllllll} \boldsymbol{E}'_0 \cdot \boldsymbol{E}_0 & = & -\gamma & \qquad & \boldsymbol{E}'_0 \cdot \boldsymbol{E}_1 & = & V\gamma \\ \boldsymbol{E}'_1 \cdot \boldsymbol{E}_0 & = & -V\gamma & \qquad & \boldsymbol{E}'_1 \cdot \boldsymbol{E}_1 & = & \gamma. \end{array}\right\} \tag{2.23}$$

In practice these relations are very useful.

The scalar product between the two *timelike* unit basis vectors is general and has a physical expression: $\boldsymbol{E}'_0 \cdot \boldsymbol{E}_0 = -\gamma = -(1 - V^2)^{-\frac{1}{2}} = -\cosh\phi$, where $\tanh\phi = V$, the relative speed of the two frames.

## 2.3 Classification of Vectors

Spacetime $\mathcal{M}$ is a four-dimensional continuum and the translation vectors $\mathcal{V}$ represent displacements. But the different directions for displacements are not at all equivalent. The translation vectors can be separated, with the help of the scalar product, into 6 classes which have different physical significance. The classification of vectors is closely related to the invariant relations between points of the previous chapter. Any vector $\boldsymbol{\Delta}$ can be illustrated by the specific translation vector $\overrightarrow{OP}$, with fixed $O$ (Figure 2.8), for which

$$\boldsymbol{\Delta} = \overrightarrow{OP} = \Delta^\mu \boldsymbol{E}_\mu = \Delta'^\mu \boldsymbol{E}'_\mu, \tag{2.24}$$

where $\Delta^\mu$ and $\Delta'^\mu$ are the coordinates of $P$ with respect to two frames with the same spacetime origin at $O$ but different basis vectors $\boldsymbol{E}_\mu$ and $\boldsymbol{E}'_\mu$, respectively.

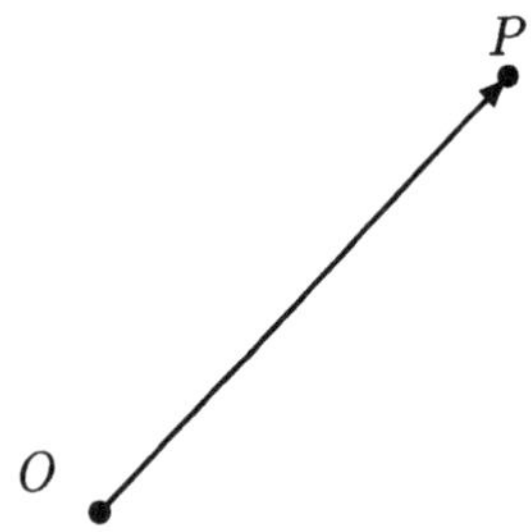


Figure 2.8

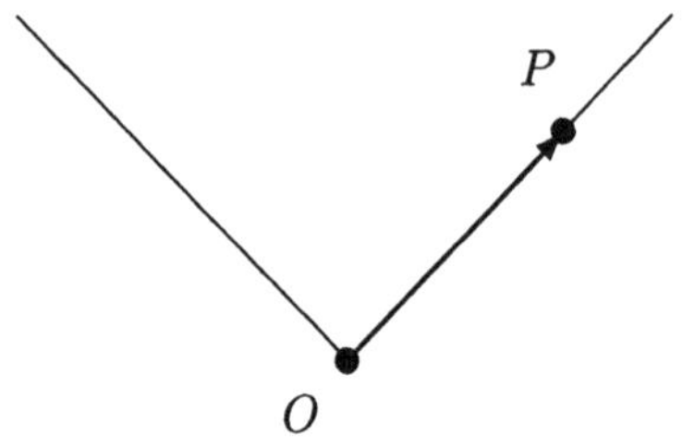


Figure 2.9

### 2.3.1 Future-pointing Lightlike Vectors

The vector $\boldsymbol{\Delta} = \overrightarrow{OP}$ satisfies $\boldsymbol{\Delta} \cdot \boldsymbol{\Delta} = 0$, with $\Delta^0 > 0$ (Figure 2.9). Because spacetime is oriented, and correspondingly the Lorentz transformations are orthochronous, the condition $\Delta^0 > 0$ is frame-independent: it implies (and is implied by) the condition $\Delta'^0 > 0$ in terms of any other frame.

All events $P$ for which the condition is satisfied constitute the *future lightcone* with vertex $O$. One abbreviates: $\boldsymbol{\Delta}$ is FPLL. A light source instantaneously at $O$ can send light rays to all such $P$. The point $P$ is inevitably later than $O$. We recall from the corresponding case of invariant relations that there is no inertial frame in which $O$ and $P$ are simultaneous, nor any in which $O$ and $P$ are at the same spatial position.

### 2.3.2 Past-pointing Lightlike Vectors

Here $\boldsymbol{\Delta} = \overrightarrow{OP}$ satisfies $\boldsymbol{\Delta} \cdot \boldsymbol{\Delta} = 0$, with $\Delta^0 < 0$ (Figure 2.10). Again the condition $\Delta^0 < 0$ implies $\Delta'^0 < 0$. The set of $P$ now form the backward lightcone with vertex $O$. All such $P$ can send light signals to $O$, and no other points can. All such $P$ are earlier than $O$. For short we say: $\boldsymbol{\Delta} = \overrightarrow{OP}$ is PPLL. If $\boldsymbol{\Delta}$ is PPLL, then $-\boldsymbol{\Delta}$ is FPLL.

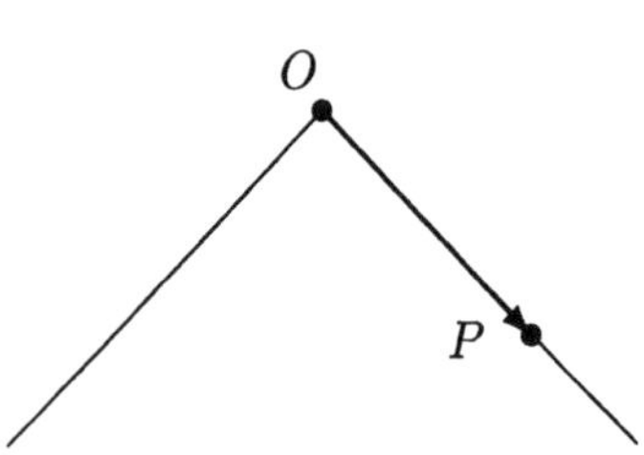


Figure 2.10

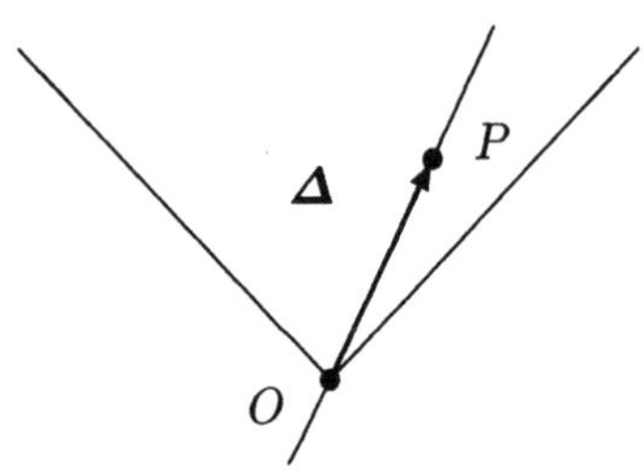


Figure 2.11

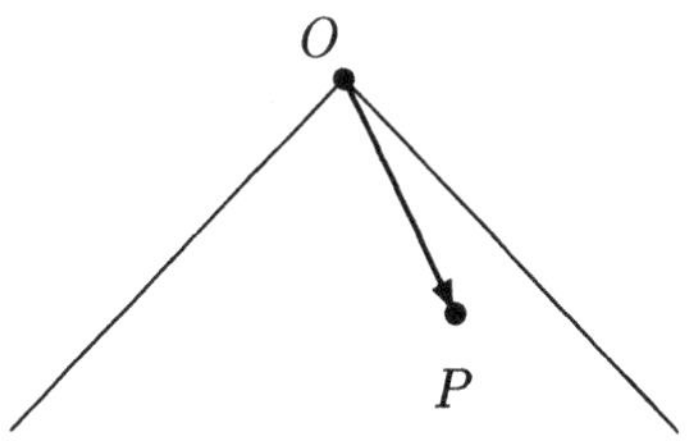

**Figure 2.12**

### 2.3.3 Future-pointing Timelike Vectors

The vector $\boldsymbol{\Delta} = \overrightarrow{OP}$ satisfies $\boldsymbol{\Delta} \cdot \boldsymbol{\Delta} < 0$, with $\Delta^0 > 0$ (Figure 2.11). The points $P$ are within the future lightcone with vertex $O$. All $P$ are later than $O$ (a frame-independent statement). In terms of components with respect to any frame the condition on $\boldsymbol{\Delta}$ is given by $\boldsymbol{\Delta} \cdot \boldsymbol{\Delta} = -(\Delta^0)^2 + \Delta^m \Delta^m < 0, \ \Delta^0 > 0$.

For any $P$ satisfying the conditions there is a special frame $\tilde{K}$ (actually many frames, related by spatial rotations) such that $\boldsymbol{\Delta} = \overrightarrow{OP} = \tilde{T}\tilde{\boldsymbol{E}}_0, \ \tilde{T} > 0$, so that in the frame $\tilde{K}$ the events $P$ and $O$ have the same spatial location but are separated by the time interval $\tilde{T}$, $\boldsymbol{\Delta} \cdot \boldsymbol{\Delta} = -\tilde{T}^2$. The two events are on the worldline of a possible free motion. The vectors $\boldsymbol{\Delta} = \overrightarrow{OP}$ are called FPTL.

### 2.3.4 Past-pointing Timelike Vectors

Here $\boldsymbol{\Delta} \cdot \boldsymbol{\Delta} < 0$, with $\Delta^0 < 0$ (Figure 2.12). Again the condition $\Delta^0 < 0$ implies $\Delta'^0 < 0$. These vectors are past-pointing timelike (PPTL) and fill the interior of the backward lightcone. The event $P$ is earlier than $O$, both lying on the worldline of a possible free motion. If $\boldsymbol{\Delta}$ is PPTL, $-\boldsymbol{\Delta}$ is FPTL.

### 2.3.5 Spacelike Vectors

If $\boldsymbol{\Delta} = \overrightarrow{OP}$ satisfies $\boldsymbol{\Delta} \cdot \boldsymbol{\Delta} > 0$, then $P$ is outside the complete lightcone with vertex $O$. The vector $\boldsymbol{\Delta}$ is spacelike (SL). There is an inertial frame $K^\dagger$, with spacetime origin $O$, such that $\boldsymbol{\Delta} = \overrightarrow{OP} = X^\dagger \boldsymbol{E}_1^\dagger$ for some $X^\dagger > 0$ (Figure 2.13). In such a frame events $O$ and $P$ are simultaneous ($\Delta^{\dagger 0} = 0$), and in that frame the distance between $O$ and $P$ is the positive square root of $\boldsymbol{\Delta} \cdot \boldsymbol{\Delta}$.

There is an inertial frame $K''$ (Figure 2.14) , for which $\Delta''^0 > 0$. In this frame and an infinite number of other such frames $P$ is later than $O$. Similarly (Figure 2.15), there is a frame $K'$ (and so an infinite number of such frames) such that $\Delta'^0 < 0$, and in which, therefore, $P$ is earlier than $O$. For a spacelike vector $\boldsymbol{\Delta}$, the sign of $\Delta^0$ depends on the frame.

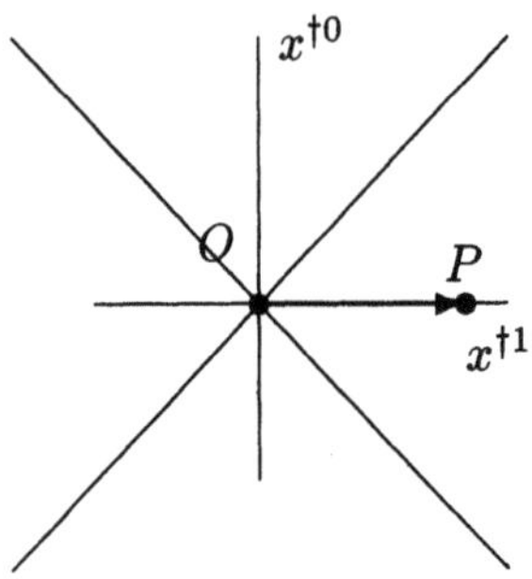


**Figure 2.13**

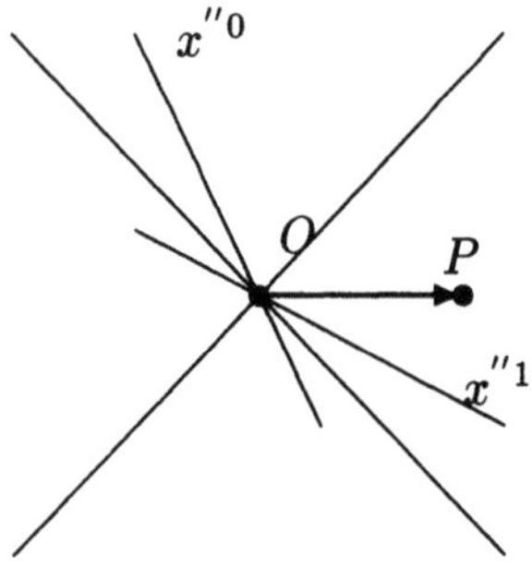


**Figure 2.14**

### 2.3.6 Zero Vector

The zero vector $\mathbf{0} = \overrightarrow{OO}$ is unique. But beware the property of an indefinite scalar product: although $\mathbf{0} \cdot \mathbf{0} = 0$, the scalar product has the same value for all *lightlike* vectors $\boldsymbol{\Delta} \cdot \boldsymbol{\Delta} = 0$.

## 2.4 The Famous Kinematical Effects

In his first paper on special relativity (translated into English in the collection of reprints cited in the previous chapter) Einstein discussed three curious consequences of his new spacetime structure: time dilation, length contraction and addition of velocities. Each is inconsistent with the corresponding Newtonian assumptions and together they highlight the need for a critical

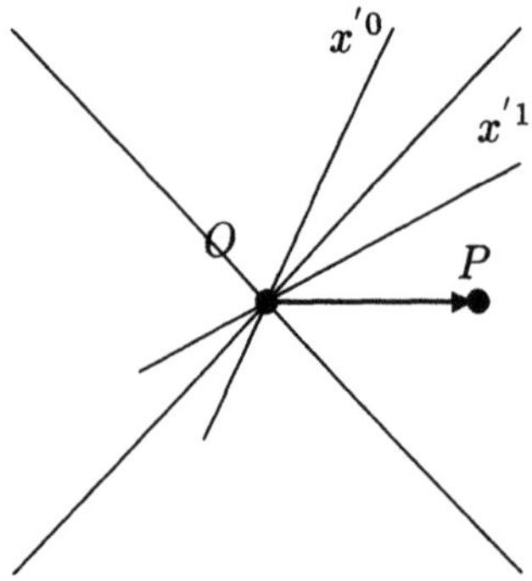


**Figure 2.15**

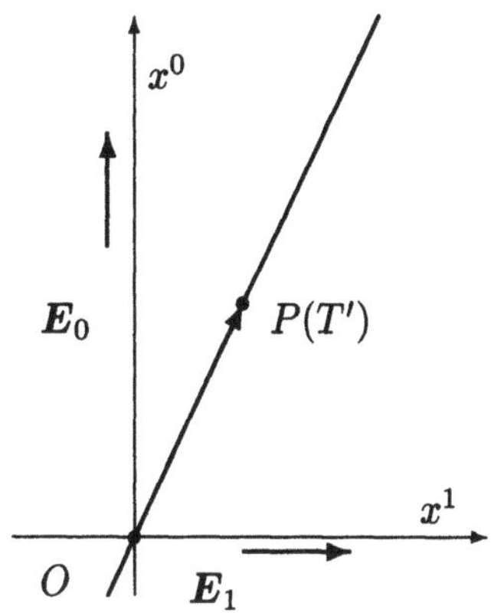


**Figure 2.16**

review of kinematics so that incorrect Newtonian preconceptions do not lead to errors.

### 2.4.1 Time Dilation

The essence of this effect is that when the time on a single, freely moving clock is compared with the time *system* of an inertial frame of reference, that is, with the times on the clocks of the inertial frame that, from moment to moment, the original clock coincides with, it appears to be running slowly.

Without real loss of generality we may consider the freely moving clock to be the master clock of an inertial frame $K'$ whose spacetime origin $O$ coincides with that of an inertial frame $K$ and whose coordinates and basis vectors are related by (1.22) and (2.12) (Figure 2.16). From the latter, if the time on the clock is $T'$ at $P(T')$,

$$\begin{aligned} \overrightarrow{OP(T')} = T'\boldsymbol{E}_0' &= T'\gamma(\boldsymbol{E}_0 + V\boldsymbol{E}_1) \\ &= T\boldsymbol{E}_0 + X\boldsymbol{E}_1, \end{aligned}$$

where $T$ and $X$ are the $K$-coordinates of $P(T')$ as in (2.5), and the speed of the moving clock, with respect to $K$, is $V$. The coordinates not only determine the two-step translation $T\boldsymbol{E}_0$ then $X\boldsymbol{E}_1$, each parallel to a $K$-axis, from $O$ to $P$. They are also the results of *physical* measurements: $T$ is the time on the $K$-clock at $P(T')$, and $X$ is on the $K$-spatial tag for the spatial vertex at the same event.

Therefore, identifying the coefficients of $\boldsymbol{E}_0$ (or taking the scalar product with $\boldsymbol{E}_0$),

$$T = \gamma T' = T'/(1 - V^2)^{\frac{1}{2}}. \tag{2.25}$$

The time interval in the system $K$ between events $O$ and $P$ (for which the time interval on the freely moving clock is $T'$) is $T = \gamma T' > T'$. The time interval between two timelike-related events is *least*, among the inertial frames, for the frame in which the events are at the same spatial position.

From the nonrelativistic point of view, when speeds are low, $|V| \ll 1$, there is little difference between the two times, $T \approx T'$. But the energies of cosmic rays (particles ejected in supernova explosions of stars in our galaxy) can be extremely high, in which case $|V|$ can be very near 1 and $\gamma$ can be very large (a value of 50,000 is not uncommon). This effect has been dramatically confirmed by muon counts at sea level. Muons are created high in the earth's atmosphere. Their average lifetime, in their own rest frame, is about $10^{-9}$secs, yet many reach sea level, a trip requiring an earth time of $10^{-8}$ secs, even at the speed of light.

When time dilation is described verbally (and inaccurately) as a comparison between two clocks, it may appear that there is a symmetry between the comparison of one with respect to the other and vice versa, hence that $T' = \gamma T$ should hold also. This is a result of careless language: we have in each case one clock observed by a *system* of clocks. There is a symmetry, but it is more delicate and leads to a relation $t' = \gamma t$ for a *different* physical situation.

In the situation described above, in which the time on the $K'$-clock is compared with the time in the $K$-system, we may say that we are comparing times on the $K$- and $K'$-clocks using the hyperplanes of $K$-simultaneity to define the association (Figure 2.17). Then $\overrightarrow{OP} = T'\boldsymbol{E}'_0$ and the $K$-time interval is $T = \gamma T'$, the same as for $OR$.

Alternatively, if we compare the master $K$-clock with the time system for $K'$, the geometrical association between points on the two worldlines is defined by the hyperplanes of $K'$-simultaneity. With $\overrightarrow{OQ} = t\boldsymbol{E}_0$, the $K'$-time interval is $t' = \gamma t$, the same as for $OS$ (Figure 2.18).

***Warning.*** One should *not make judgements of size from spacetime diagrams* interpreted as if they were Euclidean. The diagrams were not constructed to do this and such judgements *would often be wrong.* In particular, in the two diagrams above, the length of intervals on the paper (the Euclidean interpretation) does not reproduce the Minkowskian relations.

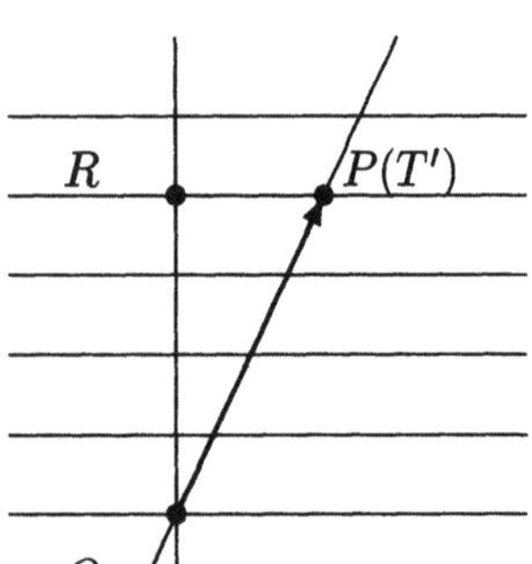


**Figure 2.17**

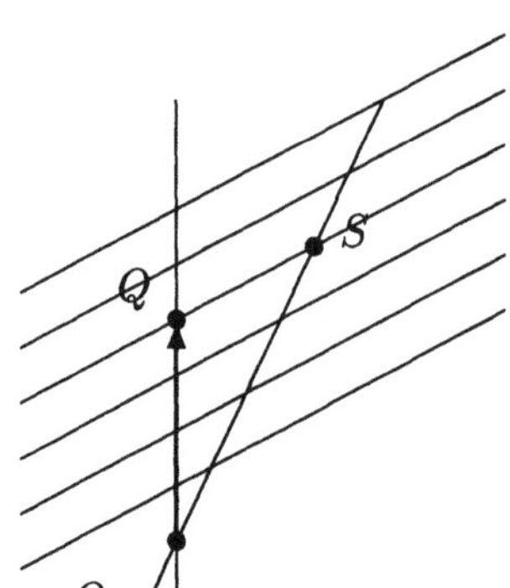


**Figure 2.18**

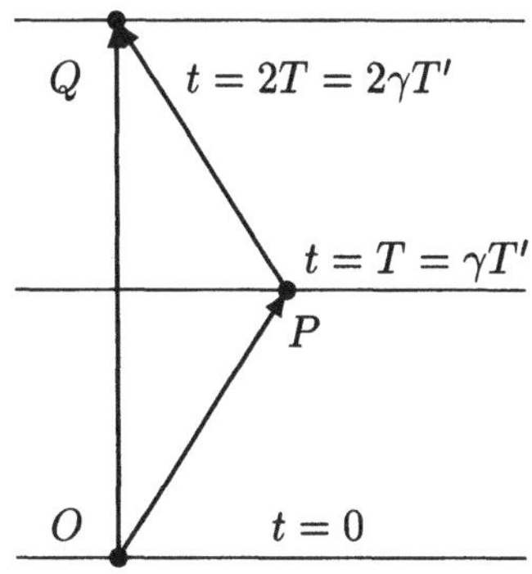


**Figure 2.19**

Several new aspects of relativistic time dilation were first revealed in a study of the twin paradox.

### 2.4.2 The Twin Paradox

The "twin paradox" is first a surprise, and second an interesting example of relativistic kinematics. Since the death of Professor Dingle[2] it has no longer been regarded as a paradox.

Two twins, Pozzo and Ulysses, live at the spatial origin of an inertial frame $K$. Ulysses takes a trip, $O$ to $P$, travelling at speed $V$ relative to $K$, in the $\boldsymbol{E}_1$-direction, for a time interval $T'$ according to his clock. Then he returns, $P$ to $Q$, again at speed $V$ relative to $K$, but in the $-\boldsymbol{E}_1$-direction. This section also takes a time interval $T'$ according to Ulysses' clock. In Figure 2.19, three hyperplanes of constant $K$-time are shown; the readings on Ulysses' clock where it intersects the three hyperplanes, that is, at $O$, $P$ and $Q$ are 0, $T'$ and $2T'$, respectively. At $P$, the fact that Ulysses' time $T'$ differs from Pozzo's $\gamma T'$ (the time in his *system*) occasions no particular surprise because the twins are physically separated. But at $Q$ the twins are together, and while Pozzo has aged $2\gamma T'$ since $O$, Ulysses has aged only $2T'$. (Recall that $\gamma > 1$, and that, in principle, $\gamma$ can be indefinitely large.) This is the surprise, since the result seems so contrary to all our intuition. But our intuition has been built up in circumstances wholly inappropriate for coming to terms with this phenomenon. In all the examples that contribute to our

[2] Herbert Dingle, 1890-1978. Almost single-handedly he kept alive a controversy about the twin paradox. He was an historian of science, influenced more by verbal descriptions than by mathematics. It would appear that he believed in absolute time. In his later years his public platform was restricted to the letters column of the Listener. His book, Science at the Crossroads (Martin Brian and O'Keefe, London, 1972), beautifully written and rationally argued, describes his struggles to maintain an outlet for his point of view.

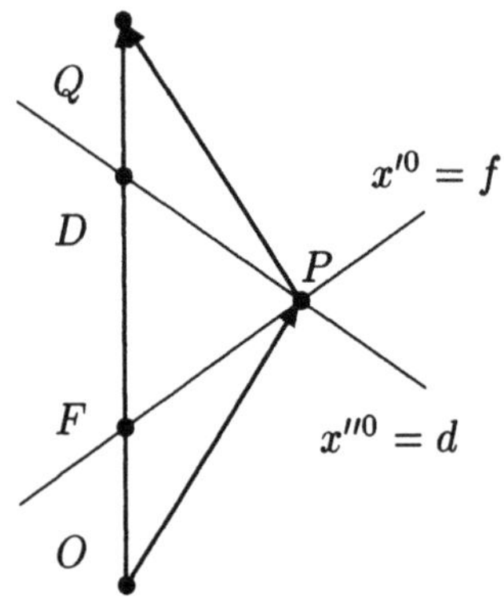


**Figure 2.20**

intuition the speeds are negligible and $\gamma$ is indistinguishable from 1. So we must admit that our surprise is not relevant to the truth of the phenomenon.

The situation in the twin paradox illustrates an interesting kinematical effect. The time interval between two events, $O$ and $Q$, measured along the timelike worldlines joining them, is *greatest* in the inertial frame in which the events are at the same place. According to a clock on any other path the time interval will be less. (This is just the *opposite* of what we learn about Euclidean distance: a straight line is the shortest distance between two points. The change is due to the different signs for time and space in the expression for the spacetime interval.)

We have noted that the time dilation effect occurs when a clock is compared with a system of clocks, and that this situation is not symmetrical. In the case of the twin paradox it appears at first glance that there is a symmetry: there are two clocks, and we may say of each that it is moving at speed $V$ with respect to the other. The situation is symmetrical when we think of Ulysses either on his outward trip or on his inward trip, but when we bear both in mind, there is no longer a symmetry. Pozzo remains at rest in an inertial frame but Ulysses does not.

It *is* possible for each twin to describe, in his own personal reference frame, the motion of the other, but this requires a generalisation, so far as Ulysses is concerned, to a noninertial frame, since Ulysses is accelerating; also, Ulysses' worldline must be made differentiable so the unphysical kink at $P$ is smoothed. If the kink (a discontinuity of velocity) is not removed, there is a section of Pozzo's worldline whose points do not appear on any of Ulysses' hyperplanes of simultaneity. To understand this we try to make a comparison of time in the discontinuous case.

In Figure 2.20, the hyperplane $x'^0 = f$, containing $F$ and $P$, is a plane of simultaneity for Ulysses as he arrives at $P$. The unit basis vector in the time direction for this rest system on the outward trip is $\boldsymbol{E}'_0 = \gamma(\boldsymbol{E}_0 + V\boldsymbol{E}_1)$. In *this system* the time intervals between $O$ and $P$, and between $O$ and $F$ are the same, $T'$.

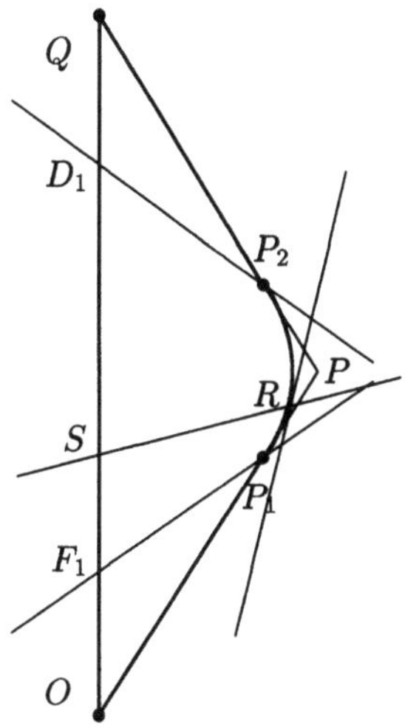


**Figure 2.21**

Similarly, $x''^0 = d$ is a hyperplane of simultaneity for Ulysses on his trip home, when the time basis vector in his rest frame is $\boldsymbol{E}''_0 = \gamma(\boldsymbol{E}_0 - V\boldsymbol{E}_1)$. In this second rest frame the time interval between $D$ and $Q$ is $T'$. The intervals $OF$ and $DQ$ account for the time interval $2T'$ in Ulysses' system, *but Ulysses does not have any temporal record* of the interval $FD$ on Pozzo's worldline. So if we try to calculate the time interval for Pozzo's stretch $OQ$, according to Ulysses'clock, it proves to be impossible, although it is clearly greater than $2T'$.

A simple way to construct a *smooth* version of Ulysses' worldline is to replace the corner at $P$ with a smoothly fitting quadratic (in $K$-time) between $P_1$ and $P_2$ (Figure 2.21). The worldline remains timelike and becomes differentiable. Using $K$-coordinates, with origin at $O$, the spatial distance to $P$ is $L$, and to $P_1$ and $P_2$ is $L - \delta$, so

$$\overrightarrow{OP} = \frac{L}{V}\boldsymbol{E}_0 + L\boldsymbol{E}_1, \quad \overrightarrow{OP_1} = \frac{L-\delta}{V}\boldsymbol{E}_0 + (L-\delta)\boldsymbol{E}_1.$$

As $\delta \to 0$, the original (unphysical) worldline is recovered. The smoothed section of the worldline, parameterised with $K$-time, $\frac{L-\delta}{V} \leq t \leq \frac{L+\delta}{V}$, is

$$z(t) = \overrightarrow{OR}(t) = t\boldsymbol{E}_0 + x(t)\boldsymbol{E}_1, \tag{2.26}$$

where

$$x(t) = (L-\delta) + \frac{\delta}{2}[1 - \frac{V^2}{\delta^2}(t - \frac{L}{V})^2]. \tag{2.27}$$

It is clear that $x(\frac{L\pm\delta}{V}) = L - \delta$ and $\frac{dx}{dt}(\frac{L\pm\delta}{V}) = \mp V$ so the quadratic does fit smoothly.

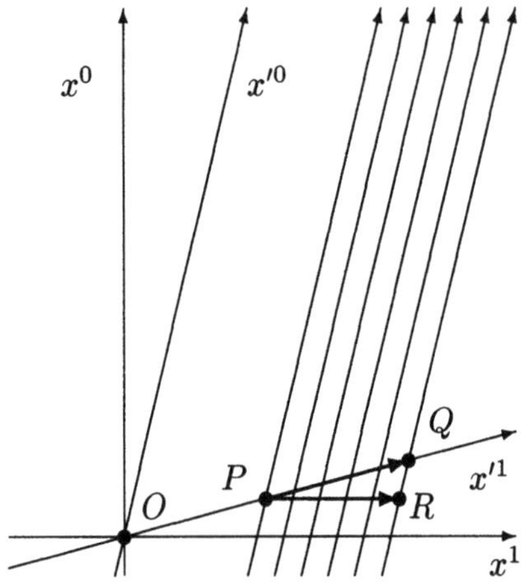


**Figure 2.22**

With the smooth worldline, every point on Pozzo's worldline is simultaneous (according to Ulysses' gradually evolving personal system of time) with a point on Ulysses' worldline; for example, in the diagram, point $S$ on Pozzo's worldline is Ulysses-simultaneous with $R$ on Ulysses' worldline.

### 2.4.3 Length Contraction

The idea of length contraction preceded Einstein's discovery of the relativity of simultaneity; the concept emerged when time was still held to be absolute and the aether was still supposed to define a privileged state of rest. At that time Fitzgerald and Lorentz imagined that motion relative to the aether brought about real physical changes that shortened bodies compared with their rest lengths. None of this is accepted today. "Length contraction" as it is now understood simply arises from measurements with respect to *different inertial frames* according to a definition of length requiring it to be made, for each frame, *at a fixed time in that frame.* There is no such thing as an absolute length that could be measured at an instant of absolute time by any observer.

In special relativity we may consider bodies as rigid if they have an inertial rest frame. Suppose we have a rigid rod, at rest in $K'$, of rest length $L_0$. Without loss of generality we may suppose the rod stretched out along the $x'^1$-axis from $x'^1 = a$ to $x'^1 = a + L_0$.

The rest length of the rod is the length of a vector $\boldsymbol{L}'$, spatial with respect to $K'$, reaching from a point $P$ on one end of the rod to a point $Q$ on the other end, $K'$-simultaneous with $P$. *As a matter of definition,* the $K$-length of the rod is the length of a vector $\boldsymbol{L}$, spatial with respect to $K$, reaching from some $P$ to a point $R$ on the other end, $K$-simultaneous with $P$ (Figure 2.22).

From the triangle $PRQ$ we have $\overrightarrow{PR} + \overrightarrow{RQ} = \overrightarrow{PQ}$ and so

$$L\boldsymbol{E}_1 + \overrightarrow{RQ} = L_0\boldsymbol{E}_1'.$$

Since $\overrightarrow{RQ}$ has the direction of $\boldsymbol{E}_0'$ (it is part of the worldline of a particle at rest in $K'$), if we take the scalar product with $\boldsymbol{E}_1'$ the term will be eliminated and what remains, using (2.23),

$$L(\boldsymbol{E}_1 \cdot \boldsymbol{E}_1') = \gamma L = L_0(\boldsymbol{E}_1' \cdot \boldsymbol{E}_1') = L_0, \tag{2.28}$$

relates the two concepts of length. Since $\gamma > 1$, the length $L$ as measured in $K$ is less than the length $L_0$ as measured in the rest frame $K'$. (A repeated warning not to make judgements of size from spacetime diagrams interpreted as if they were Euclidean: in the diagram above $L$ does appear to be contracted with respect to $L_0$, but this is just an accident.)

A significant confirmation of length contraction comes from the well verified theory of electromagnetism. Maxwell's equations are geometrical (this is discussed in detail in Chapter 5) and they can be decomposed with respect to different inertial frames of reference. One aspect is the charge density. If the density of charge in its rest frame is $\rho_{(0)}$, then in a frame in which the charge moves with speed $V$ its density is $\rho = \gamma\rho_{(0)}$, the transformation of a density being just the reciprocal of the transformation of a volume. This transformation is precisely what Maxwell's equations require.

### 2.4.4 Addition of Velocities

In Newtonian kinematics there is no upper limit to the speed at which a material particle can travel. In Einsteinian kinematics the upper limit for the speed of a particle with respect to an inertial frame is $c = 1$, the speed of light, and this bound cannot be attained by a particle with mass. (This statement is equivalent to the truth of the inequality (1.7).) It is clear, therefore, that the Newtonian rule for the addition of velocities must be replaced because it allows speeds of any magnitude to be built up by addition from lower ones.

To fit the present problem into the available theoretical framework, suppose a particle moves with constant speed $U$ with respect to an inertial frame $K'$, along its $x'^1$-axis ($U$ is the component along $\boldsymbol{E}_1'$; it is possible that $U < 0$, although diagrams are drawn for positive $U$). And suppose that $K'$ moves at $V$ along the $\boldsymbol{E}_1$-axis of another inertial frame $K$, the two frames having a common spacetime origin and basis vectors related in the standard way for a boost by (2.12). Let $K''$ be an inertial frame based on the free motion of the particle, sharing a spacetime origin with $K'$, and related to it by a boost. Then

$$\boldsymbol{E}_0'' = \gamma_U(\boldsymbol{E}_0' + U\boldsymbol{E}_1'), \qquad \boldsymbol{E}_1'' = \gamma_U(\boldsymbol{E}_1' + U\boldsymbol{E}_0').$$

with the unchanged directions in the plane

$$\boldsymbol{E}_2'' = \boldsymbol{E}_2' = \boldsymbol{E}_2, \qquad \boldsymbol{E}_3'' = \boldsymbol{E}_3' = \boldsymbol{E}_3 \quad \text{(Figure 2.23)}.$$

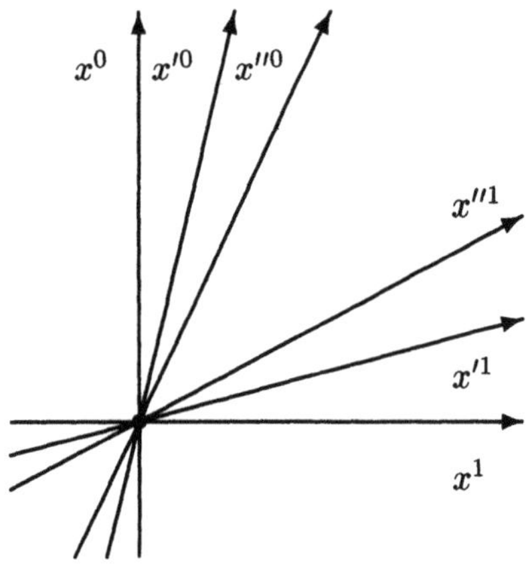


**Figure 2.23**

Substituting for the $K'$-basis vectors from (2.12) (with $\gamma$ now written as $\gamma_V$ because several velocities are involved), we get the relation between $K$- and $K''$-basis vectors:

$$\boldsymbol{E}_0'' = \gamma_U\gamma_V(\boldsymbol{E}_0 + V\boldsymbol{E}_1 + U\boldsymbol{E}_1 + UV\boldsymbol{E}_0) = \gamma_W(\boldsymbol{E}_0 + W\boldsymbol{E}_1) \tag{2.29}$$

$$\boldsymbol{E}_1'' = \gamma_U\gamma_V(\boldsymbol{E}_1 + V\boldsymbol{E}_0 + U\boldsymbol{E}_0 + UV\boldsymbol{E}_1) = \gamma_W(\boldsymbol{E}_1 + W\boldsymbol{E}_0), \tag{2.30}$$

where we can now read off the formula

$$W \equiv \frac{U+V}{1+UV} \tag{2.31}$$

for the speed of $K''$ (that is, for the speed of the particle) with respect to $K$. The derivation of (2.29) uses the shorthand

$$\gamma_W = \gamma_U\gamma_V(1+UV), \tag{2.32}$$

but since $\boldsymbol{E}_0''$ by construction is a unit vector, $\boldsymbol{E}_0'' \cdot \boldsymbol{E}_0'' = -1$ may be used to deduce the standard connection, $(1-W^2)^{-\frac{1}{2}} = \gamma_W$, a relation which can also be verified by direct calculation with (2.31). It is easy to check that $|W| < 1$.

An experiment which verifies the addition of velocities formula (2.31) was performed by Fizeau in 1851. Light travels at speed $U$ through water which itself moves at $V$ with respect to the earth. Since $V \ll U < 1$, the speed of the light with respect to the earth is well approximated by

$$W = \frac{U+V}{1+UV} \approx U + V(1-U^2).$$

The experiment confirmed this equation, although for a time the popular interpretation was in terms of partial aether drag. (Since $c = 1$ the index of refraction is $n = 1/U$.)

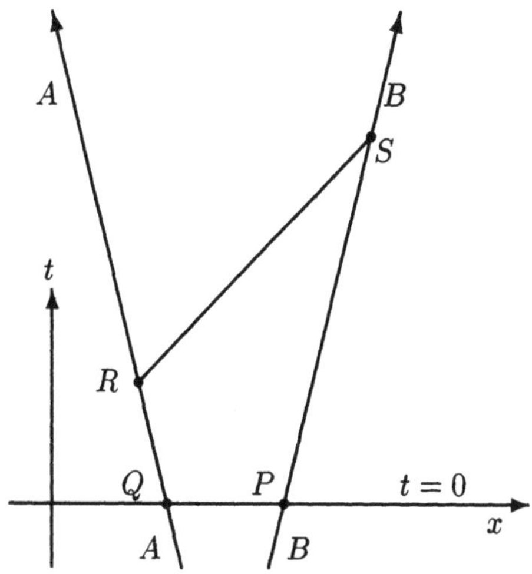


Figure 2.24

### 2.4.5 Two Moon Rockets

In applications it is usually simplest to deal with problems *ab initio*, using spacetime diagrams and vector algebra, rather than to attempt to recognise special instances of time dilation, length contraction or addition of velocities. To illustrate, consider the following problem, phrased in the natural language of examination questions:

> *Just above the surface of the moon two rockets, A and B, fly horizontally away from one another. With respect to the moon, they fly at equal speeds V along the same straight line but in opposite directions. When the moon's clocks read zero, the rockets are separated by a moon-distance L and the clocks in the two rockets read zero too. At time T according to A's clock a light signal is sent from rocket A. At what time according to B's clock does the signal arrive at rocket B?*

***The first step is always a clear spacetime diagram.***

The horizontal line in Figure 2.24 represents the $t = 0$ hyperplane of simultaneity on the moon, and specifically the diagram contains the moon's $x$-axis. The $AA$-line represents rocket $A$'s worldline and the future-pointing unit vector along it is

$$\boldsymbol{E}_A = \gamma(\boldsymbol{E}_0 - V\boldsymbol{E}_1),$$

using moon basis vectors, and noting that rocket $A$ moves to the left so its relative velocity component along $\boldsymbol{E}_1$ is $-V$. The clocks on the rockets, assumed to be standard, could be the master clocks of inertial frames of reference. Therefore, since the time interval on the $A$-clock, between $Q$ and

$R$, is $T$, $\overrightarrow{QR} = T\boldsymbol{E}_A$. Similarly, a future-pointing unit vector in the direction of the rightward-moving rocket $B$'s worldline is

$$\boldsymbol{E}_B = \gamma(\boldsymbol{E}_0 + V\boldsymbol{E}_1),$$

and $\overrightarrow{PS} = T'\boldsymbol{E}_B$, where $T'$ is the (unknown) time on $B$'s clock at $S$, when the signal arrives.

Vector addition allows us to write

$$\overrightarrow{PS} = \overrightarrow{PQ} + \overrightarrow{QR} + \overrightarrow{RS}. \tag{2.33}$$

Since $P$ and $Q$ are on the simultaneity hyperplane $t = 0$, their moon-separation is $L$, and according to the relative orientation of the diagram $\overrightarrow{PQ} = -L\boldsymbol{E}_1$. The light signal $\overrightarrow{RS}$ is proportional to $\boldsymbol{E}_0 + \boldsymbol{E}_1$, so $\overrightarrow{RS} \cdot (\boldsymbol{E}_0 + \boldsymbol{E}_1) = 0$. Using this equation and the fundamental $\boldsymbol{E}_\mu \cdot \boldsymbol{E}_\nu = \eta_{\mu\nu}$ when we take the scalar product of (2.33) with $\boldsymbol{E}_0 + \boldsymbol{E}_1$ gives immediately the result

$$T' = \frac{L}{\gamma(1-V)} + \frac{T(1+V)}{1-V} = L\sqrt{\frac{1+V}{1-V}} + \frac{T(1+V)}{1-V}.$$

Note that for the case $V = 0$, the expression reduces to $T' = L + T$, and that as $V$ increases $T'$ increases. The result comes very much more quickly and clearly using the vectorial method than it would using coordinates of events and relative velocities. The main advantage of the vectorial method is that it is coordinate-independent; although coordinates are used, different systems can be used *simultaneously* in different parts of a problem.

A second problem, again illustrating the advantages of using several inertial frames simultaneously, is as follows.

### 2.4.6 The Problem of Crashing Mirrors

*Two plane mirrors, labelled $A$ and $B$, which face each other, remain parallel and move towards each other, each with a constant high speed $V$ in the laboratory. At laboratory time $t = 0$ they are separated by a laboratory distance $2L$.*

*At $t = 0$ a light pulse leaves mirror $A$ and moves directly towards mirror $B$, and it then bounces back and forth between the mirrors until they meet. Determine the total distance travelled by the pulse as measured in the frame of reference in which mirror $B$ is at rest.*

***The first step is, as always, a clear spacetime diagram (Figure 2.25).***

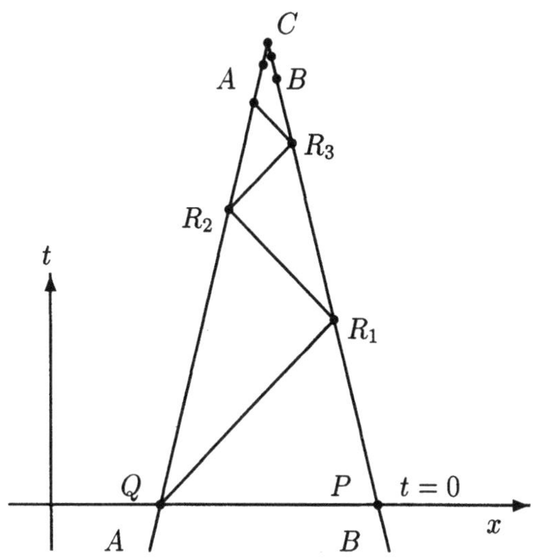


**Figure 2.25**

The worldlines of the mirrors are $AA$ and $BB$, and unit vectors along them are

$$\boldsymbol{E}_A = \gamma(\boldsymbol{E}_0 + V\boldsymbol{E}_1), \quad \boldsymbol{E}_B = \gamma(\boldsymbol{E}_0 - V\boldsymbol{E}_1),$$

decomposed with laboratory basis vectors. The pulse of radiation starts at $Q$ in the $t = 0$ hyperplane for the laboratory, then bounces $R_1, R_2, R_3, \ldots$ till the mirrors crash at $C$. Since the speed of light in any inertial reference frame is one, the distance the pulse travels with respect to a given frame equals the time elapsed with respect to the *frame*. For the problem at hand we therefore want the time interval, as measured in the $B$-frame, for the displacement $\overrightarrow{QC}$. This time interval is

$$-\boldsymbol{E}_B \cdot \overrightarrow{QC} = -\gamma(\boldsymbol{E}_0 - V\boldsymbol{E}_1) \cdot (L\boldsymbol{E}_1 + \frac{L}{V}\boldsymbol{E}_0) = L\gamma(\frac{1}{V} + V),$$

where the laboratory expression for $\overrightarrow{QC}$ is used, which represents a distance $L$ traversed in time $L/V$.

## 2.5 The Generalised Vector Space $\mathcal{V}$

The vectors that have been developed up to this point are translation vectors, in particular, simple mappings of spacetime onto itself which get their definitions and their importance from the special role of the inertial frames. Many other physical concepts have their mathematical expression in the idea of vectors. These other physical concepts, velocity, momentum, electric fields, and so on, are not translation vectors, yet they are closely related to them.

Since the spacetime directions in the real physical world, which may be expressed by translation vectors in spacetime $\mathcal{M}$, are the same for all vectorial quantities, we expect that the different spaces of vectors have a very simple relation to each other. The most elementary expectation, but one which is sufficient to determine everything, is that for each inertial frame of reference the unit basis vectors for the different spaces may be identified with each other. If this is so, then all vectors in one system have identified counterparts in the other systems by comparing components with respect to any inertial frame. Mathematically, there is an isomorphism between the two spaces.

From the mathematical point of view, a vector space is a very simple system. The vectors belonging to two different spaces of the same dimension may be put in a 1-1 relation that respects the algebraic operations in the spaces just by identifying bases and linear combinations of basis vectors. For example, if $\boldsymbol{E}_\mu$ and $F^\dagger_\mu$ (say) are inertial basis vectors in the two spaces associated with a frame $K$ (for example, $\boldsymbol{E}_m$ is a translation by unit coordinate distance in the $x^m$-direction, and $F^\dagger_m$ is unit momentum, dependent on units, in the same direction) then in each space, by (2.9), we get the inertial basis vectors for the frame $K'$, $\boldsymbol{E}'_\mu = a^{-1\nu}{}_\mu \boldsymbol{E}_\nu$ and $F'^\dagger_\mu = a^{-1\nu}{}_\mu F^\dagger_\nu$. By identifying the $K$-bases

$$\boldsymbol{E}_\mu \longleftrightarrow F^\dagger_\mu \tag{2.34}$$

and their linear combinations

$$A^\mu \boldsymbol{E}_\mu \longleftrightarrow A^\mu F^\dagger_\mu, \tag{2.35}$$

all other inertial frame bases are automatically identified, as are the same linear combinations with respect to them. Spaces related in this way differ only in name and notation, not in any algebraic property.

To take advantage of the fact that the many different vector spaces can be put into correspondence by means of their agreement on spacetime directions, it is sufficient to work with one space only. Its name, $\mathcal{V}$, and the notation for vectors $\boldsymbol{E}_\mu$, $\boldsymbol{A}$, will be taken from the translation vectors, but we think of it as a *new* space with a specially defined correspondence with the translation vectors. The other vector spaces are identified with the new space $\mathcal{V}$ as described above.

The upshot is now that all vector spaces are related in a natural and physically meaningful way. There is no longer any need to distinguish the different spaces. We agree to use the notation already developed for translations, but to extend it to all the other spaces. The space $\mathcal{V}$ becomes an all-purpose vector space, serving much the same function for vectors as the real numbers $\mathcal{R}$ do for scalars.

## 2.6 Proper Time and Concepts of Velocity

The worldline of a particle with mass is the set of point events occupied by the particle in spacetime. This is the overall spacetime description. If one

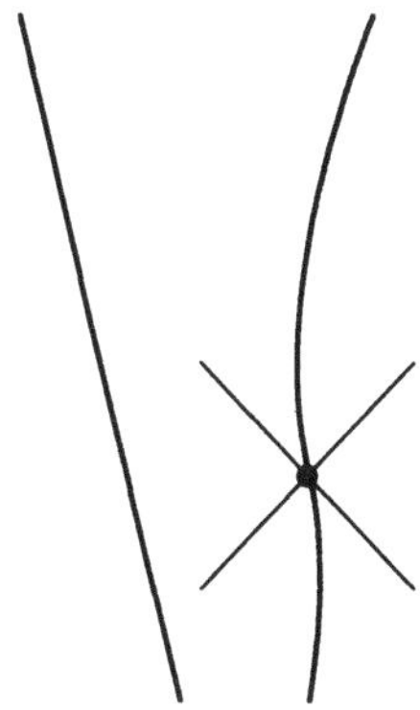

**Figure 2.26**

says "the set of points *successively* occupied by the particle" one is implicitly taking the relative point of view of an observer, perhaps an inertial observer, for whom time is passing. A straight line represents a free motion, the motion of a particle acted on by no forces. A curved worldline, on the other hand, represents a worldline for an accelerated particle on which forces do act (Figure 2.26). In both cases, successive points on the worldline are timelike displaced; for each point on the worldline a lightcone with vertex at the point contains within it all the nearby points of the worldline (and in fact *all* the points of the whole worldline). This is simply another expression of the fact that massive particles travel more slowly than light.

The spacetime velocity is a velocity concept associated with the overall spacetime description; it is a rate of change of position in *spacetime*. The relative velocity, however, is specific to a given inertial frame; it is a rate of change of *spatial* position with respect to that frame.

Although the worldline is a continuous set of points, it is very often represented in parametric form. There are many possibilities: the time $t$ in any inertial frame, a general parameter $\lambda$ with no geometrical interpretation, or proper time $\tau$, which is rather like the arc-length parameter for curves in Euclidean space. For nearby points on a worldline the change in proper time is the time interval between the points in an inertial frame in which they are at the same place (and so in which the particle is instantaneously at rest). For an accelerating particle this inertial frame that is used to help measure proper time, the instantaneous rest frame, is therefore continuously changing.

### 2.6.1 Spacetime Velocity

We start with the geometrically more fundamental concept of spacetime velocity; this is new, it is *not* the concept brought to us from prerelativistic physics, which is the relative idea, relative to some arbitrary inertial frame. The spacetime velocity, for the case of a straight worldline, is by definition

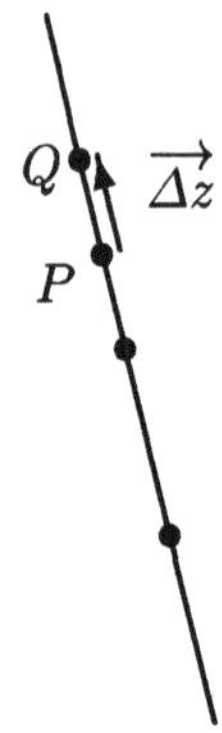

**Figure 2.27**

$$\boldsymbol{v} = \frac{\overrightarrow{\Delta z}}{\Delta \tau}, \tag{2.36}$$

where $\overrightarrow{\Delta z}$ is the translation vector between two events $P$ and $Q$ on the worldline, and $\Delta\tau$ is the time interval between the events in any inertial frame whose time basis vector has the direction of the displacement $\overrightarrow{\Delta z}$, that is, in any inertial frame in which $P$ and $Q$ have the same spatial position (Figure 2.27). Since the spatial position does not change, such an inertial frame is a *rest frame* for the particle; our standard notation for a rest frame basis is $\tilde{\boldsymbol{E}}_\mu$. Then, $\overrightarrow{\Delta z} = \tilde{\boldsymbol{E}}_0 \Delta\tau = \boldsymbol{v}\Delta\tau$. *The spacetime velocity vector* $\boldsymbol{v}$ *is a future-pointing unit timelike vector.* To find an expression for the proper time interval $\Delta\tau$, one may use the general formula, in terms of coordinate differences in any frame, $\overrightarrow{\Delta z} \cdot \overrightarrow{\Delta z} = \Delta z^\mu \Delta z^\nu \eta_{\mu\nu}$. It is the same number no matter which frame is used, and, evaluated for the rest frame, it reduces to $\overrightarrow{\Delta z} \cdot \overrightarrow{\Delta z} = \overrightarrow{PQ} \cdot \overrightarrow{PQ} = -(\Delta\tau)^2$ because there is no spatial contribution. For *any* two points on the straight worldline, one gets the same ratio $\boldsymbol{v}$.

For an accelerating worldline the geometrical idea is the same but at each point a limit must be taken:

$$\boldsymbol{v}(P) = \lim_{Q\to P} \frac{\overrightarrow{PQ}}{\delta\tau} = \lim_{Q\to P} \frac{\overrightarrow{PQ}}{\sqrt{-\overrightarrow{PQ} \cdot \overrightarrow{PQ}}}. \tag{2.37}$$

Again, $\boldsymbol{v}$ is a unit timelike vector, and indeed, $\boldsymbol{v}(P)$ is the time basis vector $\tilde{\boldsymbol{E}}_0$ for the *instantaneous* rest frame at $P$ (Figure 2.28).

Equation (2.37) is a good geometrical definition of the spacetime velocity, but for analytical purposes it is useful to be able to express it using a parametric form and coordinate functions to describe the worldline.

Suppose the worldline is parameterised by an arbitrary parameter $\lambda$, and the points on it are located, with reference to a fixed point $O$, by a vector function $\boldsymbol{z}(\lambda)$ (Figure 2.29). In terms of the coordinates of any inertial frame with spacetime origin $O$,

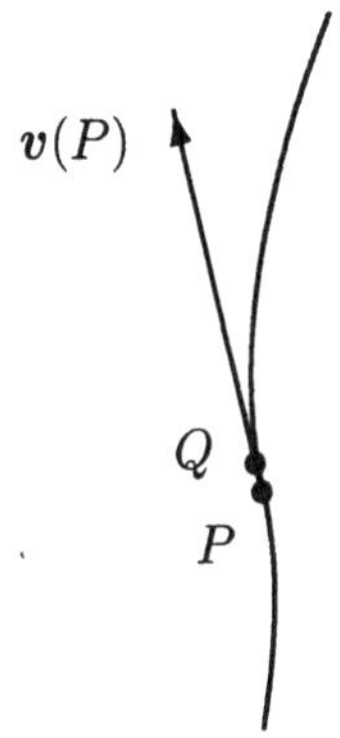


Figure 2.28

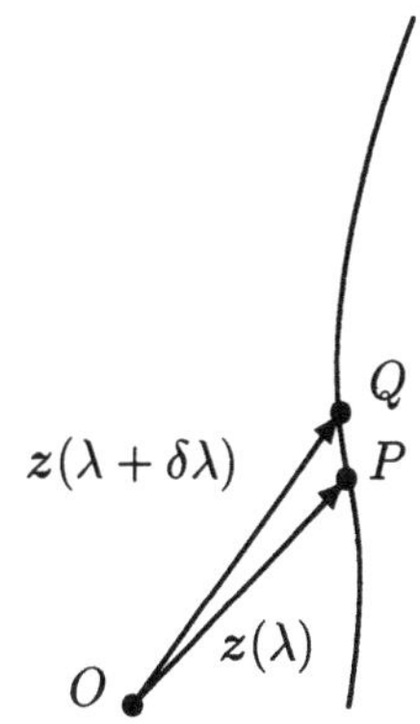


Figure 2.29

$$z(\lambda) = z^\mu(\lambda)\boldsymbol{E}_\mu, \tag{2.38}$$

and the displacement from $P$ to $Q$ is the difference

$$\delta \boldsymbol{z} = \delta z^\mu \boldsymbol{E}_\mu \simeq \delta\lambda \frac{dz^\mu}{d\lambda}\boldsymbol{E}_\mu. \tag{2.39}$$

Expressing the change in proper time from $P$ to $Q$ in terms of coordinate differences gives $\overrightarrow{PQ}\cdot\overrightarrow{PQ} = \delta\boldsymbol{z}\cdot\delta\boldsymbol{z} = \eta_{\mu\nu}\delta z^\mu \delta z^\nu = -\delta\tau^2$, and so the relation between changes in proper time $\tau$ and changes in the parameter $\lambda$ is given by

$$\delta\tau = \delta\lambda\sqrt{-\eta_{\mu\nu}\frac{dz^\mu}{d\lambda}\frac{dz^\nu}{d\lambda}}. \tag{2.40}$$

Hence, by (2.37),

$$\boldsymbol{v}(P) = \frac{dz^\rho}{d\lambda}\boldsymbol{E}_\rho\left(-\eta_{\mu\nu}\frac{dz^\mu}{d\lambda}\frac{dz^\nu}{d\lambda}\right)^{-\frac{1}{2}}. \tag{2.41}$$

### 2.6.2 Proper Time

Up to this point, proper time, for an accelerating worldline, has appeared only as an infinitesimal time interval in an instantaneous rest frame. It is very convenient to introduce an *accumulated* proper time, the integrated version of (2.40), by

$$\tau(\lambda) \equiv \int_{\lambda_0}^{\lambda}\left(-\eta_{\mu\nu}\frac{dz^\mu}{d\lambda'}\frac{dz^\nu}{d\lambda'}\right)^{\frac{1}{2}} d\lambda'. \tag{2.42}$$

The proper time, so defined, has an arbitrary zero point, when $\lambda = \lambda_0$. Its scale is fixed by the conventional units for time. A properly constructed

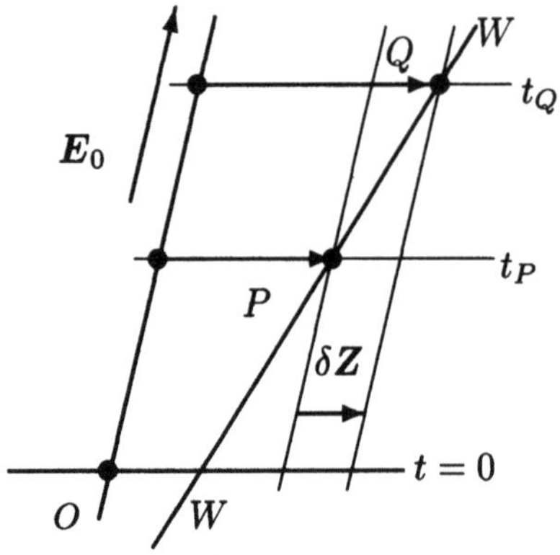


**Figure 2.30**

physical clock, if it followed an accelerating worldline, would read proper time. Only in the special case of a straight worldline, for a free motion, does the concept of proper time reduce to the concept of time in an inertial frame.

Equation (2.42), defining a continuous proper time in terms of $\lambda$, allows us to introduce $\tau$ as a new parameter for the worldline. One writes [3] $\boldsymbol{z}(\tau) = \boldsymbol{z}(\tau(\lambda)) = \boldsymbol{z}(\lambda)$ and since

$$\frac{d\tau}{d\lambda} = \left(-\eta_{\mu\nu}\frac{dz^\mu}{d\lambda}\frac{dz^\nu}{d\lambda}\right)^{\frac{1}{2}},$$

the equation (2.41) may be re-expressed in the neater form

$$\boldsymbol{v} = \frac{dz^\mu}{d\tau}\boldsymbol{E}_\mu = \frac{dz^\mu}{d\lambda}\frac{d\lambda}{d\tau}\boldsymbol{E}_\mu = \frac{d\boldsymbol{z}}{d\tau}. \tag{2.43}$$

The defining characteristic of the parameterisation $\boldsymbol{z}(\tau)$, with proper time, is that the derivative, $\boldsymbol{v} = d\boldsymbol{z}(\tau)/d\tau$, is a unit vector (albeit timelike). In this respect it behaves like the arc-length parameterisation of curves in Euclidean spaces.

### 2.6.3 Relative Velocity with Respect to an Inertial Frame

The *relative* velocity with respect to an inertial frame $K$ is the rate, per $K$-time interval, of *spatial* change with respect to $K$.

In Figure 2.30, the scale is blown up sufficiently that the worldline, $WW$, appears straight in the portion shown. We write for point $P$

[3] One should not regard $\boldsymbol{z}$ narrowly as a vector function of the real numbers, but rather as a geometrical object, $\boldsymbol{z} = \overrightarrow{OP}$, parameterised by $\tau$, or by $\lambda$. A similar remark applies to the components of $\boldsymbol{z}$ : each $z^\mu$ is the name of a physically significant quantity, not a pure mathematical function.

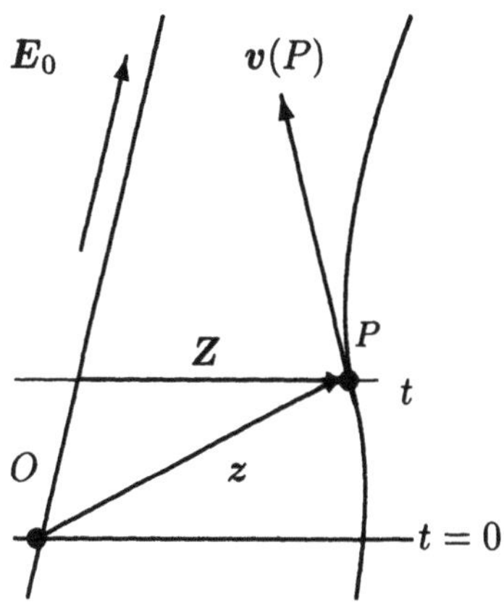


**Figure 2.31**

$$\overrightarrow{OP} = t_P \boldsymbol{E}_0 + \boldsymbol{Z}(t_P), \tag{2.44}$$

and similarly for point $Q$. With $\delta t = t_Q - t_P$, and $\delta \boldsymbol{Z} = \boldsymbol{Z}(t_Q) - \boldsymbol{Z}(t_P)$, the relative velocity with respect to the frame $K$ is defined by

$$\boldsymbol{V} = \frac{\delta \boldsymbol{Z}}{\delta t}, \tag{2.45}$$

the ratio of the spatial displacement with respect to $K$ and the $K$-time interval. For comparison, the spacetime velocity in this situation is

$$\begin{aligned} \boldsymbol{v} = \frac{\overrightarrow{PQ}}{\sqrt{-\overrightarrow{PQ} \cdot \overrightarrow{PQ}}} &= \frac{\delta t \boldsymbol{E}_0 + \delta \boldsymbol{Z}}{(\delta t^2 - \delta \boldsymbol{Z} \cdot \delta \boldsymbol{Z})^{\frac{1}{2}}} \\ &= \frac{\delta t}{\delta \tau} \boldsymbol{E}_0 + \frac{\delta \boldsymbol{Z}}{\delta \tau} = \frac{\delta t}{\delta \tau} (\boldsymbol{E}_0 + \frac{\delta \boldsymbol{Z}}{\delta t}), \end{aligned} \tag{2.46}$$

where $\delta \tau = (\delta t^2 - \delta \boldsymbol{Z} \cdot \delta \boldsymbol{Z})^{\frac{1}{2}}$ is the proper time for the interval $PQ$. Hence the relation between spacetime velocity $\boldsymbol{v}$ and relative velocity $\boldsymbol{V}$ is

$$\boldsymbol{v} = \frac{\delta t}{\delta \tau} (\boldsymbol{E}_0 + \boldsymbol{V}) = \frac{1}{(1 - \boldsymbol{V} \cdot \boldsymbol{V})^{\frac{1}{2}}} (\boldsymbol{E}_0 + \boldsymbol{V}). \tag{2.47}$$

The spacetime velocity is absolute and geometrical, depending only on the worldline, while the relative velocity also depends on the relation of the worldline to an inertial frame.

For the general accelerating worldline we may take over the formula (2.43) with the $K$-time $t$ as the parameter $\lambda$ (Figure 2.31). Then, with the decomposition

$$\overrightarrow{OP} = t \boldsymbol{E}_0 + \boldsymbol{Z}(t) \equiv \boldsymbol{z}(t), \tag{2.48}$$

and its $K$-time derivative

$$\frac{d\overrightarrow{OP}(t)}{dt} = \boldsymbol{E}_0 + \frac{d\boldsymbol{Z}}{dt} = \boldsymbol{E}_0 + \boldsymbol{V} = \frac{d\boldsymbol{z}}{dt}, \tag{2.49}$$

we get

$$\boldsymbol{v} = \frac{d\boldsymbol{z}}{d\tau} = \frac{dt}{d\tau}\frac{d\boldsymbol{z}}{dt} = \frac{dt}{d\tau}(\boldsymbol{E}_0 + \boldsymbol{V}). \tag{2.50}$$

The ratio $dt/d\tau$ is positive, and since $\boldsymbol{v}$ is a *unit* future-pointing timelike vector, by taking the scalar product of (2.50) with itself, we get its value

$$\frac{dt}{d\tau} = \frac{1}{\sqrt{1 - \boldsymbol{V}\cdot\boldsymbol{V}}} \equiv \gamma. \tag{2.51}$$

Hence,

$$\boldsymbol{v} = \gamma(\boldsymbol{E}_0 + \boldsymbol{V}). \tag{2.52}$$

The factor $\gamma$, which depends on the magnitude of the $K$-spatial relative velocity $\boldsymbol{V}$, is a measure of the relative speed of two frames with time axes along $\boldsymbol{v}$ and along $\boldsymbol{E}_0$ (a rest frame for the particle and a general frame $K$). This measure is given by the relation $\gamma = -\boldsymbol{v}\cdot\boldsymbol{E}_0$.

The spacetime velocity vector $\boldsymbol{v}$ determines the orientation in $\mathcal{M}$ of the particle's worldline. Given the time basis vector $\boldsymbol{E}_0$ for an inertial frame $K$ as well, the relative velocity with respect to $K$ is given by the unique decomposition

$$\begin{aligned}\boldsymbol{v} &= (v^0)\boldsymbol{E}_0 + (\perp \boldsymbol{E}_0)\\ &= (-\boldsymbol{v}\cdot\boldsymbol{E}_0)\boldsymbol{E}_0 + (\perp \boldsymbol{E}_0) = \gamma\boldsymbol{E}_0 + (\boldsymbol{v} - \gamma\boldsymbol{E}_0)\\ &\equiv \gamma\boldsymbol{E}_0 + \gamma\boldsymbol{V}.\end{aligned} \tag{2.53}$$

This simple relation provides an algebraic definition of relative velocity $\boldsymbol{V}$.

### 2.6.4 General Addition of Velocities Formula

In the classic addition of velocities formula, (2.31), the particle (or the frame $K''$) and the frame $K$ both move along the same axis with respect to $K'$. For the present case we suppose that $K$ and $K'$ are related by a boost in the standard way, but that the particle moves arbitrarily with respect to $K'$. If necessary after a rotation of basis vectors around the $x^1$-axis we may suppose that the relative velocity of the particle with respect to $K'$ is

$$\boldsymbol{U} = U(\cos\vartheta'\boldsymbol{E}'_1 + \sin\vartheta'\boldsymbol{E}'_2). \tag{2.54}$$

The previous case, leading to (2.31), is $\vartheta' = 0$. In order to find the relative velocity with respect to $K$, one may construct the coordinate-independent spacetime velocity,

$$\boldsymbol{v} = \gamma_U(\boldsymbol{E}_0' + \boldsymbol{U}), \tag{2.55}$$

and then decompose it with respect to the $K$-basis:

$$\boldsymbol{v} = \gamma_W(\boldsymbol{E}_0 + \boldsymbol{W}), \qquad \boldsymbol{E}_0 \cdot \boldsymbol{W} = 0. \tag{2.56}$$

Substituting (2.54) into (2.55) and converting to the $K$-basis gives

$$\begin{aligned} \boldsymbol{v} &= \gamma_U\gamma_V(1 + UV\cos\vartheta') \\ &\times \left[\boldsymbol{E}_0 + \frac{(V + U\cos\vartheta')\boldsymbol{E}_1 + (U\sin\vartheta'/\gamma_V)\boldsymbol{E}_2}{1 + UV\cos\vartheta'}\right]. \end{aligned} \tag{2.57}$$

The relative velocity with respect to $K$ may be read off directly. In particular, the angle of motion with respect to the positive $x^1$-direction is $\vartheta$, where

$$\tan\vartheta = \frac{U\sin\vartheta'}{\gamma_V(V + U\cos\vartheta')}. \tag{2.58}$$

### 2.6.5 Acceleration

When the worldline is parameterised by proper time, the spacetime velocity is given by (2.43), as a derivative with respect to proper time. Differentiating again with respect to proper time we get a frame-independent vector called the *spacetime acceleration*

$$\boldsymbol{a} \equiv \frac{d\boldsymbol{v}}{d\tau} = \frac{d^2\boldsymbol{z}}{d\tau^2}. \tag{2.59}$$

Because $\boldsymbol{v}$ is of constant magnitude, by construction a unit vector, $\boldsymbol{v}\cdot\boldsymbol{v} = -1$. By differentiation of this relation with respect to $\tau$,

$$\boldsymbol{v} \cdot \frac{d\boldsymbol{v}}{d\tau} = 0 = \boldsymbol{v} \cdot \boldsymbol{a}. \tag{2.60}$$

The spacetime acceleration is orthogonal to the spacetime velocity. Since the velocity is timelike, the acceleration must be spacelike (if it does not vanish).

The spacetime acceleration is easily related to the *relative* acceleration with respect to an inertial frame. Differentiating (2.52) with respect to $\tau$, but expressing the contribution from $\boldsymbol{V}$ in terms of its derivative with respect to $K$-time $t$,

$$\begin{aligned} \boldsymbol{a} &= \frac{d\boldsymbol{v}}{d\tau} = \frac{d\gamma}{d\tau}(\boldsymbol{E}_0 + \boldsymbol{V}) + \gamma\frac{d\boldsymbol{V}}{dt}\frac{dt}{d\tau} \\ &= \dot\gamma(\boldsymbol{E}_0 + \boldsymbol{V}) + \gamma^2\boldsymbol{A}. \end{aligned} \tag{2.61}$$

In the context of special relativity a dot is often used to indicate a derivative with respect to proper time. The relative acceleration is $\boldsymbol{A} = d\boldsymbol{V}/dt$, the derivative of relative velocity with respect to relative time. By construction $\boldsymbol{A}$ is spatial with respect to $K$ (and therefore $\boldsymbol{A}\cdot\boldsymbol{E}_0 = 0$). Since $\gamma = -\boldsymbol{v}\cdot\boldsymbol{E}_0$, $\dot{\gamma} = -\boldsymbol{a}\cdot\boldsymbol{E}_0$. Also, from $\boldsymbol{v}\cdot\boldsymbol{a} = 0$ and (2.61) follows $0 = -\dot{\gamma}/\gamma + \gamma^3\boldsymbol{A}\cdot\boldsymbol{V}$, so

$$\boldsymbol{a} = \gamma^4\boldsymbol{A}\cdot\boldsymbol{V}(\boldsymbol{E}_0 + \boldsymbol{V}) + \gamma^2\boldsymbol{A}. \tag{2.62}$$

The right-hand side is a complicated expression, but it is complicated because of the involvement in a simple geometrical vector $\boldsymbol{a}$ of a dynamically irrelevant frame $K$. For the magnitude of $\boldsymbol{a}$, one gets, expressed in terms of the relative velocity and acceleration,

$$\boldsymbol{a}\cdot\boldsymbol{a} = \gamma^4\boldsymbol{A}\cdot\boldsymbol{A} + \gamma^6(\boldsymbol{A}\cdot\boldsymbol{V})^2, \tag{2.63}$$

after a short calculation.

## 2.7 Light Rays

Light can be treated in special relativity at several different levels of detail. Here we deal primarily with light rays, but with a little added structure to provide an account of the frequency of harmonic light and its wave length. A light ray is just a directed lightlike line in spacetime. To get a geometrical picture of a harmonic light ray, a light ray with frequency, something more is needed. The frequency itself is not an absolute or geometrical idea, for it depends on the frame of reference (the Doppler effect).

The simplest way of getting a model of light with frequency is to think of a light source, the spatial origin of an inertial frame perhaps, which emits light rays at a steady rate (given by the frequency $f$ or period $T$ with respect to the inertial frame). The consequent pattern of light rays can be described from the point of view of any inertial frame. This construction is sufficient to enable us to define a lightlike momentum vector, parallel to the light ray, for a photon. The pattern of light rays provided by the steady source can be "explained" in terms of harmonic solutions of a free, massless scalar field theory. It is then found that the lightlike vector parallel to the light rays has a second rôle.

The previous model has given no account of polarisation or the spin properties of light. For those something more elaborate is needed. The full classical theory of light is co-extensive with Maxwell's theory of electromagnetism. *That* theory will be developed in relativistic form in Chapter 5.

### 2.7.1 Lightlike Vectors

A basic assumption in the derivation of the Lorentz transformation was that if a light ray can travel from $P$ to $Q$ (no matter what the state of motion of the source or receiver) then

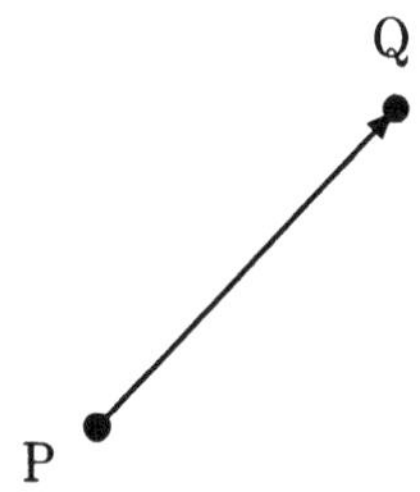


**Figure 2.32**

$$0 = \eta_{\mu\nu} \Delta^\mu \Delta^\nu = -(\Delta^0)^2 + \Delta^m \Delta^m, \qquad \Delta^0 > 0$$

where $\Delta^\mu = x^\mu(Q) - x^\mu(P)$, the coordinate differences for any inertial frame. Putting $\overrightarrow{PQ} = \Delta^\mu \boldsymbol{E}_\mu$, the main condition is $\overrightarrow{PQ} \cdot \overrightarrow{PQ} = 0$, that the vector $\overrightarrow{PQ}$ is lightlike (Figure 2.32).

If we consider light travelling, with respect to a frame $K$, in the direction of the $+x^1$-axis, then the lightlike displacement $\overrightarrow{PQ}$ (Figure 2.33) may be expressed as

$$\overrightarrow{PQ} = \Delta^0(\boldsymbol{E}_0 + \boldsymbol{E}_1).$$

The formula is simple because the spatial direction is a coordinate direction and the lightlike condition reduces to $\Delta^0 = \Delta^1$.

Now describe the same ray $\overrightarrow{PQ}$ in terms of the frame $K'$ moving at $V$ with respect to $K$ in the $+x^1$-direction (Figure 2.34). With respect to $K'$ the ray travels in the $x'^1$-direction and so

$$\overrightarrow{PQ} = \Delta'^0(\boldsymbol{E}'_0 + \boldsymbol{E}'_1).$$

The Lorentz transformation gives the relation between $\Delta^0$ and $\Delta'^0$. Alternatively, writing the $\boldsymbol{E}'_\mu$ vectors in terms of the $\boldsymbol{E}_\mu$ vectors, by (2.12),

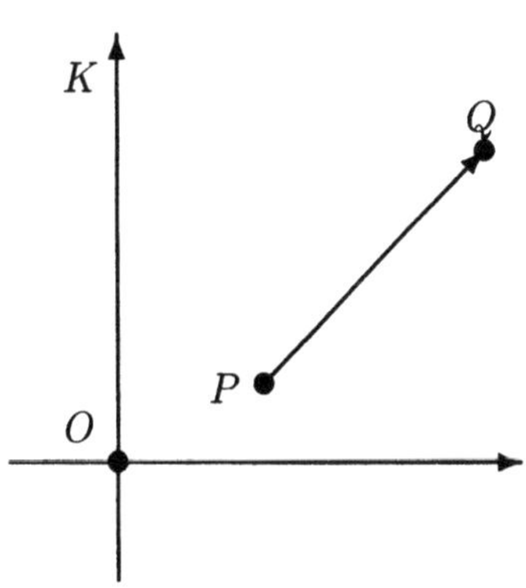


**Figure 2.33**

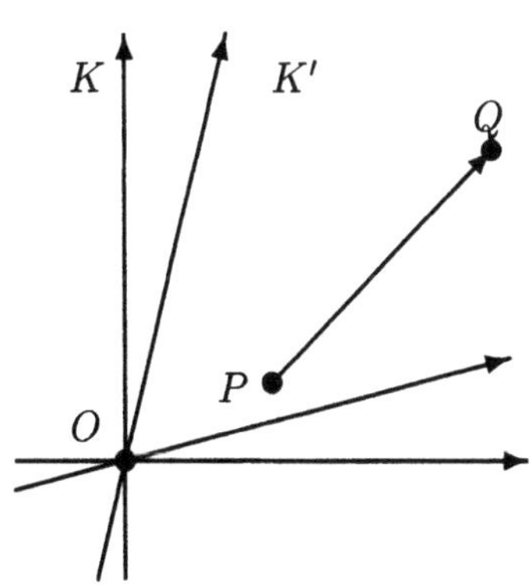


**Figure 2.34**

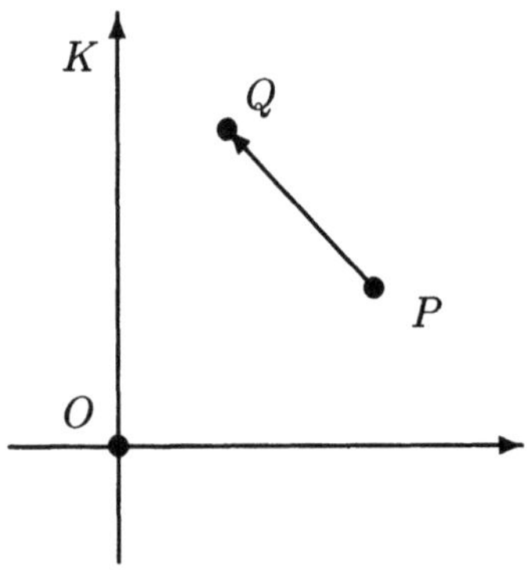


**Figure 2.35**

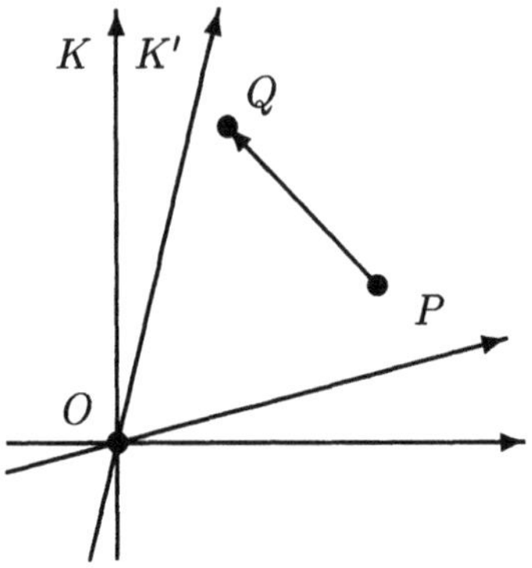


**Figure 2.36**

$$\begin{aligned}\overrightarrow{PQ} &= \Delta'^0\gamma(\boldsymbol{E}_0 + V\boldsymbol{E}_1 + \boldsymbol{E}_1 + V\boldsymbol{E}_0) \\ &= \Delta'^0\gamma(1+V)(\boldsymbol{E}_0 + \boldsymbol{E}_1)\end{aligned}$$

so

$$\Delta^0 = \Delta'^0\gamma(1+V) = \Delta'^0\sqrt{\frac{1+V}{1-V}}. \tag{2.64}$$

On the other hand, if the ray travels, with respect to $K$, in the $-x^1$-direction, as in Figure 2.35, then

$$\overrightarrow{PQ} = \Delta^0(\boldsymbol{E}_0 - \boldsymbol{E}_1),$$

with $\Delta^0 > 0$. Decomposed with respect to $K'$ (Figure 2.36),

$$\begin{aligned}\overrightarrow{PQ} &= \Delta'^0(\boldsymbol{E}'_0 - \boldsymbol{E}'_1). \\ &= \Delta'^0\gamma(1-V)(\boldsymbol{E}_0 - \boldsymbol{E}_1)\end{aligned}$$

so

$$\Delta^0 = \Delta'^0\gamma(1-V) = \Delta'^0\sqrt{\frac{1-V}{1+V}}. \tag{2.65}$$

***Relative Descriptions of a General Ray***

For a general ray travelling from $P$ to $Q$ in the $\boldsymbol{S}$-direction with respect to $K$ (where $\boldsymbol{S}$ is a unit spatial vector in $K$, $\boldsymbol{S}\cdot\boldsymbol{S} = 1$ and $\boldsymbol{E}_0\cdot\boldsymbol{S} = 0$), the form is

$$\overrightarrow{PQ} = \Delta^0(\boldsymbol{E}_0 + \boldsymbol{S}). \tag{2.66}$$

This expresses fully the description of the light ray $\overrightarrow{PQ}$ with respect to $K$. We again look for the description of $\overrightarrow{PQ}$ in a frame $K'$ boosted with respect to $K$.

After a rotation of the coordinate system (if necessary) the most general $K$-direction is $\boldsymbol{S} = \cos\vartheta \boldsymbol{E}_1 + \sin\vartheta \boldsymbol{E}_2$. Re-express $\overrightarrow{PQ}$ using the $K'$-basis by (2.12) to get

$$\begin{aligned}\overrightarrow{PQ} &= \Delta^0\gamma(\boldsymbol{E}'_0 - V\boldsymbol{E}'_1 + \cos\vartheta[\boldsymbol{E}'_1 - V\boldsymbol{E}'_0] + \frac{\sin\vartheta}{\gamma}\boldsymbol{E}'_2) \\ &= \Delta^0\gamma(1 - V\cos\vartheta) \\ &\quad \times \left[\boldsymbol{E}'_0 + \frac{(\cos\vartheta - V)\boldsymbol{E}'_1 + (\sin\vartheta/\gamma)\boldsymbol{E}'_2}{1 - V\cos\vartheta}\right]. \end{aligned} \tag{2.67}$$

This has the form

$$\overrightarrow{PQ} = \Delta'^0(\boldsymbol{E}'_0 + \boldsymbol{S}'), \qquad \boldsymbol{S}'\cdot\boldsymbol{S}' = 1$$

since

$$(\cos\vartheta - V)^2 + \gamma^{-2}\sin^2\vartheta = (1 - V\cos\vartheta)^2.$$

Therefore

$$\Delta'^0 = \Delta^0\gamma(1 - V\cos\vartheta)$$

and

$$\boldsymbol{S}' = \cos\vartheta'\boldsymbol{E}'_1 + \sin\vartheta'\boldsymbol{E}'_2, \qquad \boldsymbol{S}'\cdot\boldsymbol{E}'_0 = 0$$

with

$$\cos\vartheta' = \frac{\cos\vartheta - V}{1 - V\cos\vartheta}, \tag{2.68}$$

giving the direction of travel for the light ray with respect to the frame $K'$.

***Headlight Effect***

Inverting (2.68) gives

$$\cos\vartheta = \frac{\cos\vartheta' + V}{1 + V\cos\vartheta'}, \tag{2.69}$$

the same as replacing $V \to -V$, $\vartheta \leftrightarrow \vartheta'$. If we now consider the spray of light rays having directions with respect to $K'$ given by $0 \le \vartheta' \le \frac{\pi}{2}$, then the corresponding range of directions with respect to $K$ is, by (2.69), $V \le \cos\vartheta \le 1$, or

$$0 \le \vartheta \le \cos^{-1} V.$$

As $V \to 1$ the allowed directions in $K$ narrow to a small cone centred on the$+x^1$-direction. This has been called the "headlight effect". It can be very dramatic. Consider the radiation from an accelerating electron which is emitted into the hemisphere of its rest frame in the direction of its motion with respect to the lab. Then in the laboratory, for a 1BeV electron (say), whose rest mass is 0.5 MeV, $\gamma = 2000$ and the radiation is confined to a cone of angle $\vartheta = 0.03°$. In the lab the radiation occurs only in a very narrow pencil.

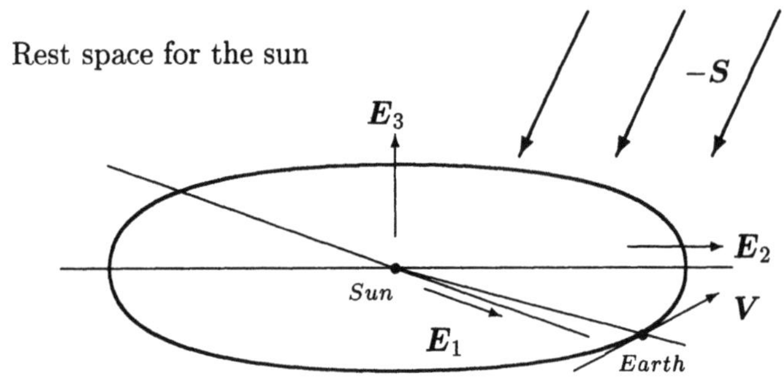


**Figure 2.37**

***Stellar Aberration***

The name "stellar aberration" was given to the first discovery, by James Bradley, of different relative descriptions of the parallel light rays from distant stars. In the early eighteenth century attempts were made to measure the distances to stars by triangulation from opposite ends of a diameter of the earth's orbit around the sun, an effect of parallax. As it turned out, the parallax effect was too small, even for the closest stars, to be measureable with the accuracy available (about half an arc second for angular measurements in the most favourable circumstances). Instead, a different effect was discovered, one which depended on the velocity of the earth rather than its position. The pattern of observations was quite different from the pattern that would be expected from parallax. The light from a distant star, received on earth anywhere in its orbit, arrives as light rays parallel to $\boldsymbol{E}_0 - \boldsymbol{S}$, where $\boldsymbol{E}_0$ is the time basis vector of an inertial rest frame for the sun, and $-\boldsymbol{S}$ is the spatial unit vector giving the direction with re-spect to the same frame (Figure 2.37). (To the extent that there was a slight nonparallel divergence of the light received over the earth's orbit, it was not measureable.) For convenience, choose a basis for the solar rest frame so that, for the star being studied,

$$\boldsymbol{S} = \cos\theta \boldsymbol{E}_2 + \sin\theta \boldsymbol{E}_3.$$

For measurements made from earth, it is appropriate to decompose $\boldsymbol{E}_0 - \boldsymbol{S}$ with respect to an instantaneous rest basis for the earth in order to get the direction of the star with respect to earth. (Although in principle one should consider the rest frame of a telescope on the rotating earth, the speed of rotation is negligible compared with the speed of orbital motion.) To simplify the calculation of a rest frame, an intermediate, rotated frame, $\boldsymbol{E}''_\mu$, in which the relative velocity vector of the earth is parallel to $\boldsymbol{E}''_2$, is useful:

$$\boldsymbol{E}''_1 = \cos\phi \boldsymbol{E}_1 + \sin\phi \boldsymbol{E}_2, \quad \boldsymbol{E}''_2 = -\sin\phi \boldsymbol{E}_1 + \cos\phi \boldsymbol{E}_2,$$

$$\boldsymbol{E}''_3 = \boldsymbol{E}_3, \quad \boldsymbol{E}''_0 = \boldsymbol{E}_0.$$

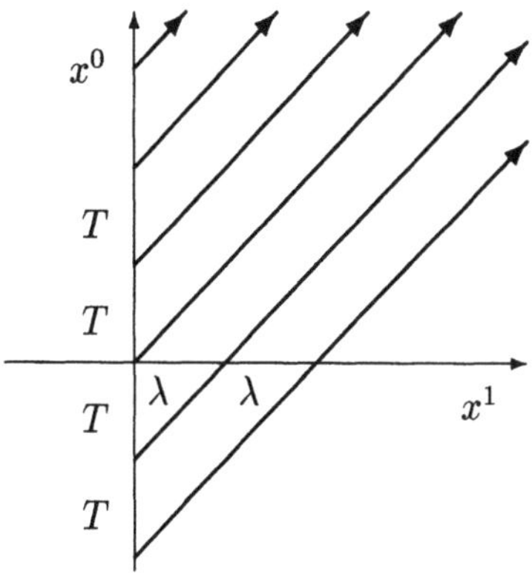


**Figure 2.38**

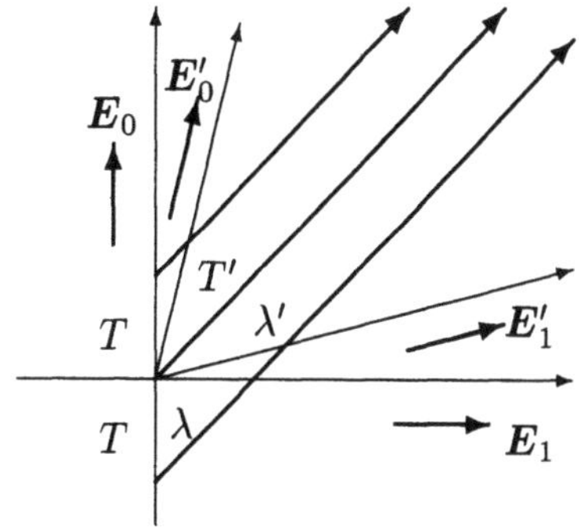


**Figure 2.39**

Neglecting the slight eccentricity of the earth's orbit, its relative velocity is $V\boldsymbol{E}_2''$ when the radius vector of the earth makes an angle $\phi$ with $\boldsymbol{E}_1$. A rest frame for the earth is now given by the boost

$$\boldsymbol{E}_0' = \gamma(\boldsymbol{E}_0'' + V\boldsymbol{E}_2''), \quad \boldsymbol{E}_2' = \gamma(\boldsymbol{E}_2'' + V\boldsymbol{E}_0''), \quad \boldsymbol{E}_1' = \boldsymbol{E}_1'', \quad \boldsymbol{E}_3' = \boldsymbol{E}_3''.$$

By expressing $\boldsymbol{E}_0 - \boldsymbol{S}$ in the $\boldsymbol{E}_\mu'$ basis, it takes the form

$$\boldsymbol{E}_0 - \boldsymbol{S} = \gamma(1 + V\cos\theta\cos\phi)(\boldsymbol{E}_0' - \boldsymbol{S}'),$$

where the unit vector $-\boldsymbol{S}'$, spatial in the earth's rest frame, is given by

$$\boldsymbol{S}' = \frac{\sin\theta\boldsymbol{E}_3' + \cos\theta\sin\phi\boldsymbol{E}_1' + \gamma(V + \cos\theta\cos\phi)\boldsymbol{E}_2'}{\gamma(1 + V\cos\theta\cos\phi)}.$$

The angle made by $\boldsymbol{S}'$ with the normal to the earth's orbit is given by $\sin\theta' = \sin\theta/\gamma(1 + V\cos\theta\cos\phi)$. The orbital speed of the earth with respect to the sun is about 30 km per second, so $V \approx 10^{-4}$ and $\gamma = 1$ is an excellent approximation. Calling the small angle of aberration $\alpha$, such that $\theta' = \theta + \alpha$, one finds to a close approximation $\alpha = -V\sin\theta\cos\phi$. For the earth at opposite ends of a diameter, as for $\phi = 0$ and $\phi = \pi$, the deviation $\alpha$ has opposite signs; its scale is set by the size of $V$, which as an angle is about 20 seconds of arc.[4]

### 2.7.2 Harmonic Light

Harmonic light is light with a frequency. In a classical theory of light it is difficult to think of a mechanism by which a frequency could be attributed

[4] There is an enjoyable account of the discovery of stellar aberration written by A.B. Stewart in Scientific American **210**, No.3 (1964) 100-108. Readers today should mentally substitute "position with respect to the rest frame of the sun" for the author's "true position".

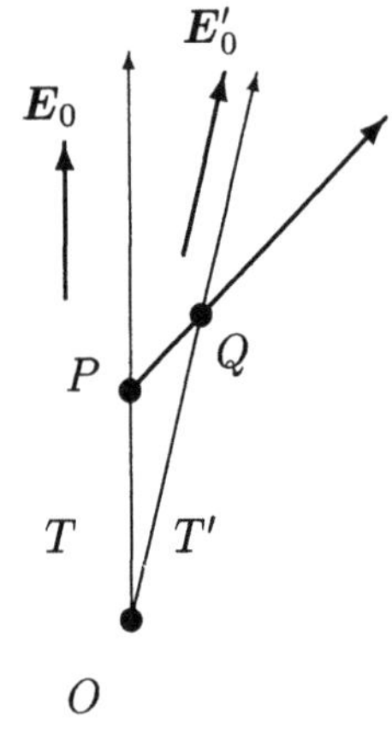


**Figure 2.40**

to a single ray of light. (In quantum theory it is difficult too, but there one is permitted not to ask the question.)

To get a provisional understanding of the frequency of light, and its relation to different reference frames, imagine a source of light at the spatial origin of a frame $K$ periodically emitting spherically symmetric signals, one signal in each $K$-time interval $T$ (Figure 2.38). The tracks of these signals on the $x^2 = x^3 = 0$ plane, for $x^1 \geq 0$, are shown in the diagram. The pattern is periodic in $K$-time $x^0$ with period $T$, and periodic in $x^1$ with period (wavelength) $\lambda = T$ (recall that the speed of light is unity). The diagram would be the same if light rays were emitted periodically in the $x^1$-direction.

When we consider the description of the pattern from the point of view of a frame $K'$ moving at $V$ in the $+x^1$-direction with respect to $K$ (Figure 2.39), we see that there are periodicities in $x'^0$ and $x'^1$ but with different periods $T'$ and $\lambda' = T'$.

In the triangle $OPQ$ (Figure 2.40), with $P$ and $Q$ on the first light ray after $O$,

$$\overrightarrow{OP} + \overrightarrow{PQ} = \overrightarrow{OQ}$$

or

$$T\boldsymbol{E}_0 + \Delta x^0(\boldsymbol{E}_0 + \boldsymbol{E}_1) = T'\boldsymbol{E}'_0,$$

where $\Delta x^0$ is the (unknown) time interval with respect to $K$ between $P$ and $Q$. We may eliminate $\Delta x^0$ by taking the scalar product with $(\boldsymbol{E}_0 + \boldsymbol{E}_1)$. Since $\boldsymbol{E}'_0 = \gamma(\boldsymbol{E}_0 + V\boldsymbol{E}_1)$, this gives

$$T = T'\gamma(1 - V) = T'\sqrt{\frac{1 - V}{1 + V}}. \tag{2.70}$$

In terms of frequencies, $f \equiv \frac{1}{T}$, $f' \equiv \frac{1}{T'}$,

$$f' = f\sqrt{\frac{1-V}{1+V}}. \tag{2.71}$$

The frequency for an observer at the $K'$-spatial origin is reduced compared to $f$ as he recedes from the source: this is the simplest instance of the *Doppler effect* for light.

***Momentum Vector for a Harmonic Light Ray***

Suppose that there is a geometrical, "absolute" vector representing a physical property of a light ray and parallel to it. With respect to $K$, possibly after a rotation it has the form $U(\boldsymbol{E}_0 + \boldsymbol{E}_1)$. With respect to $K'$, reached by a simple boost, there will be a similar expression and the two must be equal if the vector is indeed geometrical:

$$U(\boldsymbol{E}_0 + \boldsymbol{E}_1) = U'(\boldsymbol{E}_0' + \boldsymbol{E}_1').$$

Taking the scalar product with $\boldsymbol{E}_0$ immediately gives

$$U = U'\gamma(1+V) = U'\sqrt{\frac{1+V}{1-V}}. \tag{2.72}$$

We see from (2.71) that $U = f = 1/T$ satisfies the relation. Except for a multiplicative constant (Planck's constant, $h$) this gives the momentum vector for a photon of frequency $f$ (see Eq. (3.13)).

### 2.7.3 Scalar Field Theory for Light

We can approach a little closer to Maxwell's field theory of light, but still without having to deal with the complications of polarisation, by considering a scalar theory of light. We suppose that light is described by a scalar field $\phi$, which we regard as "real" (that is, a physically significant, coordinate-independent number at each point in spacetime). The field $\phi$ will give rise to lightlike rays if it satisfies, with respect to each frame $K$, the wave equation

$$\nabla^2\phi - \frac{\partial^2\phi}{\partial t^2} = 0 = \eta^{\mu\nu}\frac{\partial}{\partial x^\mu}\frac{\partial}{\partial x^\nu}\phi. \tag{2.73}$$

(Since $\phi$ is coordinate-independent,[5] if it satisfies this equation using the coordinates of $K$, it will satisfy the equation in the coordinates of any other inertial frame.)

[5] To be absolutely explicit mathematically, one distinguishes between the (geometrical, coordinate-independent) function $\phi(P)$ of points $P \in \mathcal{M}$ and the numerically equal, but coordinate-dependent, expression of it in coordinates $\phi(P) = \phi_K(x^\mu(P))$, with $x^\mu(P) \in \mathcal{R}^4$.

The plane wave solutions of (2.73) have speed one, as do the spherical solutions

$$\phi = \frac{g(r-t)}{r},$$

where $g$ is an arbitrary function and $r$ is the distance from the spatial origin.

We can regard the periodic signals with which we began our discussion of harmonic light as the phase maxima for the special case

$$\phi = \frac{A\cos\frac{2\pi}{T}(r-t)}{r}.$$

The intersections of the phase maxima with the plane $y = z = 0$ satisfy

$$x - t = x^1 - x^0 = nT, \qquad n = 0, \pm 1, \pm 2, \cdots.$$

Since $\phi$ is coordinate-independent, to get the description of this physical situation with respect to $K'$, moving at $V$ relative to $K$, we need only evaluate $\phi$ in the new coordinates. Regarding $r = \sqrt{x^2+y^2+z^2}$ and $t$ as functions of $t', x', y', z'$ according to the Lorentz transformation,

$$r = \sqrt{\gamma^2(x'+Vt')^2 + y'^2 + z'^2}, \quad t = \gamma(t'+Vx').$$

The phase maxima in the $y = z = 0$ plane (the $y' = z' = 0$ plane) are given by

$$\frac{1}{T}[\gamma(x'+Vt') - \gamma(t'+Vx')] = n,$$

or

$$x' - t' = nT/[\gamma(1-V)] = nT',$$

as before with the new, $K'$-period $T'$ by equation (2.70).

The harmonic plane wave solutions of (2.73) are interesting. They may be written

$$\phi = A\cos(\boldsymbol{h}\cdot\boldsymbol{x}), \tag{2.74}$$

with $\boldsymbol{x} = \overrightarrow{OP} = x^\mu \boldsymbol{E}_\mu$, for each $\boldsymbol{h}$ such that $\boldsymbol{h}\cdot\boldsymbol{h} = 0$. The expression (2.74) is written, appropriately, in coordinate-independent form. To express $\phi$ using (say) $K$-coordinates, the vectors $\boldsymbol{x}$ and $\boldsymbol{h}$ must be expanded in the $K$-basis, $\boldsymbol{x} = x^\mu \boldsymbol{E}_\mu, \;\; \boldsymbol{h} = h^\mu \boldsymbol{E}_\mu$, and then $\boldsymbol{h}\cdot\boldsymbol{x} = -h^0x^0 + h^m x^m$.

What is the significance of the vector $\boldsymbol{h}$? The phase maxima of $\phi$ (or any constant phase) satisfy

$$\boldsymbol{h}\cdot\boldsymbol{x} = \text{constant} = C, \quad \text{say}.$$

This defines a three-dimensional hyperplane in spacetime. If $P_0$ given by $\boldsymbol{x}_0 = \overrightarrow{OP_0}$ is on the hyperplane, that is, $\boldsymbol{h}\cdot\boldsymbol{x}_0 = C$, then other points on the hyperplane are given by $\boldsymbol{x} = \boldsymbol{x}_0 + \boldsymbol{y}$, whenever $\boldsymbol{h}\cdot\boldsymbol{y} = 0$. One solution of this equation is just $\boldsymbol{y} = \boldsymbol{h}$, or a multiple of $\boldsymbol{h}$, since $\boldsymbol{h}$ is lightlike. The two other independent solutions for $\boldsymbol{y}$ are spacelike. So the hyperplane contains one lightlike direction and two independent spacelike directions.

As an example, we choose $\boldsymbol{h}$ to reproduce the pattern of phase maxima in the $y = z = 0$ plane with which we started (but now with a plane wave, not a spherical wave):

$$\boldsymbol{h} = \frac{2\pi}{T}(\boldsymbol{E}_0 + \boldsymbol{E}_1), \quad \boldsymbol{h} \cdot \boldsymbol{x} = \frac{2\pi}{T}(x^1 - x^0).$$

Apart from a factor $\hbar \equiv h/2\pi$, the vector $\boldsymbol{h}$ is the four-momentum vector for the photons of which the scalar field is composed. The phase maximum through the spatial origin has $C = 0$, and $\boldsymbol{x}_0 = \boldsymbol{0}$ gives $P_0 = O$. The solutions $\boldsymbol{y}$ are given by

$$\boldsymbol{y} = \alpha(\boldsymbol{E}_0 + \boldsymbol{E}_1) + \beta\boldsymbol{E}_2 + \gamma\boldsymbol{E}_3. \tag{2.75}$$

The diagram of phase maxima showed only the points for which $\beta = \gamma = 0$. It is therefore to some extent misleading in suggesting (wrongly) that the planes of constant phase contain only light rays.

## 2.8 Description of Uniformly Moving Objects

We consider objects which have some inertial frame $\tilde{K}$ as a rest frame. Then if $\tilde{K}$ moves at a relative velocity $\boldsymbol{V}$ with respect to another frame $K$, every particle of the object moves at $\boldsymbol{V}$ with respect to $K$. This is what is meant by "uniformly moving". It is a severe limitation. We do not consider any sort of accelerated motion in this context. This does not mean that bodies do not accelerate; of course they do, but describing them is immensely complicated because they cease to be rigid.

To describe an object in its rest frame $\tilde{K}$ it is sufficient to describe its spatial configuration at a fixed $\tilde{K}$-time, say $\tilde{x}^0 = 0$. If the particles of the object are at spacetime positions $P, Q, \cdots$ at $\tilde{x}^0 = 0$, then at a later $\tilde{K}$-time $\tilde{x}^0$ they will each have been displaced in spacetime by $\tilde{x}^0\tilde{\boldsymbol{E}}_0$. Since the worldlines of the uniformly moving particles are straight, this determines them completely.

- To describe a straight rod at rest in $\tilde{K}$ it is sufficient to give the position of one end at $\tilde{x}^0 = 0$ and a spatial vector $\tilde{\boldsymbol{L}}$ $(\tilde{\boldsymbol{L}} \cdot \tilde{\boldsymbol{E}}_0 = 0)$ taking us to the other end.
- For a plane lamina in the shape of a parallelogram, the geometry is determined by one vertex at the rest time $\tilde{x}^0 = 0$ and an ordered pair of spatial vectors $\tilde{\boldsymbol{L}}$, $\tilde{\boldsymbol{M}}$. The order of the vectors determines the orientation of the parallelogram (which here gives a way of specifying which side of the lamina is to be called the top side).

Each edge of an object at rest in $\tilde{K}$, or each rod at rest in $\tilde{K}$, determines a sheet of worldlines in spacetime, all parallel to $\tilde{\boldsymbol{E}}_0$ (Figure 2.41).

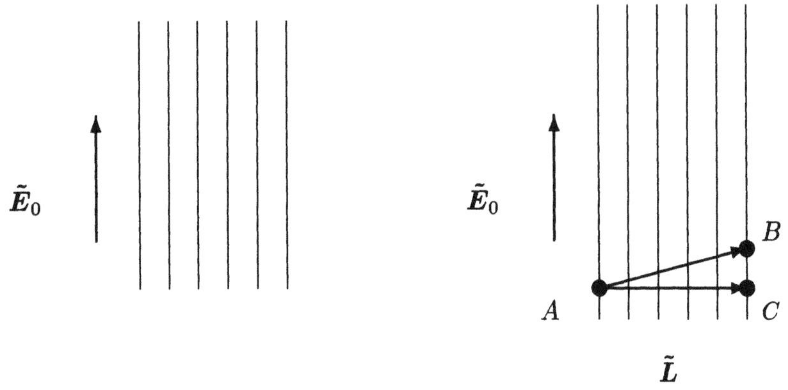


**Figure 2.41**

For any event $A$ on the worldline of one end and any event $B$ on the worldline of the other end, the vector $\overrightarrow{AB}$ has the form

$$\overrightarrow{AB} = \overrightarrow{AC} + \overrightarrow{CB} = \tilde{\boldsymbol{L}} + \lambda\tilde{\boldsymbol{E}}_0, \tag{2.76}$$

where $C$ is $\tilde{K}$-simultaneous with $A$, and $\lambda$ is the $\tilde{K}$-time interval between $C$ and $B$ (or $A$ and $B$).

Suppose $K$ is another inertial frame, with time basis vector $\boldsymbol{E}_0$, and we are interested in the description of the object with respect to $K$. Put

$$\gamma \equiv -\boldsymbol{E}_0 \cdot \tilde{\boldsymbol{E}}_0. \tag{2.77}$$

Then, calling $\boldsymbol{V}$ the relative velocity of $\tilde{K}$ with respect to $K$, and $\tilde{\boldsymbol{V}}$ the relative velocity of $K$ with respect to $\tilde{K}$, we may decompose

$$\tilde{\boldsymbol{E}}_0 = \gamma(\boldsymbol{E}_0 + \boldsymbol{V}), \tag{2.78}$$

$$\boldsymbol{E}_0 = \gamma(\tilde{\boldsymbol{E}}_0 + \tilde{\boldsymbol{V}}). \tag{2.79}$$

These could be solved to give $\boldsymbol{V}$ and $\tilde{\boldsymbol{V}}$ in terms of $\boldsymbol{E}_0$ and $\tilde{\boldsymbol{E}}_0$.

To describe with respect to $K$ an object which is at rest in $\tilde{K}$, it is convenient to specify the $K$-spatial location of the object at a fixed $K$-time, $x^0 = 0$, say. Every particle of the object has relative velocity $\boldsymbol{V}$ with respect to $K$ so at a later $K$-time $x^0$ each will have been displaced in spacetime by $x^0(\boldsymbol{E}_0 + \boldsymbol{V})$ from its location at $x^0 = 0$.

Consider again the sheet of worldlines for an edge or rod at rest in $\tilde{K}$ (Figure 2.42) Choose $A$ at the intersection of the worldline for one end with the hyperplane $x^0 = 0$. Then $B$ is specified by the condition that it is on the same hyperplane, $\overrightarrow{AB} \cdot \boldsymbol{E}_0 = 0$, or, by (2.76),

$$(\tilde{\boldsymbol{L}} + \lambda\tilde{\boldsymbol{E}}_0) \cdot \boldsymbol{E}_0 = 0. \tag{2.80}$$

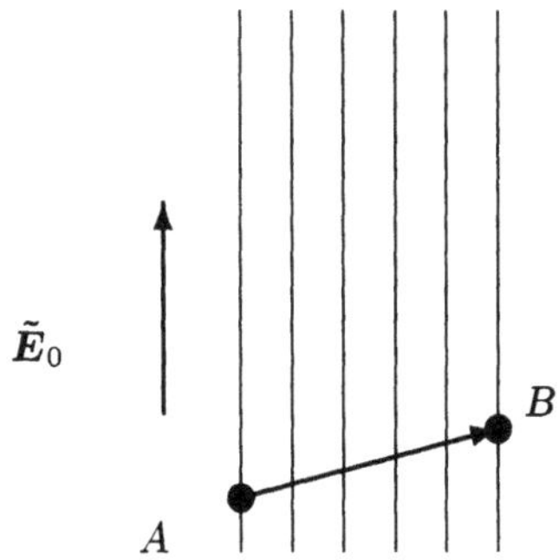


**Figure 2.42**

This gives

$$\gamma\tilde{\boldsymbol{L}}\cdot\tilde{\boldsymbol{V}} - \lambda\gamma = 0,$$

so that

$$\overrightarrow{AB} \equiv \boldsymbol{L} = \tilde{\boldsymbol{L}} + (\tilde{\boldsymbol{L}}\cdot\tilde{\boldsymbol{V}})\tilde{\boldsymbol{E}}_0 \tag{2.81}$$

determines the position with respect to $K$ of the moving rod. $\overrightarrow{AB} = \boldsymbol{L}$ is the vector taking us from one end of the rod to the other at the same $K$-time.

For further calculations it is helpful to have arranged that the spatial bases of $K$ and $\tilde{K}$ are chosen so that

$$\boldsymbol{V} = V\boldsymbol{E}_1, \quad \tilde{\boldsymbol{V}} = -V\tilde{\boldsymbol{E}}_1 \quad \text{and} \quad \boldsymbol{E}_2 = \tilde{\boldsymbol{E}}_2, \quad \boldsymbol{E}_3 = \tilde{\boldsymbol{E}}_3.$$

This makes the connection between the bases as simple as possible. They are in the form of a standard boost.

With the conveniently related bases, (2.81) takes the simple form

$$\begin{aligned} \boldsymbol{L} &= \tilde{\boldsymbol{L}} - V\tilde{L}_1\gamma(\boldsymbol{E}_0 + V\boldsymbol{E}_1) \\ &= \tilde{L}_2\boldsymbol{E}_2 + \tilde{L}_3\boldsymbol{E}_3 + \tilde{L}_1\left[\gamma(\boldsymbol{E}_1 + V\boldsymbol{E}_0) - V\gamma(\boldsymbol{E}_0 + V\boldsymbol{E}_1)\right] \\ &= \tilde{L}_2\boldsymbol{E}_2 + \tilde{L}_3\boldsymbol{E}_3 + (\tilde{L}_1/\gamma)\boldsymbol{E}_1. \end{aligned} \tag{2.82}$$

This must be an expected result: the components perpendicular to the plane $(\boldsymbol{E}_0,\ \tilde{\boldsymbol{E}}_0)$ of relative motion are unchanged, while the $\tilde{\boldsymbol{E}}_1$-component of $\tilde{\boldsymbol{L}}$ becomes, after shortening by a factor $\gamma^{-1}$, the $\boldsymbol{E}_1$-component of $\boldsymbol{L}$.

### 2.8.1 Example: Rod Lying in the Direction of Motion

In this example, and the following ones, the conveniently related bases and equation (2.82) are used.

A rod at rest in $\tilde{K}$, lying along the $\tilde{\boldsymbol{E}}_1$-axis, and of rest length $\tilde{L}$, is described in $K$ by

$$\boldsymbol{L} = (\tilde{L}/\gamma)\boldsymbol{E}_1.$$

If we agree to use the word "appearance" in $K$ for the description of the rod on the simultaneity hyperplane $x^0 = 0$, then it appears[6] to have the direction $\boldsymbol{E}_1$ and to be shortened by the factor $\gamma^{-1} = \sqrt{1 - V^2}$. This is again the Lorentz-Fitzgerald contraction.

### 2.8.2 Example: Rod at an Angle to the Direction of Motion

A rod at rest in $\tilde{K}$, of $\tilde{K}$-length $\tilde{L}$, lying in the $(\tilde{\boldsymbol{E}}_1, \tilde{\boldsymbol{E}}_2)$ plane, inclined at an angle $\tilde{\vartheta}$ to the $\tilde{\boldsymbol{E}}_1$-axis, is described in $\tilde{K}$ by the $\tilde{K}$-spatial vector

$$\tilde{\boldsymbol{L}} = \tilde{L}(\cos\tilde{\vartheta}\tilde{\boldsymbol{E}}_1 + \sin\tilde{\vartheta}\tilde{\boldsymbol{E}}_2),$$

and in $K$ on a hyperplane of constant $x^0$ by

$$\boldsymbol{L} = \tilde{L}(\frac{\cos\tilde{\vartheta}}{\gamma}\boldsymbol{E}_1 + \sin\tilde{\vartheta}\boldsymbol{E}_2).$$

In $K$ it "appears" to be inclined at an angle $\vartheta$ given by

$$\tan\vartheta = \gamma\tan\tilde{\vartheta} \tag{2.83}$$

to the $\boldsymbol{E}_1$-axis. Its length in $K$ is

$$L = \tilde{L}\sqrt{\sin^2\tilde{\vartheta} + \cos^2\tilde{\vartheta}(1 - V^2)}, \tag{2.84}$$

as would be expected if only the $\boldsymbol{E}_1$-component were shortened.

### 2.8.3 Example: Parallelogram at Rest in $\tilde{K}$

A parallelogram at rest in $\tilde{K}$ and described by edges

$$\tilde{\boldsymbol{L}} = \tilde{L}_i\tilde{\boldsymbol{E}}_i, \qquad \tilde{\boldsymbol{M}} = \tilde{M}_i\tilde{\boldsymbol{E}}_i,$$

is described on a hyperplane of simultaneity for $K$ by

$$\boldsymbol{L} = (\tilde{L}_1/\gamma)\boldsymbol{E}_1 + \tilde{L}_2\boldsymbol{E}_2 + \tilde{L}_3\boldsymbol{E}_3 = (\tilde{L}_1/\gamma)\boldsymbol{E}_1 + \boldsymbol{L}_\perp \tag{2.85}$$

$$\boldsymbol{M} = (\tilde{M}_1/\gamma)\boldsymbol{E}_1 + \tilde{M}_2\boldsymbol{E}_2 + \tilde{M}_3\boldsymbol{E}_3 = (\tilde{M}_1/\gamma)\boldsymbol{E}_1 + \boldsymbol{M}_\perp. \tag{2.86}$$

The geometry of the parallelogram as it "appears" in $K$ is completely determined by these two vectors.

---

[6] This is a new and convenient definition; the word is not to be interpreted literally. What the eye actually sees is quite different and requires a study of the light rays from the object that enter the eye. See R. Penrose, The apparent shape of a relativistically moving sphere, Proc. Camb. Phil. Soc. **55** (1959) 137-9, and also the discussion in Rindler's book.

### 2.8.4 Example: Parallelepiped at Rest in $\tilde{K}$

For a parallelepiped at rest in $\tilde{K}$, determined by a parallelogram and a third edge $\tilde{\boldsymbol{N}} = \tilde{N}_i \tilde{\boldsymbol{E}}_i$, the appearance in $K$ is given by $\boldsymbol{L}$, $\boldsymbol{M}$ and

$$\boldsymbol{N} = (\tilde{N}_1/\gamma)\boldsymbol{E}_1 + \tilde{N}_2\boldsymbol{E}_2 + \tilde{N}_3\boldsymbol{E}_3 = (\tilde{N}_1/\gamma)\boldsymbol{E}_1 + \boldsymbol{N}_\perp. \tag{2.87}$$

The $K$-volume on a hyperplane of constant $x^0$ is given, using the vector product (2.20), by

$$(K\text{-volume}) = \boldsymbol{L} \times \boldsymbol{M} \cdot \boldsymbol{N} = \frac{1}{\gamma}\tilde{\boldsymbol{L}} \times \tilde{\boldsymbol{M}} \cdot \tilde{\boldsymbol{N}} = \frac{1}{\gamma}(\tilde{K}\text{-volume}). \tag{2.88}$$

(The cross-product $\times$ that appears on the left involves only $K$-spatial vectors while on the right only $\tilde{K}$-spatial vectors are present; the connection is made with the components in the "convenient" bases.)

### 2.8.5 Example: A Uniformly Moving Rod Can Appear to Dip

In some situations the "convenient" basis we have used above is, ironically, not so very convenient after all and a discussion from first principles proves to be simpler.

A rod $\tilde{\boldsymbol{L}} = \tilde{L}\tilde{\boldsymbol{E}}_1$ is at rest in $\tilde{K}$. The frame $\tilde{K}$ moves at speed $U$ in the $\boldsymbol{E}'_2$-direction with respect to $K'$, and is boost-related to $K'$. The frame $K'$, as usual, moves at relative velocity $V\boldsymbol{E}_1$ with respect to $K$, and is boost-related to it. We are interested in getting a description of the rod from the point of view of the frame $K$. Surprisingly, it turns out that the rod, with respect to $K$, is not parallel to the $x$-axis.

Equations (2.12), with $\gamma_V$ written in place of $\gamma$, relate the bases for $K$ and $K'$. Similarly, the bases for $K'$ and $\tilde{K}$ are related by

$$\left.\begin{array}{rcl} \tilde{\boldsymbol{E}}_0 & = & \gamma_U(\boldsymbol{E}'_0 + U\boldsymbol{E}'_2) \\ \tilde{\boldsymbol{E}}_2 & = & \gamma_U(\boldsymbol{E}'_2 + U\boldsymbol{E}'_0) \\ \tilde{\boldsymbol{E}}_1 & = & \boldsymbol{E}'_1, \quad \tilde{\boldsymbol{E}}_3 = \boldsymbol{E}'_3. \end{array}\right\} \tag{2.89}$$

Therefore,

$$\tilde{\boldsymbol{E}}_0 = \gamma_U\gamma_V\left[\boldsymbol{E}_0 + V\boldsymbol{E}_1 + \frac{U}{\gamma_V}\boldsymbol{E}_2\right]$$

and so the relative velocity of $\tilde{K}$ with respect to $K$ is

$$\boldsymbol{W} = V\boldsymbol{E}_1 + \frac{U}{\gamma_V}\boldsymbol{E}_2. \tag{2.90}$$

For the application to (2.81) we need the relative velocity of $K$ with respect to $\tilde{K}$. An identical calculation gives for it

$$\tilde{\boldsymbol{W}} = -U\tilde{\boldsymbol{E}}_2 - \frac{V}{\gamma_U}\tilde{\boldsymbol{E}}_1. \tag{2.91}$$

By (2.81) the description of the rod in $K$ is given by the $K$-spatial vector

$$\boldsymbol{L} = \tilde{\boldsymbol{L}} + \tilde{\boldsymbol{E}}_0(\tilde{\boldsymbol{L}} \cdot \tilde{\boldsymbol{W}}). \tag{2.92}$$

Using $\tilde{\boldsymbol{L}} = \tilde{L}\tilde{\boldsymbol{E}}_1$ to express this vector in the $K$-basis we get

$$\boldsymbol{L} = (\tilde{L}/\gamma_V)\boldsymbol{E}_1 - (\tilde{L})UV\boldsymbol{E}_2 = (\tilde{L}/\gamma_V)\left[\boldsymbol{E}_1 - \gamma_V UV\boldsymbol{E}_2\right]. \tag{2.93}$$

We see the effect of contraction, but also we see that the rod appears to dip with respect to the $x$-axis by an angle $\tan^{-1}\gamma_V UV$ in the direction of $-\boldsymbol{E}_2$.

## 2.9 Problems

1. From the fundamental definition of vectors, show that $\boldsymbol{A} + \boldsymbol{B} = \boldsymbol{B} + \boldsymbol{A}$ and $\boldsymbol{A} + (\boldsymbol{B} + \boldsymbol{C}) = (\boldsymbol{A} + \boldsymbol{B}) + \boldsymbol{C}$. Illustrate the results with diagrams.

2. Derive the transformation law for inertial basis vectors,

$$\boldsymbol{E}_\nu = a^\mu{}_\nu \boldsymbol{E}'_\mu,$$

   by expressing the definition (2.4) in $K'$-coordinates, and using vector addition and multiplication by scalars.

3. If $\boldsymbol{\Delta} = 2\boldsymbol{E}_0 + \boldsymbol{E}_1 - \boldsymbol{E}_2$, check that $\boldsymbol{\Delta}$ is FPTL and find an inertial basis $\boldsymbol{E}'_\mu$ such that $\boldsymbol{\Delta} = T\boldsymbol{E}'_0$, where $T$ is a number which should be found.

4. If $\boldsymbol{\Delta} = \boldsymbol{E}_0 + 2\boldsymbol{E}_1 - 2\boldsymbol{E}_2$, check that $\boldsymbol{\Delta}$ is spacelike and find an orthonormal basis $\boldsymbol{E}'_\mu$ such that $\boldsymbol{\Delta} = X\boldsymbol{E}'_1$. If $\boldsymbol{\Delta} = \overrightarrow{OP}$, describe two other frames with respect to one of which $P$ is earlier than $O$, with respect to the other of which $P$ is later than $O$.

5. Let $\boldsymbol{A}$ and $\boldsymbol{B}$ be two different, spacelike, unit vectors, $\boldsymbol{A} \cdot \boldsymbol{A} = \boldsymbol{B} \cdot \boldsymbol{B} = 1$. Show that if $|\boldsymbol{A} \cdot \boldsymbol{B}| < 1$, there is a frame of reference with respect to which both $\boldsymbol{A}$ and $\boldsymbol{B}$ are spatial, but that if $|\boldsymbol{A} \cdot \boldsymbol{B}| \geq 1$ there is no such frame.

6. If $\boldsymbol{E}_\mu$ is an inertial basis for the space of vectors, then another basis (not inertial) consists of the vectors $\boldsymbol{E}_\pm = \frac{1}{\sqrt{2}}(\boldsymbol{E}_0 \pm \boldsymbol{E}_1)$, $\boldsymbol{E}_2$, $\boldsymbol{E}_3$. Write down the expansion of an arbitrary vector $\boldsymbol{A}$ in this basis, expressing the coefficients as scalar products.

7. In an inertial frame $K$, with inertial coordinates $t = x^0$, $x^m$, we can use standard spherical polar coordinates $r, \vartheta, \varphi$ in place of $x^m$. Then, if we define a vector field $\boldsymbol{x}(P) = \overrightarrow{OP}$, in terms of displacements from the spacetime origin for $K$, the inertial basis vectors are given by $\boldsymbol{E}_\mu = \partial\boldsymbol{x}/\partial x^\mu$. Define basis vectors similarly related to the new coordinates by

$$\boldsymbol{\partial}_t \equiv \frac{\partial \boldsymbol{x}}{\partial t} = \boldsymbol{E}_0, \quad \boldsymbol{\partial}_r \equiv \frac{\partial \boldsymbol{x}}{\partial r}, \quad \boldsymbol{\partial}_\vartheta \equiv \frac{\partial \boldsymbol{x}}{\partial \vartheta}, \quad \boldsymbol{\partial}_\varphi \equiv \frac{\partial \boldsymbol{x}}{\partial \varphi}.$$

Calculate these explicitly in terms of the inertial basis, and work out the matrix of scalar products, with entries $\eta_{tt} \equiv \boldsymbol{\partial}_t \cdot \boldsymbol{\partial}_t$, $\eta_{tr} \equiv \boldsymbol{\partial}_t \cdot \boldsymbol{\partial}_r$, $\eta_{t\vartheta} \equiv \boldsymbol{\partial}_t \cdot \boldsymbol{\partial}_\vartheta, \ldots$. Finally, express an arbitrary vector $\boldsymbol{A} = A^\mu \boldsymbol{E}_\mu$ as a decomposition in the new basis.

8. Verify that if $\boldsymbol{E}_\mu$ is an inertial basis, then another one is $\boldsymbol{E}'_\mu$, given by

$$\begin{aligned}
\boldsymbol{E}'_0 &= \gamma(\boldsymbol{E}_0 + V\cos\vartheta\,\boldsymbol{E}_1 + V\sin\vartheta\,\boldsymbol{E}_2), \quad \gamma^{-2} = 1 - V^2, \\
\boldsymbol{E}'_1 &= \frac{\cos\vartheta - V}{1 - V\cos\vartheta}\gamma(V\boldsymbol{E}_0 + \cos\vartheta\,\boldsymbol{E}_1 + \sin\vartheta\,\boldsymbol{E}_2) \\
&\qquad + \frac{\sin\vartheta}{\gamma(1 - V\cos\vartheta)}(\sin\vartheta\,\boldsymbol{E}_1 - \cos\vartheta\,\boldsymbol{E}_2), \\
\boldsymbol{E}'_2 &= \boldsymbol{E}_2 + \frac{V\sin\vartheta}{1 - V\cos\vartheta}(\boldsymbol{E}_0 + \boldsymbol{E}_1), \quad \boldsymbol{E}'_3 = \boldsymbol{E}_3.
\end{aligned}$$

(Do not forget to check time and space orientations.)

9. Consider the set of future-pointing timelike vectors, together with the zero vector. Does it form a vector space (with usual addition and multiplication by reals)? Explain. The set of *all* timelike vectors, future- and past-pointing, together with the zero vector: does *it* form a vector space?

10. Let $\boldsymbol{t}$ be any future-pointing, timelike unit vector, and $\boldsymbol{U}$ be any future-pointing lightlike vector. Find another future-pointing lightlike vector $\boldsymbol{D}$ so $\boldsymbol{t}$ may be written

$$\boldsymbol{t} = \lambda(\boldsymbol{U} + \boldsymbol{D})$$

for some $\lambda$. Write an expression for $\lambda$.

11. The half-life of pions, when measured in their rest frame, is $2.5\times10^{-8}$secs. What is their half-life when measured with respect to a frame in which they are moving at speed $V = \sqrt{3}/2$?

12. Resolve the confusion in the following analysis. "Two rings rotate in a plane with equal and opposite angular velocities about a common centre. Amos rides on one ring, Europa on the other. At some moment they pass each other and their clocks agree. But at the moment they pass, Europa

sees Amos's clock running more slowly (so she expects to be ahead next time they pass), while Amos sees just the reverse." What really happens?

13. A train travels uniformly at speed $V$ with respect to the earth. An arrow, of rest length $L_0$, travels at speed $W$ with respect to the train towards the front. Define and calculate the length of the arrow with respect to the earth.

14. Frames $K$ and $K'$ are related by a standard boost. If $V$ is large enough, a 20-metre pole at rest in $K'$ will appear so short in $K$ that it will fit (for a moment) in a 10-metre barn, at rest in $K$. No one any longer disputes that. But the 10-metre barn will appear, in $K'$, to be 5 metres long. Resolve the apparent paradox with spacetime diagrams.

15. A train, of rest length $L_0$, travels on a straight track at speed $V_0$ (with respect to the earth). At the back of the train a trolley is moving forward at speed $V$ (with respect to the train). A gun on the trolley fires a bullet which moves at $W$ with respect to the trolley. Ignoring gravity and air resistance, calculate the time intervals taken for the bullet to reach the front of the train in the rest frames of (a) the train, (b) the trolley, (c) the bullet.

16. Just above the (flat) surface of the moon two rockets approach each other horizontally along the same straight line. Rocket $A$ travels at speed $V$ with respect to the moon, rocket $B$ at $W$. At moon time zero the rockets are separated by a moon distance $L$, and at this moon moment the rockets' clocks read zero too. When $A$'s clock reads $T$ it fires a missile towards $B$. The missile travels at $U$ with respect to $A$. Show that the time on $B$'s clock when the missile arrives at $B$ is $T'$, where

$$T' = \frac{L\gamma(V)(1+UV)+TU}{\gamma(V)\gamma(W)(V+W+U+VWU)},$$

with the notation $\gamma(V) = (1-V^2)^{-1/2}$.

17. The inertial frame $K'$, whose $y'$ and $z'$ axes are parallel to the $y$ and $z$ axes of $K$, moves at speed $V$ with respect to $K$, in the $+x$ direction. As the spatial origin of $K'$ passes the spatial origin of $K$, their master clocks read zero, and at this event a projectile moves in the $+x'$ direction with speed $W$ with respect to $K'$. After travelling for a $K'$-time interval $T'$, the projectile passes a second projectile moving in the $-x'$ direction at speed $U$ with respect to $K'$. At what $K$-time $T_1$ does the second projectile meet the spatial origin of $K'$? At what $K$-time $T_2$ does it meet the spatial origin of $K$?

18. At the moment a missile carrier passes its base at the spatial origin of the $K$-frame, its clock reads zero. It is travelling at relative speed $V$, in

the $\boldsymbol{E}_1$-direction, with respect to $K$. After a time $T_0$ (on its clock) the missile carrier fires a time bomb at speed $U$ with respect to itself, in the same $K$-direction.

a) If the bomb is set to explode after a time interval $T_1$ (on its own clock), what is the $K$-distance to the base when the explosion occurs?
b) When light from the explosion arrives at the missile carrier, what is the time on its clock?
c) If material particles of the exploded bomb travel in all directions at speed $W(> U)$ relative to the bomb's rest frame, at what time will they hit the missile carrier (according to the latter's clock)?

19. A Cruise missile, of rest length $C$, is travelling down a street in Xanadu at speed $V$, right next to the 6th floor windows of the Intercontinental Hotel, of rest length $H$. John Simpson, at one end of the 6th floor corridor, and his cameraman, at the other end, time its passage from the moment the tip of the missile reaches the hotel until the tail just leaves. How long does it take?

20. Frames $K$ and $K'$ are related by a standard boost. Two particles travel along the $x'$-axis of $K'$ at speed $U$, with a constant distance $d$ between them. Show that the distance between them in $K$ is

$$d\sqrt{1-V^2}/(1+UV).$$

21. A massive particle travels uniformly between two spatial points fixed in an inertial frame $K$. (You may think of a muon travelling between a point at the top of the earth's atmosphere to a point at sea level.) Its speed is $V$, the time it travels is $T$ and the distance travelled is $L$, all with respect to $K$. In the rest frame of the particle, $K'$, what is the distance between the two points? (If you're thinking the answer is zero, reconsider the relativistic meaning of "length". Zero is the answer to a different question: the distance between two spatial points fixed in $K'$.) What is the $K'$-time interval between the events where the particle meets the two points?

22. From the spacetime origin of an inertial frame $K$ two particles are shot out in perpendicular directions with the same speed $V$ (with respect to $K$). Show that the speed of one particle with respect to the rest frame of the other is $V\sqrt{2-V^2}$.

23. The inertial frame $K'$ is related by a standard boost to $K$ and they share a common spacetime origin $O$. If a free particle passes through $O$ and reaches a point with spatial coordinates $(a,b,c)$ in $K'$ at $K'$-time $T$, what is its spacetime velocity? What are its relative velocities with respect to $K'$ and to $K$?

24. With respect to a frame $K$ a particle has relative velocity $\boldsymbol{V}_1$. A second particle has relative velocity $\boldsymbol{V}_2$ with respect to $K$. Show that the relative velocity $\boldsymbol{V}_{21}$ of the first particle with respect to a rest frame for the second has magnitude given by

$$V_{21}^2 = 1 - (1 - V_1^2)(1 - V_2^2)/(1 - \boldsymbol{V}_1 \cdot \boldsymbol{V}_2)^2.$$

25. Two particles with mass have spacetime velocity vectors $\boldsymbol{v}_1$ and $\boldsymbol{v}_2$. Their relative velocities, with respect to a frame $K$ whose time basis vector is $\boldsymbol{E}_0 = \boldsymbol{v}$, are $\boldsymbol{V}_1$ and $\boldsymbol{V}_2$. Verify that

$$\boldsymbol{V}_1 - \boldsymbol{V}_2 = \frac{(-\boldsymbol{v} \cdot \boldsymbol{v}_2)\boldsymbol{v}_1 - (-\boldsymbol{v} \cdot \boldsymbol{v}_1)\boldsymbol{v}_2}{(-\boldsymbol{v} \cdot \boldsymbol{v}_2)(-\boldsymbol{v} \cdot \boldsymbol{v}_1)},$$

and describe in words the interpretation of this vector.

26. The straight worldlines of two observers, moving at speed $V$ relative to each other, cross at $O$. One observer, $K$, ejects particles at a $K$-time interval $T$ at speed $W$ relative to himself. Show that they are received by the other observer, $K'$, at a $K'$-time interval $T' = WT/[\gamma_V(W - V)]$. If the worldlines cross but $K$ moves with velocity $\boldsymbol{V}_0$ relative to a frame $K_0$, and $K'$ moves at velocity $\boldsymbol{V}_0'$ relative to $K_0$, and $K$ emits particles at a $K_0$-interval $T_0$ at speed $W$ with respect to himself, what is the $K_0$-interval at which they are received by $K'$?

27. In the laboratory a particle travels in a circle of radius $R$ at uniform angular velocity $\Omega$. The circle is in the $z = 0$ plane of the lab, centre at the origin. The particle travels anticlockwise with respect to the $z$-axis, and at lab time $t = 0$ it is at $x = R$, $y = 0$ (spatial point $A$). Use the lab's basis vectors $\boldsymbol{E}_\mu$ to write down an expression for the worldline of the particle as a displacement vector from the spacetime origin, $\boldsymbol{OP} = \boldsymbol{z}(t)$. With $\tau$ the proper time, calculate explicitly the vectors

$$\boldsymbol{v} = \frac{d\boldsymbol{z}}{d\tau}, \qquad \boldsymbol{a} = \frac{d\boldsymbol{v}}{d\tau}, \qquad \boldsymbol{b} = \frac{d\boldsymbol{a}}{d\tau},$$

and verify $\boldsymbol{v} \cdot \boldsymbol{a} = 0$, $\boldsymbol{a} \cdot \boldsymbol{a} + \boldsymbol{v} \cdot \boldsymbol{b} = 0$. Find the basis vectors for an instantaneous rest frame for the particle whenever it returns to $A$.

28. Two rockets are travelling uniformly and in the same direction along the positive $x$-axis of an inertial frame $K$. Rocket $A$ (whose speed is $V$ with respect to $K$) is well behind rocket $B$ (speed $W$ with respect to $K$), but $V > W$. At time zero according to a clock on $A$ a radar pulse is sent by $A$ in the direction of $B$. Having been reflected by $B$ the signal returns to $A$ at time $T_1$. Immediately, another radar pulse is sent to $B$ and it returns at $T_1 + T_2$. Show that $A$ will overtake $B$ at $A$-time $(T_1)^2/(T_1 - T_2)$.

29. A light source $S$, with frequency $f_0$ in its own rest frame, moves with constant speed $V$ (with respect to $K$) along the $+x$-axis of $K$. A receiver $R$ recedes from $S$, moving at speed $V$ (with respect to $K$) in the opposite direction, along $K$'s negative $x$-axis. What is the frequency of the light in the rest frame of $R$?

30. Frames $K$ and $K'$ are related by a standard boost, as are $K$ and $K''$ moving at $U$ with respect to $K$ in the $+x$-direction. A rod, inclined to the $x''$-axis, is at rest in $K''$. Show that its slope $m$ relative to the $x$-axis in $K$ is related to its slope $m'$ relative to the $x'$-axis in $K'$ by

$$m' = m\gamma(V)(1 - UV), \qquad \gamma(V) = 1/\sqrt{1 - V^2}.$$

[Note: slope=tangent of the angle made with the axis.]

31. Frames $K$ and $K'$ are boost-related in the standard way. A "rod" is instantaneously at rest at time $t' = 0$ in frame $K'$, stretched along the $x'$-axis from $x' = 0$ to $x' = L$. Each particle of the rod has relative acceleration $\boldsymbol{A}' = A\boldsymbol{E}'_2$ with respect to the $K'$-frame. Show that the configuration of the particles of the rod with respect to the $K$-frame, at the $K$-instant $t = 0$, is along the parabola $y = \frac{1}{2}AV^2\gamma^2x^2$.

32. Frames $K$ and $K'$ are related by a standard boost. A cylinder, with axis along the $x'$-axis, rotates uniformly with respect to $K'$ at constant angular velocity $\omega$. It does not translate with respect to $K'$. Show that, with respect to $K$, the cylinder is twisted through an angle $\omega V\gamma$ per unit $K$-length.

33. When a light ray is reflected by a mirror the law of reflection is most simply stated in a rest frame for the mirror: the ray is reflected at the same angle with respect to the vertical and in the same plane. With respect to the standard frames $K, K'$, a mirror is at rest with respect to $K'$ in the $x' = 0$ plane and a light ray approaches from the positive side at an angle $\theta$ with respect to the $x$-axis of $K$. Show that after reflection the angle it makes with the $x$-axis is $\theta_R$ given by

$$\cos\theta_R = \frac{2V + (1 + V^2)\cos\theta}{1 + V^2 + 2V\cos\theta}.$$

34. If $P$ and $Q$ are two different points on a timelike worldline, then the vector $\overrightarrow{PQ}$ is itself timelike. (The worldline given by $z(\lambda)$ is timelike if $dz/d\lambda$ is timelike for all $\lambda$.)

If $P$ and $Q$ are two different points on a backward lightcone with vertex at $O$, then either $\overrightarrow{PQ}$ is spacelike or it is lightlike, in which case $P, Q$, and $O$ are on a single light ray. (We get the same alternative with a forward lightcone, of course.)

Combining the two results we see that a particle cannot cross a lightcone (either forward or backward) twice. It's once or never.

35. Consider a future lightcone with vertex at some event $O$, and the timelike worldline of a particle described by $\overrightarrow{OP}(\tau) = \boldsymbol{z}(\tau)$. Suppose that for $\tau > 0$ (any value would do) the particle moves freely. Show that it crosses the lightcone exactly once. (We get the same result for a backward lightcone and free motion for $\tau < 0$.)

36. From the result in the previous question one might wonder if it is even possible for a timelike worldline to avoid crossing a lightcone. The worldlines that manage the trick are called "runaways" and the simplest of these are the so-called hyperbolic orbits. Consider a particle motion which, with respect to some inertial frame, is along a fixed spatial axis. Then the spacetime velocity vector has the form $\boldsymbol{v}(\tau) = \gamma(\tau)(\boldsymbol{E}_0 + V(\tau)\boldsymbol{E}_1)$. A $\tau$-dependent rest frame for the particle is given by $\boldsymbol{E}_0' = \boldsymbol{v}$ and $\boldsymbol{E}_1' = \gamma(\boldsymbol{E}_1 + V\boldsymbol{E}_0)$. Show that the spacetime acceleration for such a motion is given by $\boldsymbol{a}(\tau) = d\boldsymbol{v}/d\tau = (\gamma^2 dV/d\tau)\boldsymbol{E}_1'$. The hyperbolic motion is the one for which $\gamma^2 dV/d\tau = a$, a constant. In this case, with a convenient choice for the zero of propertime, $V = \tanh a\tau$, $\gamma = \cosh a\tau$, $V\gamma = \sinh a\tau$. The hyperbolic motion requires a spacetime acceleration of constant magnitude in the instantaneous rest frame (we do not try to explain how this could physically be achieved). For this motion the equation $d\boldsymbol{a}/d\tau = (\boldsymbol{a}\cdot\boldsymbol{a})\boldsymbol{v}$ is satisfied. With a convenient choice for the initial position of the particle, the whole worldline is given by
$$\boldsymbol{z}(\tau) = \frac{1}{a}(\sinh a\tau \boldsymbol{E}_0 + \cosh a\tau \boldsymbol{E}_1).$$
Since $\boldsymbol{z}\cdot\boldsymbol{z} = \frac{1}{a^2} > 0$ for all $\tau$, this worldline, always timelike, remains spacelike with respect to $O$ and avoids both forward and backward lightcones with vertex $O$.

37. For the smoothed worldline (2.26) for the travelling twin, Ulysses, calculate the proper time $\tau(t)$ as a function of $K$-time $t$ with origin $O$. Show that
$$\tau(\frac{L}{V}) = \frac{L-\delta}{V\gamma_V} + \frac{\delta}{2V^2}[\frac{V}{\gamma_V} + \sin^{-1} V].$$
As the smoothing is removed, $\delta \to 0$, and the previous result for the proper time at the turning point $P$ is recovered, $L/(V\gamma_V)$.

38. Event $P$ is timelike displaced with respect to event $Q$. Show that the timelike curve connecting the two events with the greatest possible proper time is the straight line (no acceleration). For spacelike displaced events the straight line has the least proper distance.

39. Let $E$ and $O$ be any two events in spacetime. Let $l$ be a straight timelike worldline through $O$. Exactly two light rays through $E$ meet $l$: at events $P_1$ and $P_2$ which occur at times $T_1$ and $T_2$ according to a standard clock with worldline $l$ and origin (zero) $O$. Show that

$$\overrightarrow{OE} \cdot \overrightarrow{OE} = -T_1 T_2.$$

40. From an orthonormal inertial basis $\boldsymbol{E}_\mu$ construct a basis consisting of future-pointing lightlike vectors. Can the same be done with timelike vectors? With spacelike vectors?

41. If the Poincaré transformation between inertial frames $K$ and $K'$ is $x'^\mu = a^\mu + a^\mu{}_\nu x^\nu$, what are the relative velocities of $K$ with respect to $K'$ and of $K'$ with respect to $K$?

42. The Poincaré transformation that relates the coordinates of an event with respect to two inertial frames $K$ and $K'$ is $x'^\mu = a^\mu + a^\mu{}_\nu x^\nu$. A freely moving massive particle has a straight timelike worldline $l$. If the components of the relative velocities of the particle with respect to $K$ and to $K'$ are $V^m$ and $V'^m$ respectively, show that

$$V'^m = (a^m{}_0 + V^n a^m{}_n)/(a^0{}_0 + V^p a^0{}_p).$$

Suppose that the worldline $l$ intersects the $K$-simultaneous hyperplane $t = 0$ at $P$ with spatial $K$-coordinates $z^m$, and intersects the $K'$-simultaneous hyperplane $t' = 0$ at $P'$ with spatial $K'$-coordinates $z'^m$. Show that

$$z'^n = a^n + a^n{}_m z^m - \frac{(a^n{}_0 + V^p a^n{}_p)}{(a^0{}_0 + V^r a^0{}_r)}(a^0 + a^0{}_m z^m).$$

If we have a third inertial frame, $K''$, for which $x''^\mu = b^\mu + b^\mu{}_\nu x'^\nu = c^\mu + c^\mu{}_\nu x^\nu$, where $c^\mu \equiv b^\mu + b^\mu{}_\nu a^\nu$, and $c^\mu{}_\nu \equiv b^\mu{}_\lambda a^\lambda{}_\nu$, we may also use the intersection of $l$ with the hyperplane $t'' = 0$ at $P''$ with spatial $K''$-coordinates $z''^m$. Show that

$$\begin{aligned} V''^m &= (b^m{}_0 + V'^n b^m{}_n)/(b^0{}_0 + V'^r b^0{}_r) \\ &= (c^m{}_0 + V^n c^m{}_n)/(c^0{}_0 + V^r c^0{}_r), \end{aligned}$$

and

$$\begin{aligned} z''^n &= b^n + b^n{}_m z'^m - V''^n (b^0 + b^0{}_m z'^m) \\ &= c^n + c^n{}_m z^m - V''^n (c^0 + c^0{}_m z^m). \end{aligned}$$

This provides a 6-d nonlinear representation of the Poincaré group.

## References

A.P. French, Special Relativity (Nelson, 1968)
C.W. Misner, K.S. Thorne and J.A. Wheeler, Gravitation (Freeman, 1973)
W. Rindler, Introduction to Special Relativity (OUP, 1982)
B.F. Schutz, A First Course in General Relativity (CUP, 1985)
J.L.Synge, Relativity: the Special Theory, Second Edition (North-Holland, 1964)

# 3. Asymptotic Momentum Conservation

*Primitive Theory, Significant Consequences*

That momentum should satisfy a conservation law is, after relativistic invariance, the most important property of a physical theory in special relativity. (*Four*-momentum is meant here, as will always be understood when the word is not qualified.) In many theories of importance today the physical system is decomposed into simpler subsystems of particles and fields. Momentum is attributed to each subsystem, and if the subsystems interact momentum can be transferred between them, but overall it must be conserved.

Energy and three-momentum are relative concepts, dependent on the choice of an inertial frame. Four-momentum is a spacetime vector and energy is its temporal component, three-momentum its spatial part. Momentum conservation, expressed with vectors belonging to $\mathcal{V}$, is an example of an absolute, geometrical law. The conservation of energy (with respect to a given frame) and of three-momentum are the relative aspects of the law. The experimental content is the same but the geometrical formulation, being a single vector equation, is simpler and more elegant, and it *explains* the relative version, which is an infinite collection of equalities of components connected by transformation relations.

In the most detailed theories conservation of momentum is a local affair, holding in every spacetime volume no matter how small. In the present chapter, however, we consider a more primitive version. The physical processes now considered can usefully be regarded as a transformation of an incoming set of particles to an outgoing set of particles. Fields, as such, are neglected. Conservation of momentum in this context is both total and asymptotic, referring to the sum of momenta for all incoming or outgoing particles, in the distant past or far future, respectively. Conservation simply means that the two sums are equal.

## 3.1 Particle Momenta

In a classical theory a particle is specified by its mass and its worldline, timelike if the mass is positive and lightlike if the mass is zero. In the simplest

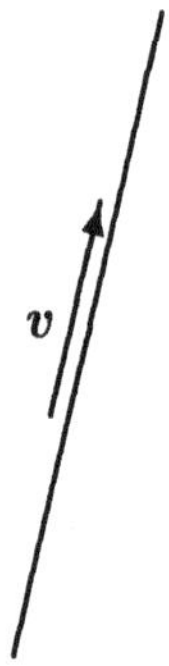


**Figure 3.1**

case it has no other attributes, but at the microscopic level its spin must be considered. In the present chapter we ignore the possible influence of spin on particle momentum.

The concept of mass which appeared in the last paragraph was taken over uncritically from Newtonian dynamics. In the present context it is perhaps better to regard mass as a secondary concept: for a particle it is the least value of the energy in the set of possible inertial frames. For a particle with positive mass a frame in which the minimum is attained is a rest frame. For a particle with zero mass the least value must be interpreted as the greatest lower bound, which cannot actually be attained. It is particularly helpful here to use units with $c = 1$ since it reduces the temptation to introduce redundant physical concepts. Energy, three-momentum and mass all have the same units.

### 3.1.1 Massive Particles

A noninteracting classical particle has a straight worldline. For a particle with a timelike worldline (which will have positive mass), if the the four-momentum is to have the direction of the worldline it will simply be a positive numerical multiple of the spacetime velocity vector (a future-pointing, unit timelike vector) (Figure 3.1).

$$\boldsymbol{p} = m\boldsymbol{v} \tag{3.1}$$

This is an absolute, or geometrical concept. The actual value of the multiple (the mass) $m$ depends on a choice of units and on experiments which implement momentum conservation. The mass value is characteristic of the type of particle.

In a first reading of special relativity the usual practice is to construct the four-momentum from the separate Newtonian ideas of mass, kinetic energy and three-momentum. Such a treatment requires somewhere a conceptual leap. In the present discussion, driven by the four-dimensional geometry, the leap occurs when theequation (3.1) is set down. The Newtonian ideas then re-emerge when the equation is decomposed with respect to an inertial frame.

For an interacting, accelerating particle, with a curved worldline, the definition is the same but $\boldsymbol{v}$, the unit tangent vector to the curve, is then a changing function of proper time. The definition (3.1) in special relativity is the simplest spacetime vector which reduces to the nonrelativistic forms for energy and three-momentum at low speeds (low speeds in the frame in which the energy and three-momentum are measured). This momentum appears in the relativistic generalisation of Newton's equations for a charged particle in an electromagnetic field. This same momentum appears in the theoretical explanation of the Compton effect and a myriad of other phenomena in high energy physics.

If we decompose the spacetime velocity vector $\boldsymbol{v}$ as a combination of the basis vectors for some inertial frame $K$, we get from (2.52) the form

$$\boldsymbol{p} = m\gamma(\boldsymbol{E}_0 + \boldsymbol{V}) = m\gamma\boldsymbol{E}_0 + m\gamma\boldsymbol{V}, \tag{3.2}$$

where $\boldsymbol{V} = V^m \boldsymbol{E}_m$, showing the geometrical four-momentum $\boldsymbol{p}$ decomposed into a unit temporal vector times the relative energy and a spatial vector, the relative three-momentum. These are new definitions of frame-dependent concepts. The coefficient of the time basis vector $\boldsymbol{E}_0$ is called the "energy with respect to $K$," or $K$-energy,

$$m\gamma \equiv E = -\boldsymbol{p} \cdot \boldsymbol{E}_0. \tag{3.3}$$

The $K$-spatial component is called the "three-momentum with respect to $K$"

$$m\gamma\boldsymbol{V} \equiv \boldsymbol{P} = \boldsymbol{p} - E\boldsymbol{E}_0. \tag{3.4}$$

In a rest frame for the particle the time basis vector is $\tilde{\boldsymbol{E}}_0 = \boldsymbol{v}$ and so $\boldsymbol{p} = m\tilde{\boldsymbol{E}}_0$. With respect to *this* frame the particle has only energy (its minimum energy, its rest energy or mass $m$), but no three-momentum.

The energy and the three-momentum are *relative* ideas, dependent for their definitions onsome inertial frame of reference. The same is inevitably true for any concept (except a Poincaré invariant) which has been inherited from Newtonian dynamics. For the energy the relation between the special relativistic (SR) and nonrelativistic (NR) concepts is given by

$$\begin{aligned} E = m\gamma &= m\frac{1}{\sqrt{1-V^2}} = m\left(1 + \frac{1}{2}V^2 + \frac{3}{8}V^4 + \ldots\right) && (3.5)\\ &= m(\text{rest mass}) + (\text{NR kinetic energy}) && (3.6)\\ &\qquad + (\text{SR corrections}). \end{aligned}$$

The first term on the right, $m$, is the celebrated $mc^2$ expressed in units such that $c = 1$. It is the value of the energy in a frame in which the particle is at rest ($\boldsymbol{V} = \boldsymbol{0}$). It is sometimes thought that the rest energy $m$ need not have a physical significance because "energy can be redefined by an additive constant." Such an argument is not correct; the form of the energies of individual particles is completely determined by the geometry, and the relation between them is determined by conservation of four-momentum. (It is true that the *total* momentum could be altered by a constant additive vector without changing the content of the conservation law, but this is clearly a trivial change without physical consequences.)

In any frame other than the rest frame the additional energy $T$ is called "kinetic." In this context the definition of kinetic energy is new, made to conform to nonrelativistic usage:

$$T \equiv E - m = (\text{NR kinetic energy}) + (\text{SR corrections}). \tag{3.7}$$

As $V^2 \to 0$, so $\gamma \to 1$. In a frame in which the particle moves slowly, the special relativistic three-momentum, $m\gamma\boldsymbol{V}$, differs insignificantly from $m\boldsymbol{V}$, the nonrelativistic expression for three-momentum.

The relative velocity of a particle of the simple type considered here is related to the (relative) three-momentum and energy by

$$\boldsymbol{V} = \frac{m\gamma\boldsymbol{V}}{m\gamma} = \frac{\boldsymbol{P}}{E}, \tag{3.8}$$

all quantities with respect to a given frame.

Since the spacetime velocity vector is a unit timelike vector, the four-momentum $\boldsymbol{p} = m\boldsymbol{v} \equiv p^\mu \boldsymbol{E}_\mu$ satisfies

$$\boldsymbol{p}\cdot\boldsymbol{p} = -m^2 = p^\mu p^\nu \eta_{\mu\nu} = -(p^0)^2 + p^m p^m. \tag{3.9}$$

The latter can be rearranged to give $p^0 = E = (m^2 + p^m p^m)^{1/2}$. Only the positive sign is relevant here for $p^0 = E = m\gamma$.

### 3.1.2 Massless Particles

If the four-momentum $\boldsymbol{k}$ of a particle with a lightlike worldline has the same direction in spacetime as the worldline, it too will be lightlike and therefore will satisfy

$$\boldsymbol{k}\cdot\boldsymbol{k} = 0. \tag{3.10}$$

By comparison of this equation with the formula $\boldsymbol{p}\cdot\boldsymbol{p} = -m^2$ for a massive particle with a timelike worldline, we are led to call the present particle massless. Indeed, its minimum energy is zero. It is easy to show that as one ranges over all inertial frames, all positive values of energy are possible. The greatest lower bound is zero. There can be no rest frame as there is for a massive particle, for if there were, the four-momentum would be a multiple of some $\tilde{\boldsymbol{E}}_0$, and this is not possible for a lightlike vector. The four-momentum

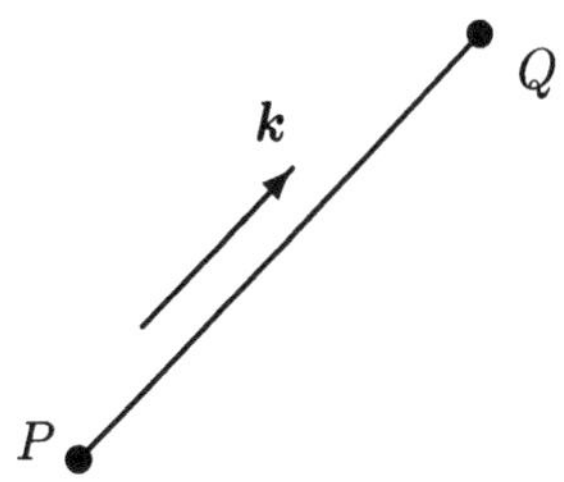


**Figure 3.2**

of the photon, the particle manifestation of light, is expected, almost by definition, to be a massless vector satisfying (3.10). The same is true for the conventional model of the neutrinos.[1] In physical applications of this theory one is usually in the quantum mechanical regime of electromagnetism. Hence photons, with sharp momenta, may not be supposed to have positions or worldlines. The existence of the vector $\boldsymbol{k}$ (Figure 3.2) is more important than any particular classical model for it; it is determined by a lightlike direction in spacetime, together with an appropriately transforming frequency for each inertial frame.

With respect to the basis $\boldsymbol{E}_\mu$ for the inertial frame $K$, the decomposition

$$\boldsymbol{k} = E\boldsymbol{E}_0 + \boldsymbol{K} = E(\boldsymbol{E}_0 + \boldsymbol{S}) \tag{3.11}$$

defines $E$, the energy with respect to $K$, and $\boldsymbol{K}$, the three-momentum with respect to $K$ ($\boldsymbol{K} \cdot \boldsymbol{E}_0 = 0,\ \boldsymbol{S} \cdot \boldsymbol{E}_0 = 0$). These quantities satisfy

$$E = -\boldsymbol{k} \cdot \boldsymbol{E}_0, \qquad \boldsymbol{K} = \boldsymbol{k} - E\boldsymbol{E}_0, \tag{3.12}$$

exactly as in (3.3) and (3.4) for the massive case.

Since the four-momentum vector $\boldsymbol{k}$ is geometrical (basis-independent) and has a lightlike direction, the energies with respect to different frames satisfy $\boldsymbol{k} = E(\boldsymbol{E}_0 + \boldsymbol{S}) = E'(\boldsymbol{E}_0' + \boldsymbol{S}')$. Equation (2.72) then shows how the energies are related. If the light particle has in each frame a unique frequency $f$, satisfying the same relation, (2.71), it is inevitable that the energy $E$ should be proportional to $f$. The constant of proportionality is called Planck's constant, $h$, so

$$E = hf. \tag{3.13}$$

The relation is well verified experimentally by the photoelectric effect.

---

[1] At the time of writing, a significant experiment in the deep mines at Kamiokande in Japan has found evidence for a small mass for the neutrinos. Confirmation is awaited.

The vector $\boldsymbol{S}$ is a $K$-spatial unit vector which gives the $K$-direction of the massless particle. In fact, $\boldsymbol{S}$ is the relative velocity with respect to $K$ ($\boldsymbol{S} \cdot \boldsymbol{S} = 1$, $\boldsymbol{S} = \boldsymbol{K}/E$, the latter to be compared with (3.8)).

### 3.1.3 Energy and Three-momentum of One Particle With Respect to the Rest Frame of Another Particle

Let a particle (the p-particle) have mass $M$, spacetime velocity $\boldsymbol{v}$, and momentum $\boldsymbol{p} = M\boldsymbol{v}$. For a second particle (the q-particle), the momentum is $\boldsymbol{q} = m\boldsymbol{u}$. A rest frame for the q-particle has time basis vector $\boldsymbol{E}_0 = \boldsymbol{u}$. If we expand the p-particle momentum in such a basis,

$$\boldsymbol{p} = p^0 \boldsymbol{E}_0 + p^m \boldsymbol{E}_m = (-\boldsymbol{E}_0 \cdot \boldsymbol{p})\boldsymbol{E}_0 + \boldsymbol{P} = E\boldsymbol{u} + \boldsymbol{P}. \tag{3.14}$$

Since $\boldsymbol{P} \cdot \boldsymbol{u} = 0$, the energy of the p-particle with respect to the rest frame of the q-particle is

$$E = -\boldsymbol{u} \cdot \boldsymbol{p} = -\frac{\boldsymbol{q} \cdot \boldsymbol{p}}{m}. \tag{3.15}$$

(*Vice versa,* the energy of the q-particle with respect to the rest frame of the p-particle is $-\boldsymbol{q} \cdot \boldsymbol{p}/M$. Recall that $-\boldsymbol{q} \cdot \boldsymbol{p} = -Mm\boldsymbol{v} \cdot \boldsymbol{u} = Mm\gamma$.) From (3.14) and (3.15) the three-momentum of the p-particle in the same frame is

$$\boldsymbol{P} = \boldsymbol{p} - E\boldsymbol{u} = \boldsymbol{p} + \frac{(\boldsymbol{q} \cdot \boldsymbol{p})\boldsymbol{q}}{m^2}, \tag{3.16}$$

and its magnitude is given by

$$\boldsymbol{P} \cdot \boldsymbol{P} = -M^2 + (\boldsymbol{q} \cdot \boldsymbol{p})^2/m^2 = -M^2 + E^2, \tag{3.17}$$

as must be so by (3.9). These equations are entirely geometrical, linear vectorial equations with coefficients defined in terms of scalar products with physical significance.

The relative velocity of the p-particle with respect to the rest frame of the q-particle is the $q$-spatial vector $\boldsymbol{U}$ given by an instance of (2.52),

$$\boldsymbol{v} = \gamma(\boldsymbol{u} + \boldsymbol{U}), \qquad \frac{1}{\gamma^2} = 1 - \boldsymbol{U} \cdot \boldsymbol{U}. \tag{3.18}$$

One can therefore get $U^2$ from $\gamma = -\boldsymbol{v} \cdot \boldsymbol{u} = -\boldsymbol{p} \cdot \boldsymbol{q}/Mm$.

## 3.2 Conservation of Asymptotic Momentum

Conservation of four-momentum is such an important physical law that one can very nearly say: a physical theory is a theoretical system which provides a realisation of momentum conservation (and whatever other absolute conservation laws the system obeys). Asymptotic momentum conservation

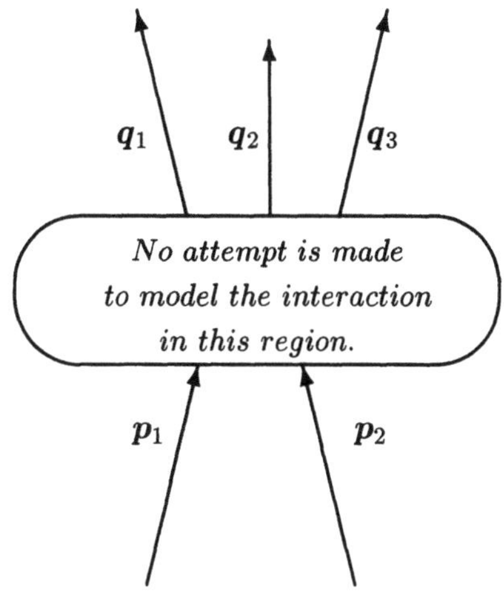


**Figure 3.3**

is a rather primitive implementation, valid in only very special, but very important, circumstances: in the distant past and in the distant future the system consists of widely separated, freely moving particles, and nothing else. For electromagnetically charged particles the conditions cannot be rigorously satisfied since the Coulomb force has infinite range. At high energies the electromagnetic forces may be negligible, or can be allowed for. Electromagnetic energy and three-momentum is included only in the form of photons (for example, gamma rays), particles of zero mass.

Figure 3.3 refers to a high energy, quantum-mechanical régime in which all four-momentum is attributable to particles, not fields, and in which long distance electromagnetic interactions between particles must be negligible. In these circumstances, four-momentum conservation is expressed by

$$\Sigma \boldsymbol{p}_i = \Sigma \boldsymbol{q}_i. \tag{3.19}$$

In general, we expect the particles participating in the process to move freely (and so have constant momentum vectors) only asymptotically, either as $t \to -\infty$ when the particles are approaching one another from infinite separations, or as $t \to +\infty$ when the final particles diverge ($t$ may be the time in any inertial frame). Later, in Chapter 5, a local form of four-momentum conservation will be formulated in terms of the divergence of the (stress-)energy(-momentum) tensor. This will only be possible when a theory is available for the interaction in finite regions.

Equation (3.19) is geometrical, or absolute, independent of particular inertial frames. Each particle is described by a single (geometrical) vector. By expanding all the momentum vectors in the basis vectors $\boldsymbol{E}_\mu$ of some inertial frame $K$, (3.19) can be rewritten as the four equations

$$\Sigma p_i^\mu = \Sigma q_i^\mu. \tag{3.20}$$

The $\mu = 0$ component expresses the equality of total $K$-energy before and after interaction; the spatial components do the same for the three-momentum in the $K$-frame. This latter version of momentum conservation is the *relative* form. If it is valid with respect to one frame, then by transforming the components it is valid with respect to all; also, it implies the geometrical form (3.19).

## 3.3 Three-particle Processes

Suppose a process $A \to B + C$ can occur physically, where the particles have masses $M$, $m$ and $\mu$, for the present all assumed to be nonzero (we will consider later the case when these may be zero), and momenta $\boldsymbol{p}$, $\boldsymbol{q}$, and $\boldsymbol{k}$, respectively (Figure 3.4). Then, by momentum conservation,

$$\boldsymbol{p} = \boldsymbol{q} + \boldsymbol{k}, \tag{3.21}$$

and $\boldsymbol{p} \cdot \boldsymbol{p} = -M^2$, $\boldsymbol{q} \cdot \boldsymbol{q} = -m^2$, $\boldsymbol{k} \cdot \boldsymbol{k} = -\mu^2$.

Taking the scalar product of $\boldsymbol{p}$ in (3.21) with itself gives

$$M^2 = m^2 + \mu^2 - 2\boldsymbol{q} \cdot \boldsymbol{k} = m^2 + \mu^2 + 2m\mu\gamma, \tag{3.22}$$

where $\gamma = -\boldsymbol{q} \cdot \boldsymbol{k}/(m\mu)$ is determined, as in (3.15) and (3.18), by the relative velocity of $B$ in the rest frame of $C$, or vice versa, by the relative velocity of $C$ in the rest frame of $B$. Since $\gamma \geq 1$, we have $M^2 \geq (m + \mu)^2$. Equality can occur only if $\gamma = 1$, that is, if the relative velocity of $C$ with respect to $B$ is zero. *If a massive particle decays into two other massive particles, the decay products are lighter than the original particle.* Exceptionally, the mass of the decay products may *equal* that of the original, but then the whole process occurs at a single spatial point in the original rest frame, like an apple gently split into two halves on the head of William Tell's son by a razor-sharp arrow.

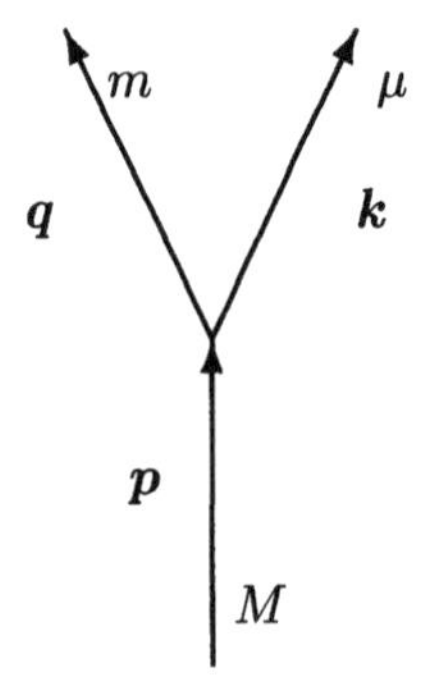


**Figure 3.4**

If the mass of $C$ is zero, $\mu = 0$, we will call the particle a photon. Equation (3.22) is replaced with

$$M^2 = m^2 - 2\boldsymbol{q} \cdot \boldsymbol{k} = m^2 + 2mE_{(B)}, \tag{3.23}$$

where, by (3.12), $E_{(B)}$ is the energy of the photon in the rest frame of $B$. Clearly, $E_{(B)} > 0$, for if $E_{(B)} = 0$ the momentum vector is zero and no particle exists. Hence, in this case, $M > m$. *A particle cannot emit a photon, or any other massless particle, without itself losing mass.* If the particle is elementary, for which only one mass is possible, it cannot emit a photon without changing to a different sort of particle, with a lower mass.

If both $m$ and $\mu$ are zero, and we express the momentum vectors with respect to an arbitrary frame $K$, $\boldsymbol{q} = E_B(\boldsymbol{E}_0 + \boldsymbol{S}_B)$, $\boldsymbol{k} = E_C(\boldsymbol{E}_0 + \boldsymbol{S}_C)$, so in place of (3.22) we get

$$M^2 = -2\boldsymbol{q} \cdot \boldsymbol{k} = 2E_B E_C(1 - \cos\vartheta), \tag{3.24}$$

where $\vartheta$ is the angle between the $K$-spatial unit vectors $\boldsymbol{S}_B$ and $\boldsymbol{S}_C$. $M$ must be positive in this case apart from the exceptional possibility of a photon decay into two lower energy photons travelling in the same direction. (If two photons travel in the same direction with respect to one frame of reference, they do so with respect to all frames, a statement that is not true for massive particles.)

The energy of particle $B$ in the rest frame of particle $A$ (which may very well be the laboratory in which the decay occurs) is, as in (3.15), $-\boldsymbol{p} \cdot \boldsymbol{q}/M$. Using (3.21) and(3.22), it may be written

$$\begin{aligned} -\frac{\boldsymbol{p}}{M} \cdot \boldsymbol{q} &= -\frac{1}{M}(-m^2 + \boldsymbol{q} \cdot \boldsymbol{k}) \\ &= \frac{1}{2M}(M^2 + m^2 - \mu^2). \end{aligned} \tag{3.25}$$

This formula may be obtained even more easily just by taking the scalar product of $\boldsymbol{k} = \boldsymbol{p} - \boldsymbol{q}$ with itself. Similarly, the energy of $C$ in the rest frame of $A$ is

$$-\frac{\boldsymbol{p}}{M} \cdot \boldsymbol{k} = \frac{1}{2M}(M^2 - m^2 + \mu^2). \tag{3.26}$$

The sum of these two energies of $B$ and $C$ is $M$, the total energy in the rest frame of $A$. The speeds of $B$ and $C$ in the decay frame can be calculated from these energies.

The algebra for this physical decay, which is all a consequence of momentum conservation (3.21), is exactly the same for the corresponding creation process, $B + C \rightarrow A$, since the equations of momentum conservation are exactly the same for the time-reversed process.

## 3.4 A Kinematical Function

There is a useful kinematical function, $\lambda(m_1^2, m_2^2, m_3^2)$, which arises whenever four-momentum is conserved at a three particle vertex. In the general case we have three timelike momentum vectors obeying

$$\boldsymbol{p}_1 + \boldsymbol{p}_2 + \boldsymbol{p}_3 = \boldsymbol{0}. \tag{3.27}$$

One or two of the momentum vectors here must be past-pointing; we may regard these as the negative of the physical momentum vectors by having masses the negative of the physical values. For each momentum vector, $\boldsymbol{p}_i = m_i \boldsymbol{v}_i$, where $\boldsymbol{v}_i$ is a timelike, future-pointing unit vector, and $m_i > 0$ if $\boldsymbol{p}_i$ is the physical vector, $m_i < 0$ if $-\boldsymbol{p}_i$ is physical; in both cases, $\boldsymbol{p}_i \cdot \boldsymbol{p}_i = -m_i^2$. In one application, $\boldsymbol{p}_1$ and $\boldsymbol{p}_2$ are physical particle momentum vectors, and $-\boldsymbol{p}_3$ is the total four momentum.

Without any real loss of generality we expand $\boldsymbol{p}_1$ in the rest frame of the number 2 particle:

$$\boldsymbol{p}_1 = \left( \frac{-\boldsymbol{p}_1 \cdot \boldsymbol{p}_2}{m_2^2} \right) \boldsymbol{p}_2 + \boldsymbol{P}_{12}. \tag{3.28}$$

The three-momentum of number 1 in the rest frame of number 2 is $\boldsymbol{P}_{12}$. Taking the scalar product of $\boldsymbol{P}_{12}$ with itself gives

$$\boldsymbol{P}_{12} \cdot \boldsymbol{P}_{12} = \left( \boldsymbol{p}_1 + \frac{\boldsymbol{p}_1 \cdot \boldsymbol{p}_2}{m_2^2} \boldsymbol{p}_2 \right) \cdot \left( \boldsymbol{p}_1 + \frac{\boldsymbol{p}_1 \cdot \boldsymbol{p}_2}{m_2^2} \boldsymbol{p}_2 \right), \tag{3.29}$$

or

$$\boldsymbol{P}_{12} \cdot \boldsymbol{P}_{12} = -m_1^2 + \left( \frac{\boldsymbol{p}_1 \cdot \boldsymbol{p}_2}{m_2^2} \right)^2 m_2^2 = \frac{4(\boldsymbol{p}_1 \cdot \boldsymbol{p}_2)^2 - 4m_1^2 m_2^2}{4m_2^2}. \tag{3.30}$$

The numerator in the final form proves to be a symmetrical function of the squares of the particle masses. Using (3.27) to get an expression for $(\boldsymbol{p}_1 \cdot \boldsymbol{p}_2)$,

$$\boldsymbol{P}_{12} \cdot \boldsymbol{P}_{12} = \frac{(m_3^2 - m_1^2 - m_2^2)^2 - 4m_1^2 m_2^2}{4m_2^2} \equiv \frac{\lambda(m_1^2, m_2^2, m_3^2)}{4m_2^2}. \tag{3.31}$$

The symmetry in the function $\lambda$ becomes obvious when the square in the definition is expanded,

$$\lambda = m_1^4 + m_2^4 + m_3^4 - 2m_1^2 m_2^2 - 2m_2^2 m_3^2 - 2m_3^2 m_1^2. \tag{3.32}$$

The symmetry leads to simple and useful kinematical relations. In a decay $M \to m + \mu$ (with particles represented by their masses), if $\boldsymbol{Q}$ is the three-momentum of the $m$-particle in the rest frame of the $M$-particle, and $\boldsymbol{K}$ is the three-momentum of the $\mu$-particle in the rest frame of the $m$-particle, then, by (3.31),

$$\lambda(M^2, m^2, \mu^2) = 4M^2 \boldsymbol{Q} \cdot \boldsymbol{Q} = 4m^2 \boldsymbol{K} \cdot \boldsymbol{K}, \tag{3.33}$$

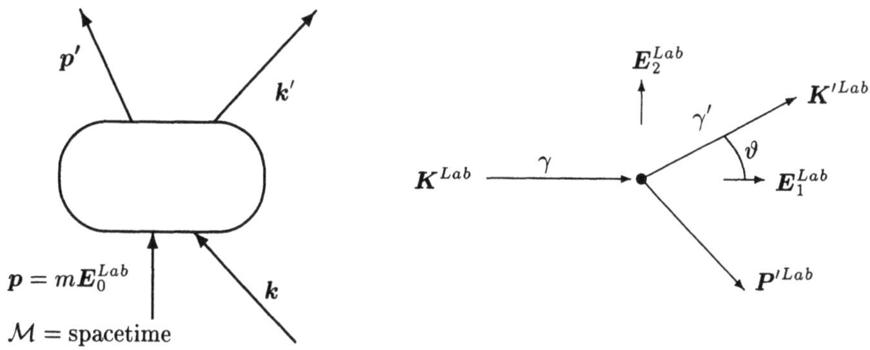


**Figure 3.5** Space in the laboratory.

a simple and significant geometrical relation.

As a second application, one is sometimes interested in comparing the magnitudes of the three-momenta of a given particle (say, number 1) in the rest frames of numbers 2 and 3. For this we have immediately the relation

$$\frac{1}{4}\lambda(m_1^2, m_2^2, m_3^2) = m_2^2 \boldsymbol{P}_{12} \cdot \boldsymbol{P}_{12} = m_3^2 \boldsymbol{P}_{13} \cdot \boldsymbol{P}_{13}. \tag{3.34}$$

## 3.5 Compton Effect

In the Compton effect (1923), a photon collides elastically with an electron, regarded as stationary in the laboratory. After the collision the (reduced) energy of the photon emerging at an angle $\vartheta$ to its original direction is measured. The significant experimental relation may be derived solely from asymptotic momentum conservation, (3.19). Historically, this great success of the "particle" model for light finally convinced the physics community as a whole of the correctness of Einstein's "photon" concept, originally put forward in 1905.

The left hand part of Figure 3.5 shows a standard two-particle elastic scattering process, and labels the asymptotic electron momenta before and after scattering $\boldsymbol{p}$ and $\boldsymbol{p}'$, for the photons $\boldsymbol{k}$ and $\boldsymbol{k}'$. The laboratory frame is the rest frame of the target electron before scattering. The right hand part shows that the $x^1$-axis is chosen along the direction of motion of the incident photon, and the $x^2$-axis is chosen to lie in the plane of motion of the photons before and after scattering. The scattering angle in the laboratory is $\vartheta$.

Asymptotic momentum conservation for the elastic scattering of a photon with an electron of mass $m$ is

$$\boldsymbol{p} + \boldsymbol{k} = \boldsymbol{p}' + \boldsymbol{k}'. \tag{3.35}$$

Since the recoil electron is not observed, and nothing is known about it except its mass, the recoil momentum is eliminated by evaluating $\boldsymbol{p}' \cdot \boldsymbol{p}' = -m^2$ with the help of momentum conservation; this gives

$$(\boldsymbol{p}+\boldsymbol{k}-\boldsymbol{k}')\cdot(\boldsymbol{p}+\boldsymbol{k}-\boldsymbol{k}')=-m^2.$$

Hence, with the laboratory decompositions

$$\begin{aligned}\boldsymbol{k} &= E(\boldsymbol{E}_0^{Lab}+\boldsymbol{E}_1^{Lab}) = E\boldsymbol{E}_0^{Lab}+\boldsymbol{K}^{Lab} = E(\boldsymbol{E}^{Lab}+\boldsymbol{S}^{Lab}),\\ \boldsymbol{k}' &= E'(\boldsymbol{E}_0^{Lab}+\cos\vartheta\boldsymbol{E}_1^{Lab}+\sin\vartheta\boldsymbol{E}_2^{Lab})\\ &= E'\boldsymbol{E}_0^{Lab}+\boldsymbol{K}'^{Lab} = E'(\boldsymbol{E}^{Lab}+\boldsymbol{S}'^{Lab})\end{aligned}$$

for the initial and final photons, we get

$$-mE+mE'+EE'(1-\cos\vartheta)=0,$$

where $\cos\vartheta = \boldsymbol{K}^{Lab}\cdot\boldsymbol{K}'^{Lab}/EE' = \boldsymbol{S}^{Lab}\cdot\boldsymbol{S}'^{Lab}$. Expressing the photon energies in terms of Planck's constant and laboratory frequencies (or wavelengths), $E=hf=h\lambda^{-1},\;\; E'=hf'=h\lambda'^{-1}$, we get

$$\frac{1}{f'}-\frac{1}{f}=\lambda'-\lambda=\frac{h}{m}(1-\cos\vartheta). \tag{3.36}$$

It is this equation which has been well verified experimentally. Note, in particular, that whenever $\vartheta\neq 0$, the photon loses energy ($f'<f$), and so gets redder ($\lambda'>\lambda$). The ratio $h/m$ is called the Compton wave length, $\lambda_c$, for the electron; it sets the scale for the shift in wavelength of the photon during scattering. The Compton wavelength for the electron is the wavelength of a photon whose energy equals the (rest) mass of the electron: $m=h(\frac{1}{\lambda_c})$. Its value is about $0.024\times10^{-8}$cm (for comparison, the wavelengths of visible light are around $5000\times10^{-8}$cm).

A curious property of the physical configuration in the Compton experiment, which follows directly from (3.35) by squaring each side, is that the frequency of the incident photon in the rest frame of the target electron (the laboratory) is equal to the frequency of the scattered photon in the rest frame of the recoil electron. This is a result of the definition (3.12) and the fact that the electron has a fixed mass.

## 3.6 Centre-of-momentum Frame

The centre-of-momentum frame is the conceptual successor to the centre-of-mass frame of Newtonian dynamics. In the Newtonian context, for each Galilean-inertial frame and at each moment of (absolute) time, a centre-of-mass point is defined by weighting particle positions with their Newtonian masses. The so-defined point, for a *closed system* in which the sum of the Newtonian forces is zero, moves at a constant velocity. Hence, in any frame moving at the same constant velocity, the centre-of-mass point will be stationary. Such a frame is the centre-of-mass frame (or centre-of-gravity frame). In such a frame the total Newtonian three-momentum is zero. The relativistic version of this frame is the centre-of-momentum frame.

The Newtonian construction of the centre-of-mass point loses interest in special relativity because the point it identifies turns out to depend on the Galilean frame used in the construction, and because it seems evident that the particle energies should be used for the weighting of particle positions instead of the masses. No other definition of a centre-of-mass point has all the attractive features that the Newtonian one does.[2]

It *is*, however, easily possible to introduce in special relativity a frame of reference in which the total three-momentum is zero: ***one just chooses a frame of reference whose time basis vector has the spacetime direction of the total four-momentum.***

The introduction of the special centre-of-momentum frame depends on a simple property of a collection of momentum vectors: A sum of any future-pointing timelike or lightlike momentum vectors,

$$\boldsymbol{p} \equiv \Sigma \boldsymbol{p}_i, \tag{3.37}$$

is future-pointing timelike. For,

$$\boldsymbol{p} \cdot \boldsymbol{p} = \boldsymbol{p}_1 \cdot \boldsymbol{p}_1 + \cdots + 2\boldsymbol{p}_1 \cdot \boldsymbol{p}_2 + \ldots .$$

The terms $\boldsymbol{p}_i \cdot \boldsymbol{p}_i = -m_i^2$ are negative (or zero for massless particles). For the cross-terms: if particle $i$ (say) is massive, then $\boldsymbol{p}_i \cdot \boldsymbol{p}_j = -m_i E_j$, where $E_j$ is the energy of particle $j$ in the rest frame of particle $i$; if both particles are massless, then $\boldsymbol{p}_i \cdot \boldsymbol{p}_j = -E_i E_j (1 - \cos\vartheta_{ij})$, where $E_i$ and $E_j$ are the energies in some inertial frame and $\vartheta_{ij}$ is the angle between their three-momenta in that frame. In all cases (ignoring the one exceptional case in which all momentum vectors are lightlike and parallel), $\boldsymbol{p} \cdot \boldsymbol{p} < 0$, so that we can introduce an effective mass $M$ and a future-pointing timelike unit vector $\boldsymbol{E}_0^{CM}$

$$M^2 \equiv -\boldsymbol{p} \cdot \boldsymbol{p}, \qquad \boldsymbol{E}_0^{CM} \equiv \frac{\boldsymbol{p}}{M}, \qquad \boldsymbol{p} = M \boldsymbol{E}_0^{CM}. \tag{3.38}$$

When we deal with a group of particles with momenta $\boldsymbol{p}_i$, any inertial frame of reference whose time basis vector is $\boldsymbol{E}_0^{CM}$ is called a centre-of-momentum frame (CM-frame) for that group. According to the definitions (3.14) of relative energy and relative three-momentum,

$$\boldsymbol{p} = \Sigma \boldsymbol{p}_i = E \boldsymbol{E}_0^{CM} + \boldsymbol{P}^{CM}. \tag{3.39}$$

So, the centre-of-momentum energy is $E = M$ and the centre-of-momentum three-momentum vanishes by construction, $\boldsymbol{P}^{CM} = \boldsymbol{0}$.

---

[2] There is a useful discussion of six alternatives for the definition of the centre-of-mass in a paper of M.H.L. Pryce.

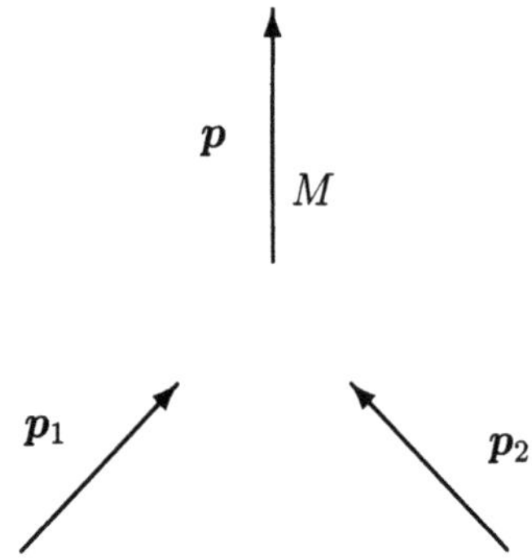


**Figure 3.6**

The concept of a "centre-of-momentum frame" does not, as we have noted, specify a *particular* inertial frame. Only the time basis vector $\boldsymbol{E}_0^{CM}$ is determined. The location of the spatial origin is completely arbitrary, as is the configuration of the orthogonal spatial axes. These circumstances are just what is wanted for many important applications in which the discussion is inevitably in the quantum mechanical régime, and in momentum space. In quantum mechanics, if the momentum is precisely known the position is completely unknown. It is appropriate too that in a quantum mechanical discussion of momenta no worldlines should appear.

### 3.6.1 Two-particle CM-frame

To see explicitly how the CM-frame can be used, consider the case when just two particles contribute to the total momentum

$$\boldsymbol{p} = \boldsymbol{p}_1 + \boldsymbol{p}_2 = M\boldsymbol{E}_0^{CM}, \qquad \boldsymbol{p}\cdot\boldsymbol{p} = -M^2 \quad \text{(Figure 3.6)}. \tag{3.40}$$

For each particle we have a CM-energy

$$\begin{aligned} E_i &= -\boldsymbol{p}_i\cdot\boldsymbol{E}_0^{CM} = -\boldsymbol{p}_i\cdot(\boldsymbol{p}_1+\boldsymbol{p}_2)/M \\ &= (m_i^2 - \boldsymbol{p}_1\cdot\boldsymbol{p}_2)/M & (3.41) \\ &= \frac{1}{2M}(2m_i^2 + M^2 - m_1^2 - m_2^2). & (3.42) \end{aligned}$$

If the two particles scatter elastically (so the masses are the same afterwards as before), then $\boldsymbol{p} = \boldsymbol{p}_1 + \boldsymbol{p}_2 = \boldsymbol{p}_1' + \boldsymbol{p}_2'$, and after squaring, $\boldsymbol{p}_1\cdot\boldsymbol{p}_2 = \boldsymbol{p}_1'\cdot\boldsymbol{p}_2'$. Hence the energy in the CM for each particle remains the same after scattering, $E_i = E_i'$, as does the magnitude of the three-momentum in the CM. Note also that

$$E_1 + E_2 = (m_1^2 + m_2^2 - 2\boldsymbol{p}_1\cdot\boldsymbol{p}_2)/M = -\boldsymbol{p}\cdot\boldsymbol{p}/M = M^2/M = M.$$

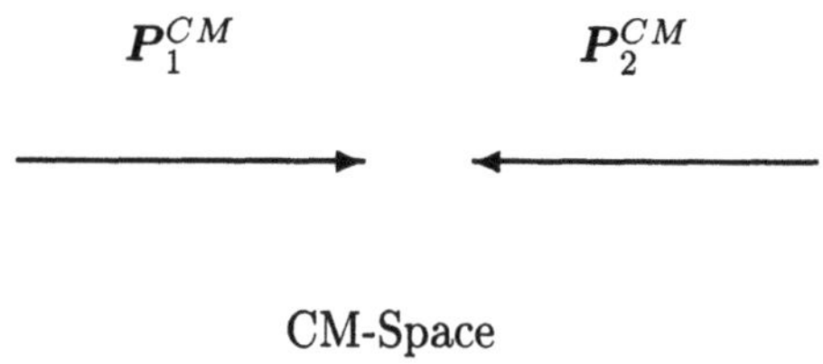


**Figure 3.7**

The velocity relative to the CM is given for each particle (massless as well as massive Figure 3.7) by the form

$$\boldsymbol{p}_i = E_i(\boldsymbol{E}_0^{CM} + \boldsymbol{V}_i) = E_i\boldsymbol{E}_0^{CM} + \boldsymbol{P}_i^{CM}, \tag{3.43}$$

where the CM-energy is given by (3.42). Hence

$$\boldsymbol{V}_i = \frac{\boldsymbol{P}_i^{CM}}{E_i} = \frac{\boldsymbol{p}_i}{E_i} - \boldsymbol{E}_0^{CM} = \frac{\boldsymbol{p}_i}{E_i} - \frac{\boldsymbol{p}}{M} = \frac{M\boldsymbol{p}_i - E_i\boldsymbol{p}}{E_iM}. \tag{3.44}$$

Note that the total three-momentum with respect to the CM vanishes:

$$\boldsymbol{P}_1^{CM} + \boldsymbol{P}_2^{CM} = E_1\boldsymbol{V}_1 + E_2\boldsymbol{V}_2 = \frac{M(\boldsymbol{p}_1 + \boldsymbol{p}_2) - (E_1 + E_2)\boldsymbol{p}}{M} = \boldsymbol{0},$$

as it must. The speed in the CM-frame may be obtained from the formula (3.31):

$$\begin{aligned}\boldsymbol{V}_i \cdot \boldsymbol{V}_i = \frac{\boldsymbol{P}_i^{CM} \cdot \boldsymbol{P}_i^{CM}}{E_i^2} &= \frac{\lambda(M^2, m_1^2, m_2^2)}{4M^2E_i^2} \\ &= \frac{\lambda(M^2, m_1^2, m_2^2)}{(2m_i^2 + M^2 - m_1^2 - m_2^2)^2}.\end{aligned} \tag{3.45}$$

## 3.7 Threshold Energy for Particle Production

In a common high energy physics experiment involving elementary particles a projectile particle $A$ (mass $M_A$) is fired at a target particle $B$ (mass $M_B$) at rest in the laboratory. Depending on the energy of $A$ various outcomes are possible. If we fix attention on a particular process, say, $A + B \to m_1 + m_2 + m_3 + m_4$ (Figure 3.8), where the new particles are labelled by their masses, we are interested in knowing the minimum laboratory energy for $A$ for which this process will go, for which it is kinematically possible.

In the laboratory, the target particle, $B$, is initially at rest. If the initial particle four-momenta are $\boldsymbol{p}_A = M_A\boldsymbol{v}_A$ and $\boldsymbol{p}_B = M_B\boldsymbol{v}_B$, then the time basis

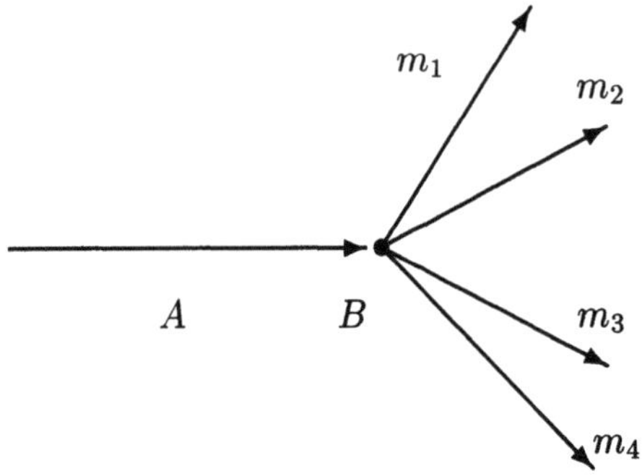


**Figure 3.8** Laboratory space

vector for the laboratory is $\boldsymbol{E}_0^{Lab} = \boldsymbol{v}_B$. If the process goes, then momentum conservation gives

$$\boldsymbol{p}_A + \boldsymbol{p}_B = \Sigma \boldsymbol{q}_k = \Sigma m_k \boldsymbol{v}_k \tag{3.46}$$

in an obvious notation. Squaring each side gives

$$\begin{aligned} -\boldsymbol{p}_A \cdot \boldsymbol{p}_A - \boldsymbol{p}_B \cdot \boldsymbol{p}_B - 2\boldsymbol{p}_A \cdot \boldsymbol{p}_B &= M_A^2 + M_B^2 + 2M_B E_A^{Lab} \\ &= -\left(\Sigma \boldsymbol{q}_k\right)^2, \end{aligned} \tag{3.47}$$

where $E_A^{Lab}$ is the laboratory energy of particle $A$.

The minimum for $E_A^{Lab}$ corresponds to the minimum for $-(\Sigma \boldsymbol{q}_k)^2$. It is easy to find the minimum that is allowed by the kinematics. Multiplying out we get

$$-(\Sigma \boldsymbol{q}_k)^2 = \Sigma_k m_k^2 - \Sigma_{k \neq j} \boldsymbol{q}_k \cdot \boldsymbol{q}_j = \Sigma_k m_k^2 + \Sigma_{k \neq j} m_k m_j \gamma_{kj}, \tag{3.48}$$

where $\gamma_{kj} = -\boldsymbol{v}_k \cdot \boldsymbol{v}_j$. The minimum occurs when all $\gamma_{kj}$ equal one; this is only possible if all the $\boldsymbol{v}_k$ are the same, $\boldsymbol{v}_k = \boldsymbol{v}$ (from (3.18), $\gamma_{kj} = 1$ implies that the relative velocities of $m_k$ with respect to $m_j$, and *vice versa,* are zero). In this case, momentum conservation at the minimum becomes

$$\boldsymbol{p}_A + \boldsymbol{p}_B = \Sigma \boldsymbol{q}_k = \Sigma m_k \boldsymbol{v}_k = (\Sigma m_k)\, \boldsymbol{v}. \tag{3.49}$$

We see that $\boldsymbol{v}$ is the time basis vector for the CM-frame at the minimum, and that the total CM-energy is $E_{CM} = (\Sigma m_k)$. At the minimum, the particles produced in the process (which is *just* possible) are relatively at rest in the CM-frame. From (3.47), the minimum laboratory energy for $A$, the laboratory threshold energy, is

$$(E_A^{Lab})_{Min} = \frac{(\Sigma m_k)^2 - M_A^2 - M_B^2}{2M_B}. \tag{3.50}$$

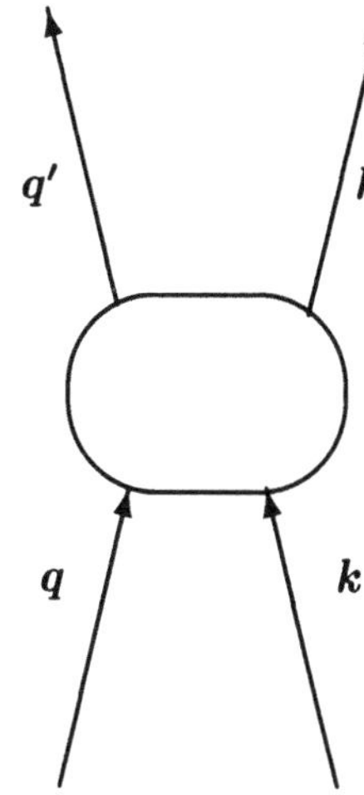


**Figure 3.9**

## 3.8 Scattering Formulæ

We consider the elastic scattering of a particle of mass $\mu$ and momentum $\boldsymbol{k}$ (the projectile) by a particle of mass $m$ and momentum $\boldsymbol{q}$ (the target) (Figure 3.9).

The traditional scattering diagram represents the process by means of a projection onto an arbitrary timelike plane (a plane containing a timelike vector, hence parallel to the time basis vector for some arbitrary inertial frame). The infinite interval of time, from past to future, is expressed explicitly in the diagram. Momentum conservation is expressed by

$$\boldsymbol{q}+\boldsymbol{k}=\boldsymbol{q}'+\boldsymbol{k}'\equiv\boldsymbol{p}=M\boldsymbol{E}_0^{CM}, \tag{3.51}$$

where $\boldsymbol{p}$ is the total momentum and $M$ is the total CM-energy. The invariant variables conventionally used to describe the scattering are

$$s \equiv -(\boldsymbol{q}+\boldsymbol{k})\cdot(\boldsymbol{q}+\boldsymbol{k})=+M^2 \tag{3.52}$$
$$t \equiv -(\boldsymbol{q}-\boldsymbol{q}')\cdot(\boldsymbol{q}-\boldsymbol{q}') \tag{3.53}$$
$$u \equiv -(\boldsymbol{q}-\boldsymbol{k}')\cdot(\boldsymbol{q}-\boldsymbol{k}'), \tag{3.54}$$

which are related by the relation

$$s+t+u=2m^2+2\mu^2. \tag{3.55}$$

The quantity $s$ is the square of the CM-energy, and $t$ is the square (negative scalar product) of the four-momentum transferred in the process from $\boldsymbol{q}$ to $\boldsymbol{q}'$. Similarly, $u$ is another (the last independent) square of a momentum transferred.

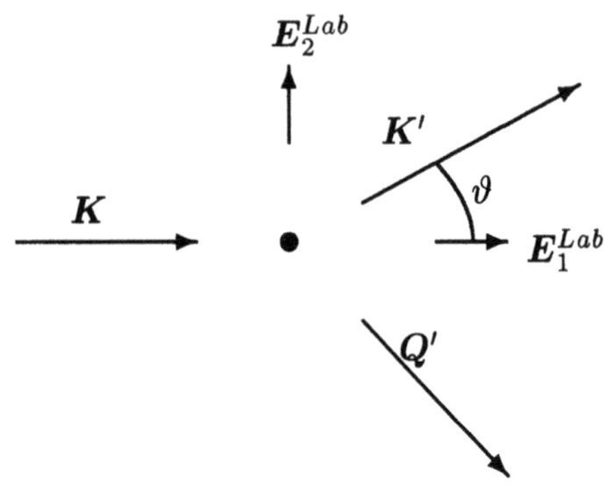


**Figure 3.10** Space in the laboratory

### 3.8.1 Laboratory frame

The laboratory is an inertial frame in which the target particle $m$ is initially at rest. Its time basis vector is therefore $\boldsymbol{E}_0^{Lab} = \boldsymbol{q}/m$.

To visualise the scattering event in the laboratory we consider a projection orthogonal to $\boldsymbol{q}$ onto the scattering plane in the lab (Figure 3.10). Time is now shown implicitly; as the process unfolds (from left to right), the tracks are preserved in the diagram. Only the three-momentum vectors with respect to the laboratory decomposition, all orthogonal to $\boldsymbol{q}$, are shown on the diagram.

The notations for the laboratory energies and three-momenta are given by

$$\boldsymbol{k} = k^0 \left(\frac{\boldsymbol{q}}{m}\right) + \boldsymbol{K}, \quad \boldsymbol{k}' = k'^0 \left(\frac{\boldsymbol{q}}{m}\right) + \boldsymbol{K}', \quad \boldsymbol{q}' = q'^0 \left(\frac{\boldsymbol{q}}{m}\right) + \boldsymbol{Q}'. \qquad (3.56)$$

When the laboratory spatial basis vectors are conveniently chosen as in the diagram, so that $\boldsymbol{K} = |\boldsymbol{K}|\boldsymbol{E}_1^{Lab}$, and $\boldsymbol{K}' = |\boldsymbol{K}'|(\cos\vartheta \boldsymbol{E}_1^{Lab} + \sin\vartheta \boldsymbol{E}_2^{Lab})$, the laboratory scattering angle satisfies

$$\tan\vartheta = \frac{\boldsymbol{K}' \cdot \boldsymbol{E}_2^{Lab}}{\boldsymbol{K}' \cdot \boldsymbol{E}_1^{Lab}}. \qquad (3.57)$$

### 3.8.2 CM-frame

An alternative visualisation of the process is achieved by projecting orthogonal to the CM-time basis vector $\boldsymbol{E}_0^{CM} = \boldsymbol{p}/M$ onto the scattering plane in a CM- frame (Figure 3.11). In this case the three-momentum vectors with respect to the CM remain after projection. The spatial basis for the CM is fixed by

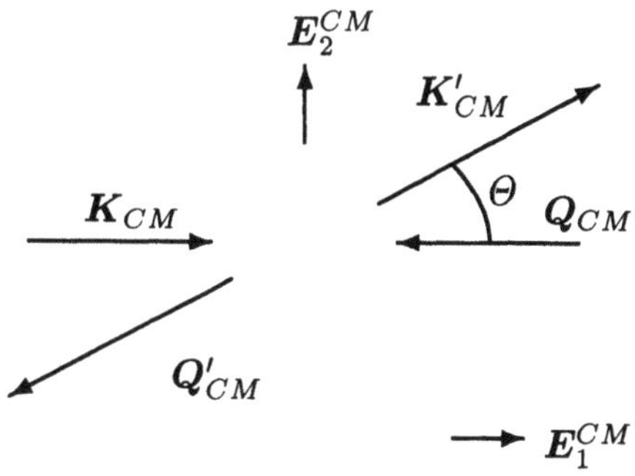


**Figure 3.11**

$$k = k^0_{CM}\left(\frac{p}{M}\right) + K_{CM} = k^0_{CM}\left(\frac{p}{M}\right) + |K_{CM}|E_1^{CM} \tag{3.58}$$

and

$$k' = k'^0_{CM}\left(\frac{p}{M}\right) + K'_{CM}, \quad K'_{CM} = |K'_{CM}|(\cos\Theta E_1^{CM} + \sin\Theta E_2^{CM}), \tag{3.59}$$

where $\Theta$ is the CM scattering angle.

To get the explicit connection between the laboratory basis and the CM basis, we have first of all

$$E_0^{CM} = \frac{p}{M} = \frac{q+k}{M} = \frac{m+k^0}{M}\left(E_0^{Lab} + \frac{|K|}{m+k^0}E_1^{Lab}\right). \tag{3.60}$$

By (3.58) we see that $E_1^{CM}$ is in the plane of $k$ and $p$, hence in the plane of $E_0^{Lab}$ and $E_1^{Lab}$; we need only write down the spatial unit vector (with the correct sense) which is orthogonal to $E_0^{CM}$ :

$$E_1^{CM} = \frac{m+k^0}{M}\left(E_1^{Lab} + \frac{|K|}{m+k^0}E_0^{Lab}\right). \tag{3.61}$$

Finally, (3.59) involves only $E_2^{CM}$, not $E_3^{CM}$, so that

$$E_2^{CM} = E_2^{Lab}, \qquad E_3^{CM} = E_3^{Lab}. \tag{3.62}$$

The bases have been arranged so that the CM is reached from the Lab by a simple boost with relative velocity $\frac{|K|}{m+k^0}E_1^{Lab}$ with respect to the laboratory.

Before considering the relation between the laboratory and the CM scattering angles we derive some useful relations for the magnitudes of various three-momenta. To get the (squared) magnitude of the CM three-momenta $Q_{CM}$ and $K_{CM}$, we use the fact that $p = ME_0^{CM} = q + k$ is of the form (3.27), and we apply (3.31) to two cases in which the rest frame has mass $M$,

$$\boldsymbol{Q}_{CM} \cdot \boldsymbol{Q}_{CM} = \boldsymbol{K}_{CM} \cdot \boldsymbol{K}_{CM} = \frac{\lambda(M^2, m^2, \mu^2)}{4M^2}. \tag{3.63}$$

By the same argument, based on $\boldsymbol{p} = \boldsymbol{q}' + \boldsymbol{k}'$, we get for $\boldsymbol{Q}'_{CM} \cdot \boldsymbol{Q}'_{CM} = \boldsymbol{K}'_{CM} \cdot \boldsymbol{K}'_{CM}$ exactly the same value (because the masses are all the same).

The other relation that is needed is a comparison of the CM magnitude $|\boldsymbol{K}_{CM}| = |\boldsymbol{K}'_{CM}|$ and the laboratory magnitude $|\boldsymbol{K}|$. In view of $\boldsymbol{p} = M\boldsymbol{E}_0^{CM} = \boldsymbol{q} + \boldsymbol{k} = m\boldsymbol{E}_0^{Lab} + \boldsymbol{k}$, we may apply (3.34) and get

$$m|\boldsymbol{K}| = M|\boldsymbol{K}_{CM}| = M|\boldsymbol{K}'_{CM}|. \tag{3.64}$$

To get the connection between the two scattering angles, equation (3.57) is rewritten using $\boldsymbol{E}_1^{Lab}$ derived from (3.60) and (3.61):

$$\begin{aligned} \tan\vartheta &= \frac{\boldsymbol{K}' \cdot \boldsymbol{E}_2^{Lab}}{\boldsymbol{K}' \cdot \boldsymbol{E}_1^{Lab}} = \frac{\boldsymbol{k}' \cdot \boldsymbol{E}_2^{Lab}}{\boldsymbol{k}' \cdot \boldsymbol{E}_1^{Lab}} = \frac{\boldsymbol{k}' \cdot \boldsymbol{E}_2^{CM}}{\boldsymbol{k}' \cdot \frac{m+k^0}{M}(\boldsymbol{E}_1^{CM} - \frac{|\boldsymbol{K}|}{m+k^0}\boldsymbol{E}_0^{CM})} \\ &= \frac{|\boldsymbol{K}'_{CM}|\sin\Theta}{\frac{m+k^0}{M}|\boldsymbol{K}'_{CM}|\cos\Theta + k'^0_{CM}\frac{|\boldsymbol{K}|}{M}} \\ &= \frac{m\sin\Theta}{k^0_{CM} + q^0_{CM}\cos\Theta}. \end{aligned} \tag{3.65}$$

(In the last step, (3.64) has been used, and $k^0_{CM} = k'^0_{CM}$, which follows from the corresponding equality of the magnitudes of the CM three-momenta before and after scattering; and lastly, $q^0_{CM} = -\boldsymbol{p} \cdot \boldsymbol{q}/M = (m^2 + mk^0)/M$.)

We may regard $\boldsymbol{p} = \boldsymbol{q} + \boldsymbol{k}$ as a three-particle "process" in which a CM-particle of mass $M$ decays into particles of masses $m$ and $\mu$. We recognise that (3.25) and (3.26) provide formulae for $q^0_{CM}$ and $k^0_{CM}$. Hence, (3.65) becomes

$$\begin{aligned} \tan\vartheta &= \frac{2mM\sin\Theta}{(M^2 + \mu^2 - m^2) + (M^2 + m^2 - \mu^2)\cos\Theta} \\ &= \frac{2mM\tan(\Theta/2)}{M^2 + (\mu^2 - m^2)\tan^2(\Theta/2)}. \end{aligned} \tag{3.66}$$

The scattering angles may be expressed easily in terms of the invariants $s$ and $t$ defined in (3.52) and (3.53). From the former, $M^2 = s$, and from the latter,

$$t = -(\boldsymbol{q} - \boldsymbol{q}') \cdot (\boldsymbol{q} - \boldsymbol{q}') = -(\boldsymbol{k} - \boldsymbol{k}') \cdot (\boldsymbol{k} - \boldsymbol{k}') = 2\boldsymbol{k} \cdot \boldsymbol{k}' + 2\mu^2.$$

Using (3.58), (3.59) and (3.64), we get

$$\begin{aligned} \frac{t}{2} &= -k^0_{CM}k'^0_{CM} + |\boldsymbol{K}_{CM}||\boldsymbol{K}'_{CM}|\cos\Theta + \mu^2 \\ &= \boldsymbol{K}_{CM} \cdot \boldsymbol{K}_{CM}(\cos\Theta - 1) = \frac{\lambda(M^2, m^2, \mu^2)}{4M^2}(\cos\Theta - 1). \end{aligned} \tag{3.67}$$

## 3.9 Problems

1. How fast, with respect to a frame $K$, must a particle move so its kinetic energy (with respect to $K$) equals its rest energy?

2. A particle of mass $m$ decays from rest (in $K$) into a particle of mass $m'$ and a photon $\gamma$. Show that the $K$-energies of the end products are $E' = (m^2 + m'^2)/2m$, and $E_\gamma = (m^2 - m'^2)/2m$.

3. If $\boldsymbol{k}$ is a lightlike momentum vector, verify that for every positive number $E$ there is at least one inertial frame $K$ such that the associated $K$-energy is $E$.

4. A particle $A$, of unit mass, decays from rest in $K$ into two particles $B$ and $C$. The speeds $V_B$ and $V_C$ of $B$ and $C$ with respect to $K$ are measured and have $\gamma$-factors $\gamma_B = 2$, $\gamma_C = 3$ ($\gamma = (1 - V^2)^{-1/2}$). What are the masses of $B$ and $C$?

5. We use the usual boost-related frames $K$ and $K'$. In $K$ there is a uniform spatial distribution of particles of mass $m \neq 0$, all moving, at speed $U$ with respect to $K$, in the direction $\cos\theta \boldsymbol{E}_1 + \sin\theta \boldsymbol{E}_2$. Show that the relation between the $K'$-number density $n'$ and the $K$-number density $n$ is
$$n' = n\gamma(1 - UV\cos\theta).$$
(Hint: since the particles are massive one can relate each density to the rest frame density.)
6. We use the usual boost-related frames $K$ and $K'$. In $K$ there is a uniform spatial distribution of photons, all moving in the $\cos\theta \boldsymbol{E}_1 + \sin\theta \boldsymbol{E}_2$ direction. Show that the relation between the $K$-number density $n$ (at $t = 0$, say) and the $K'$-number density $n'$ (at $t' = 0$) is
$$n' = n\gamma(1 - V\cos\theta),$$
where $V$ is the speed of $K'$ with respect to $K$. (This result was worked out by several people in 1968 to use in looking for a directional effect in the background cosmic radiation.)
7. An atomic transition occurs with the emission of a photon, $A \to B + \gamma$, the rest masses being $M_A$ and $M_B$.
   a) If $A$ is initially at rest in the lab, find the lab energy $E$ of the photon and the speed $V$ of recoil (with respect to the lab) of $B$.
   b) What lab energy $E'$ would a (different) photon have to have to stimulate the reverse transition $\gamma + B \to A$ if the state $B$ is initially at rest in the lab? If the photon's energy $E'' < E'$, show that the minimum speed $V''$ (with respect to the lab) with which $B$ would have to move for the transition to occur is

$$V'' = \frac{(M_A^2 - M_B^2)^2 - 4(E''M_B)^2}{(M_A^2 - M_B^2)^2 + 4(E''M_B)^2}.$$

8. If a particle of mass $m$ collides with a particle of mass $M$ to form a particle of mass $M + m$, the relative speed of the collision must be zero. (In general the mass of the compound particle determines the relative speed, and $M + m$ is the threshold value of the compound mass.)

9. A particle $A$ of mass $M$ decays from rest (with respect to frame $K$) into a particle $B$ of mass $m$ and a particle $C$ of mass $\mu$.
   a) Find expressions in terms of the masses alone for the $K$-energies of $B$ and $C$.
   b) Show that the two decay products $B$ and $C$ are separated by a $K$-distance
   $$\frac{2M^2T\sqrt{\lambda(M^2, m^2, \mu^2)}}{M^4 - (m^2 - \mu^2)^2}$$
   at a $K$-time interval $T$ after the decay event.
   c) Find the $\tilde{K}_C$-distance between $B$ and $C$ at a $\tilde{K}_C$-time $T_C$ after the decay event, where $\tilde{K}_C$ is the rest frame of $C$.

10. Frames $K$ and $K'$ are in standard boost relation. If light of frequency $\nu$ in $K$ travels in a direction inclined at an angle $\psi$ to the $+x$-axis, what is its frequency and direction with respect to $K'$? (This is the Doppler effect for the case when the relative direction of the frames is different from the direction of the light.)

11. In a Compton scattering the ingoing photon has frequency $\nu$ and the scattered photon $\nu'$, both with respect to the lab, where the target electron is at rest. If the electron recoils with lab-speed $V$ at a lab-angle $\psi$ to the direction of the incoming photon, show that
$$\frac{\nu'}{\nu} = \frac{1 - V\cos\psi}{\sqrt{1 - V^2}}.$$
(Note that in this question the photons with frequencies $\nu$ and $\nu'$ are different, whereas in the previous question there is only one sort of photon viewed from different frames $K$ and $K'$.)

12. A particle $A$ of mass $m_A$ is struck by an incident photon; the result of the collision is a particle $B$ of mass $m_B$ and a scattered photon. Show that if the frequency of the incident photon in the rest frame of $A$ is $f$, then the frequency of the scattered photon in the rest frame of $B$ is $(m_A/m_B)f$.

13. In a Compton experiment the initial photon has energy $E_\gamma$ in the lab, where the target electron, of mass $m$, is initially at rest. Show that the

CM-frame moves at speed $V = E_\gamma/(E_\gamma + m)$ with respect to the lab and find the CM-energies for the initial photon and initial electron. What is the CM-energy of the photon after scattering? If the scattering angle of the photon in the lab is $\theta$, what is it in the CM-frame?

14. A disc rotates with angular velocity $\omega$ with respect to the laboratory. On its rim, at radius $R$, is an emitter $E$, and at angular separation $\alpha$ there is an absorber $A$. The quantities $\omega, \alpha, R$ are all derived from lab measurements. A photon is emitted from $E$ at just the right angle to be absorbed by $A$ when the photon meets the rim again. Show that the frequency of the photon in the rest frame of $E$ at the moment of emission is equal to its frequency in the rest frame of $A$ at the moment of absorption.

15. Two particles, of masses $m_1$ and $m_2$ and four-momenta $\boldsymbol{p}_1$ and $\boldsymbol{p}_2$, approach each other. Define a centre-of-momentum (CM) frame for the particles. Obtain expressions for the energies $E_1$ and $E_2$ and relative velocities $\boldsymbol{V}_1$ and $\boldsymbol{V}_2$ of each particle with respect to the CM-frame. Verify that
$$\boldsymbol{V}_1 = -\boldsymbol{V}_2(m_2^2 - \boldsymbol{p}_1 \cdot \boldsymbol{p}_2)/(m_1^2 - \boldsymbol{p}_1 \cdot \boldsymbol{p}_2).$$
If the particles interact and emerge with the same masses but momenta $\boldsymbol{p}'_1, \boldsymbol{p}'_2$, show that the new CM-energies are the same as before, and the magnitudes of the new relative velocities are the same too.

16. The worldlines $L_1$ and $L_2$ of two nonintersecting, freely moving massive particles are parallel to their momentum vectors $\boldsymbol{p}_1 = m_1\boldsymbol{v}_1$ and $\boldsymbol{p}_2 = m_2\boldsymbol{v}_2$. Let $O_1$ and $O_2$ be arbitrary origins on $L_1$ and $L_2$, respectively. Other points on each worldline can then be parameterised with their proper times, $\overrightarrow{O_1P_1} = \tau_1\boldsymbol{v}_1$, $\overrightarrow{O_2P_2} = \tau_2\boldsymbol{v}_2$. Show that the point $S_2$ on $L_2$ which is simultaneous in the CM frame with $O_1$ on $L_1$ has proper time $\tau_2 = -m_2(\boldsymbol{p}\cdot\overrightarrow{O_1O_2})/(\boldsymbol{p}\cdot\boldsymbol{p}_2)$, where $\boldsymbol{p}$ is the total momentum. Show that the minimum distance in the CM frame, $|\overrightarrow{Q_1Q_2}|$, between the two worldlines, is given by
$$\overrightarrow{Q_1Q_2}\cdot\overrightarrow{Q_1Q_2} = \overrightarrow{O_1S_2}\cdot\overrightarrow{O_1S_2} - (\overrightarrow{O_1S_2}\cdot\boldsymbol{q})^2/(\boldsymbol{q}\cdot\boldsymbol{q}),$$
where $\boldsymbol{q} \equiv (\boldsymbol{p}_1 \wedge \boldsymbol{p}_2)\cdot\boldsymbol{p} \equiv \boldsymbol{p}_1(\boldsymbol{p}_2\cdot\boldsymbol{p}) - \boldsymbol{p}_2(\boldsymbol{p}_1\cdot\boldsymbol{p})$.

17. An electron-positron pair annihilates into two photons. Suppose that in the CM-frame $K$ the direction of motion of the photons is perpendicular to the direction of motion of the electron. Show that in a frame $K'$ moving at speed $V$ with respect to $K$ in the direction of the electron's motion the angle between the directions of the two photons is $\theta$ given by $\cos\theta = 2V^2 - 1$.

18. As the headlight effect exemplifies, the angle between the spatial directions of two photons depends on the frame of reference. Two photons which are moving parallel with respect to one frame, however, move parallel with respect to any other frame. This property is *not* true of massive particles. Explain and give an example.

19. In the reaction $\pi^- + p^+ \longrightarrow K^o + \Lambda^o$ the target proton is at rest in the lab frame.
    a) Using the masses $m_{\pi^-} = 0.140 Gev$, $m_p = 0.938 Gev$, $m_{K^o} = 0.498 Gev$, $m_{\Lambda^o} = 1.116 Gev$ show that the total lab energy of the pion at threshold is $0.909 Gev$.
    b) In an experiment in which the lab three-momentum of the pions has magnitude 2.50Gev the $\Lambda^o$ are observed to have three-momentum 0.60Gev at a lab angle $45^o$ with respect to the incident pions. What is the velocity of the centre-of-momentum frame in this case? What is the magnitude of the three-momentum of the $K^o$ in the lab frame, and in the CM-frame?

20. Two particles $A$ and $B$ approach each other, one with mass $m_A$ and spacetime velocity $\boldsymbol{v}_A$, the other with mass $m_B$ and spacetime velocity $\boldsymbol{v}_B$. Show that the relative velocity of the centre-of-momentum frame with respect to the rest frame of $B$ may be written

$$\frac{m_A \boldsymbol{v}_A + m_A(\boldsymbol{v}_A \cdot \boldsymbol{v}_B)\boldsymbol{v}_B}{m_B - m_A(\boldsymbol{v}_A \cdot \boldsymbol{v}_B)}.$$

21. A photon strikes a particle, of mass $m$, at rest in $K$. The products of the collision are three particles, each of mass $m$. Show that the $K$-energy of the photon is at least $4m$.
22. A particle of mass $M$, at rest in $K$, decays into a particle of mass $m$, a particle of mass $\mu$ and a photon. Show that the *maximum* energy (with respect to $K$) that the photon can have is $[M^2 - (m+\mu)^2]/(2M)$.

23. A particle of mass $m$ collides elastically with a stationary (with respect to $K$) particle of equal mass. ("Elastic" means that the end products of the process have the same masses as the original particles.) The incident particle has $K$-kinetic energy $T$ and it is scattered in the collision by a $K$-angle $\theta$. Show that its $K$-kinetic energy after the collision is

$$T' = 2mT\cos^2\theta/(2m + T\sin^2\theta).$$

24. Two particles, each of mass $m$, collide elastically. If one of the particles is at rest in $K$ before collision and the other has total $K$-energy $E$, and after the collision the two particles move (with respect to $K$) with the

same speed, show that the angle between their directions of motion is $\chi$, where

$$\cos^2(\chi/2) = (E^2 - m^2)/[(E+m)^2 - 4m^2] = (E+m)/(E+3m).$$

In the nonrelativistic limit, $\chi \to \pi/2$.

25. Two particles, of masses $\mu$ and $m$, scatter elastically. Momentum conservation is expressed by $\boldsymbol{k} + \boldsymbol{p} = \boldsymbol{k}' + \boldsymbol{p}'$. The Breit frame (or brick wall BW-frame) is defined by having its timelike basis vector $\boldsymbol{E}_0$ in the direction of $\boldsymbol{k} + \boldsymbol{k}'$. The energies and spatial momenta with respect to this frame are $\epsilon, \boldsymbol{K}; \epsilon', \boldsymbol{K}'; E, \boldsymbol{P}; E', \boldsymbol{P}'$. Show that $\epsilon = \epsilon', \boldsymbol{K} + \boldsymbol{K}' = 0, E = E', 2\boldsymbol{K} = \boldsymbol{P}' - \boldsymbol{P}$. If $t$ is defined by $t = -(\boldsymbol{k} - \boldsymbol{k}')^2$, show that $t = -4\boldsymbol{K} \cdot \boldsymbol{K}$, and that the angle $\theta_B$ between $\boldsymbol{P}$ and $\boldsymbol{P}'$ is given by $\cos\theta_B = 1 + \frac{1}{2}t/(E^2 - m^2)$.

26. Two particles of equal rest mass scatter elastically (same particle types after the collision as before). Before the collision the projectile has speed $V$ with respect to lab and $\gamma$-factor $\gamma$, while the target particle is at rest. After the collision the projectile goes off at an angle $\theta_1$ with respect to its incident direction and its new velocity has $\gamma$-factor $\gamma_1$. The target particle goes off at an angle $\theta_2$ with $\gamma$-factor $\gamma_2$. Show that for $i = 1, 2$

$$\cos^2\theta_i = \frac{\gamma_i - 1}{\gamma_i + 1}\frac{\gamma + 1}{\gamma - 1}$$

and hence that $\tan\theta_1 \tan\theta_2 = \frac{2}{\gamma+1}$.

27. Suppose a particle of mass $M$ decays into a particle of mass $m$ and a particle of mass $\mu$. If $\boldsymbol{Q}$ is the three-momentum of the $m$-particle in the rest frame of the $M$-particle, and $\boldsymbol{K}$ is the three-momentum of the $\mu$-particle in the rest frame of the $m$-particle, show that $m^2\boldsymbol{K} \cdot \boldsymbol{K} = M^2\boldsymbol{Q} \cdot \boldsymbol{Q}$.

28. Two particles, one of mass $m_1$ and momentum $\boldsymbol{p}_1$, the other $m_2$, $\boldsymbol{p}_2$, travel freely. Denote the total four-momentum by $\boldsymbol{K}$ and $\boldsymbol{K} \cdot \boldsymbol{K} = -M^2$. Show that the unit spatial vector in the CM-frame which points in the direction of the relative three-momentum (with respect to CM) of the first particle is

$$\boldsymbol{q} = \frac{M}{\sqrt{\lambda}}[\boldsymbol{p}_1 - \boldsymbol{p}_2 - \frac{(m_1^2 - m_2^2)}{M^2}\boldsymbol{K}],$$

where $\lambda$ is the standard function $\lambda(M^2, m_1^2, m_2^2) \equiv M^4 + m_1^4 + m_2^4 - 2m_1^2m_2^2 - 2M^2(m_1^2 + m_2^2)$. Write $\boldsymbol{p}_1$ and $\boldsymbol{p}_2$ in terms of $\boldsymbol{K}$ and $\boldsymbol{q}$.

29. If a particle of mass $m$ collides elastically with a particle of mass $M$, the relative speed of one particle with respect to the other is the same after the collision as before.

30. Two particles of equal mass collide head-on (the collision takes place in one straight spatial line in $K$, the rest frame of one of the particles before the collision). Show that if $V$ is the initial relative speed of the projectile with respect to $K$, then after the collision the target recoils with speed $V$ (with respect to $K$) and the projectile is reduced to rest.

31. A projectile particle, of mass $m$, collides head-on with a target particle of mass $M$, which is initially at rest in $K$. If $V$ is the initial speed of the projectile with respect to $K$, show that $U$, the recoil speed of the target (with respect to $K$), is given by

$$\frac{1}{\sqrt{1-U^2}} = \frac{(\gamma+x)^2+\gamma^2-1}{(\gamma+x)^2-\gamma^2+1},$$

where $x = M/m$ and $\gamma^{-2} = 1-V^2$. (Hint: bear in mind the results of the previous two problems, the second of which is the case of equal masses.)

32. In a high energy collider an electron and positron, of equal Lab energies, meet head on and produce a pair of heavy particles, $W^+$ and $W^-$, of equal mass $M$. In this experiment the Lab is also the CM-frame, and the total energy is determined by $s \equiv -(\boldsymbol{p}_+ + \boldsymbol{p}_-)\cdot(\boldsymbol{p}_+ + \boldsymbol{p}_-)$. Suppose the $W^+$ decays to $a+b$ and the $W^-$ decays to $c+d$. Show that, in the approximation in which all particles $a, b, c, d$ are massless, the Lab energy $E$ of any one of them is bounded: $E_- \leq E \leq E_+$, where

$$E_{\pm} = \frac{1}{4}\left(\sqrt{s} \pm \sqrt{s-4M^2}\right).$$

This bound can be used to measure the mass $M$ of the $W^{\pm}$ particles. (Note that if the $W^{\pm}$ are not assumed to mediate the process $e^+e^- \to abcd$, the bound on the energy of any of the decay products would be simply $0 \leq E \leq \sqrt{s}/2$.)

## References

A.H. Compton, Physical Review **21** (1923) 483-502

R. Hagedorn, Relativistic Kinematics (Benjamin, 1973)

M.H.L. Pryce, The mass-centre in the restricted theory of relativity, Proc. Roy. Soc. (London) **195A** (1948) 62-81

# 4. Covectors and Dyadics in Spacetime

*Generalisations of Vectors*

In Chapter 2, vectors, the elements of $\mathcal{V}$, were introduced and used to discuss many kinematical problems. The kinematical applications followed immediately from the original definition of vectors as a class of displacements in spacetime. In Chapter 3 the very slightly more general space of vectors was used to consider particle momentum.

In the present chapter we consider two new classes of geometrical objects in spacetime. One new class is based on the fact that linear mappings of a vector space themselves form vector spaces. The linear mappings of $\mathcal{V}$ to the real numbers constitute the space $\mathcal{V}^*$ of covectors, the space dual to $\mathcal{V}$. The linear mappings of $\mathcal{V}$ to itself serve to define the dyadics. These two cases are sufficient to provide the tools to construct many interesting and useful physical theories (but the process can be extended indefinitely to produce further geometrical objects).

The second new class of object is a *field,* a scalar field or vector field, for example, providing a numerical-valued or vector-valued function of position in spacetime. An accelerating cloud of dust provides physical examples: at each point $P$ of spacetime which is occupied by the dust the local velocity vector $\boldsymbol{v}(P)$ is the value of the velocity field, and the mass-density of the dust, in the local rest frame at $P$, is an example of a scalar field. Dyadic-valued fields are possible too, and they are extremely important: the electromagnetic field is an example, and the energy tensor for any physical system is another.

When fields are present we must have a calculus to analyse them. In the present chapter we consider not only the gradient of scalar fields but also the gradient of vector fields and dyadics. There are in addition some meaningful combinations, corresponding to the divergence and curl of three-dimensional vector analysis. Finally, the various fields may be integrated over various regions of spacetime; such integrals may be reduced to numerical integrals when a constant basis, derived from those of inertial frames, is introduced.

Much of the material in the present chapter may be regarded as formally "obvious", since the operational rules are familiar from vector analysis or the calculus of real functions. The chapter has been written with the idea of providing help when the rules no longer seem obvious and a more detailed development is needed. The following sections should be skimmed lightly, then returned to as required.

## 4.1 Covectors in Spacetime

Covectors $\underset{\leftarrow}{Z}$ are linear mappings of vectors (from $\mathcal{V}$) into the real numbers.

$$\underset{\leftarrow}{Z} \;:\; \mathcal{V} \to \mathcal{R} \tag{4.1}$$

$$\;:\; \boldsymbol{A} \in \mathcal{V} \mapsto \underset{\leftarrow}{Z}(\boldsymbol{A}) \in \mathcal{R} \tag{4.2}$$

(The notation for $\underset{\leftarrow}{Z}$, with an underarrow, is chosen to provide a constant visual reminder of the type of object that $\underset{\leftarrow}{Z}$ is, and specifically to distinguish it from the vectors $\boldsymbol{A}$. With increasing experience one can do without the notational assistance that the arrows provide.)

### 4.1.1 Components of a Covector

Because the map $\underset{\leftarrow}{Z}$ is *linear*, we can expand its vector argument in a basis to get

$$\underset{\leftarrow}{Z}(\boldsymbol{A}) = \underset{\leftarrow}{Z}(A^{\mu}\boldsymbol{E}_{\mu}) = A^{\mu}\underset{\leftarrow}{Z}(\boldsymbol{E}_{\mu}) \equiv A^{\mu}Z_{\mu}, \tag{4.3}$$

where the $Z_{\mu}$, the numbers which fully determine $\underset{\leftarrow}{Z}$ by its effect on the four basis vectors $\boldsymbol{E}_{\mu}$, will be seen in a moment to be the components of the covector with respect to a particular *cobasis* for the space of covectors. The cobasis for the covectors is derived from the basis $\boldsymbol{E}_{\mu}$ for the vectors, and the $Z_{\mu} = \underset{\leftarrow}{Z}(\boldsymbol{E}_{\mu})$ are the relevant *components* of the covector $\underset{\leftarrow}{Z}$. The calculation of such components can be made for any inertial basis.

### 4.1.2 Transformation Law for Components

We have already compared the vector components and vector bases generated by two different inertial frames: from $\boldsymbol{A} = A^{\nu}\boldsymbol{E}_{\nu} = A'^{\mu}\boldsymbol{E}'_{\mu} = a^{\mu}{}_{\nu}A^{\nu}\boldsymbol{E}'_{\mu}$ we have $\boldsymbol{E}_{\nu} = a^{\mu}{}_{\nu}\boldsymbol{E}'_{\mu}$ and so $a^{-1\nu}{}_{\lambda}\boldsymbol{E}_{\nu} = \boldsymbol{E}'_{\lambda}$. These relations allow us to get the connection between the components of $\underset{\leftarrow}{Z}$ with respect to different cobases,

$$Z'_{\lambda} \equiv \underset{\leftarrow}{Z}(\boldsymbol{E}'_{\lambda}) = \underset{\leftarrow}{Z}(a^{-1\nu}{}_{\lambda}\boldsymbol{E}_{\nu}) = a^{-1\nu}{}_{\lambda}Z_{\nu}; \tag{4.4}$$

we see that the components, $Z_{\mu}$, $Z'_{\mu}$ of $\underset{\leftarrow}{Z}$ transform in exactly the same way as the basis vectors $\boldsymbol{E}_{\mu}$, $\boldsymbol{E}'_{\mu}$. An aspect of the notation should now begin to make sense: quantities which transform in one way are written with subscripts, those which transform another way with superscripts; summing one type with the other type, for which scripts are always one up and one down, results in an invariant.

### 4.1.3 The Dual Space (or Cospace)

The collection of covectors becomes a vector space, called $\mathcal{V}^*$, when we have defined addition and multiplication by scalars. The definition of both is given by the single equation

$$(a\underset{\sim}{Z} + b\underset{\sim}{Y})(\boldsymbol{A}) = a\underset{\sim}{Z}(\boldsymbol{A}) + b\underset{\sim}{Y}(\boldsymbol{A}). \tag{4.5}$$

The space $\mathcal{V}^*$ is the dual space, or cospace, for $\mathcal{V}$.

### 4.1.4 Cobases and Their Transformation Law

Once we are legitimately able to form linear combinations of covectors, now known to form a vector space of their own, we may look for sets of basis covectors. We shall see that the cobasis referred to above, associated with a basis $\boldsymbol{E}_\mu$ for $\mathcal{V}$, consists of covectors $\underset{\sim}{E}^\mu$ defined by

$$\underset{\sim}{E}^\mu(\boldsymbol{E}_\nu) \equiv \delta^\mu_\nu. \tag{4.6}$$

(By linearity such a formula defines the action of $\underset{\sim}{E}^\mu$ on *any* vector.) If, for any covector $\underset{\sim}{Z}$, with components $Z_\mu$, we consider the linear combination $Z_\mu \underset{\sim}{E}^\mu$ and evaluate it on an arbitrary vector $\boldsymbol{A} = A^\mu \boldsymbol{E}_\mu$, we get

$$Z_\mu \underset{\sim}{E}^\mu(\boldsymbol{A}) = Z_\mu \underset{\sim}{E}^\mu(A^\nu \boldsymbol{E}_\nu) = Z_\mu A^\nu \delta^\mu_\nu = Z_\mu A^\mu = \underset{\sim}{Z}(\boldsymbol{A}), \tag{4.7}$$

by (4.3). This verifies that $\underset{\sim}{Z} = Z_\mu \underset{\sim}{E}^\mu$, and shows that the collection $\underset{\sim}{E}^\mu$ really does span the space $\mathcal{V}^*$. It is easy to see that the $\underset{\sim}{E}^\mu$ are linearly independent, so they form a basis. Both $\mathcal{V}$ and $\mathcal{V}^*$ have dimension four.

Equation (4.6) is the general formula defining a cobasis in terms of a basis; the cobasis $\underset{\sim}{E}'^\mu$ corresponding to $\boldsymbol{E}'_\mu$ is given by $\underset{\sim}{E}'^\mu(\boldsymbol{E}'_\nu) = \delta^\mu_\nu$. One can check easily that

$$\underset{\sim}{E}'^\mu = a^\mu{}_\nu \underset{\sim}{E}^\nu \tag{4.8}$$

by evaluating the right-hand side on $\boldsymbol{E}'_\lambda$ :

$$a^\mu{}_\nu \underset{\sim}{E}^\nu(\boldsymbol{E}'_\lambda) = a^\mu{}_\nu \underset{\sim}{E}^\nu({a^{-1}}^\rho{}_\lambda \boldsymbol{E}_\rho) = a^\mu{}_\nu {a^{-1}}^\rho{}_\lambda \delta^\nu_\rho = \delta^\mu_\lambda. \tag{4.9}$$

Note that vector components $A^\mu$ and cobasis vectors $\underset{\sim}{E}^\mu$, both with superscripts, transform in the same way, and that covector components $Z_\mu$ and basis vectors $\boldsymbol{E}_\mu$, both with subscripts, transform in the same way. The contraction (a sum with one up and one down) of one member of the first pair and one of the second pair gives an invariant (a number, a vector, a covector or a unit operator, as the case may be).

### 4.1.5 The Natural Isomorphism Between $\mathcal{V}$ and $\mathcal{V}^*$

Because there is a scalar product between vectors we get a natural 1-1 relation between vectors and covectors, a natural isomorphism between the spaces $\mathcal{V}$ and $\mathcal{V}^*$ which does not depend on any arbitrary choice of basis. To see this relationship, consider that for a fixed vector $\boldsymbol{B}$ we can define a linear mapping $\mathcal{V} \to \mathcal{R}$ by

$$\boldsymbol{A} \mapsto \boldsymbol{B} \cdot \boldsymbol{A}. \tag{4.10}$$

Since $\boldsymbol{B} \cdot \boldsymbol{A}$ is a real-valued, linear function of $\boldsymbol{A}$, the map *is*, therefore, an example of what we have called a covector; we give it the name $\underset{\sim}{B}$. In terms of components with respect to some basis and its cobasis we have, by (4.3) and (2.19),

$$\underset{\sim}{B}(\boldsymbol{A}) = B_\mu A^\mu = \boldsymbol{B} \cdot \boldsymbol{A} = B^\nu \eta_{\nu\mu} A^\mu \tag{4.11}$$

so that the relations between the components of the corresponding vector and covector (with respect to a basis and its associated cobasis) are

$$B_\mu = \eta_{\mu\nu} B^\nu \qquad \text{or} \qquad B^\nu = \eta^{\nu\mu} B_\mu. \tag{4.12}$$

Since $\mathcal{V}$ and $\mathcal{V}^*$ are finite-dimensional vector spaces, and $\eta_{\mu\nu}$ is invertible, we see that all covectors arise this way. Given any $\underset{\sim}{Z} \in \mathcal{V}^*$ there is a $\boldsymbol{Z} \in \mathcal{V}$ such that $\underset{\sim}{Z}(\boldsymbol{A}) = \boldsymbol{Z} \cdot \boldsymbol{A}$ for all $\boldsymbol{A}$ :

$$\underset{\sim}{Z} = Z_\mu \underset{\sim}{E}^\mu \Longleftrightarrow \boldsymbol{Z} = Z^\mu \boldsymbol{E}_\mu = (\eta^{\mu\nu} Z_\nu)\boldsymbol{E}_\mu. \tag{4.13}$$

In the context of special relativity the distinction between vectors and covectors is solely mathematical. One can go back and forth between the two simply for the sake of convenience. For example, the covector gradient (to be defined in the next section) has a simpler coordinate expression than the corresponding vector gradient. If the vector represents a physical quantity the covector may be regarded as providing a different representation of it, and *vice versa*.[1]

### 4.1.6 Geometrical Interpretation

There is a simple geometrical interpretation of covectors in $\mathcal{V}^*$ corresponding to the picture of vectors in $\mathcal{V}$ as arrows in spacetime. A "picture" of $\underset{\sim}{Z}$ is a geometrical realisation of the function $\underset{\sim}{Z}(\boldsymbol{x})$, for all $\boldsymbol{x}$. Let us associate the vectors $\boldsymbol{x}$ with the end points $X \in \mathcal{M}$ of translation vectors $\overrightarrow{OX} = \boldsymbol{x}$, for some fixed point $O$. Then $\underset{\sim}{Z}(\boldsymbol{x})$ becomes a real-valued function of points in

[1] The situation is somewhat different in general relativity, where the metric tensor, which connects the vectors and covectors, is itself a varying physical quantity.

spacetime $\mathcal{M}$. To get an understanding of this function it is convenient to look at its level hypersurfaces, $\underset{\sim}{Z}(\boldsymbol{x}) =$ constant. Start with the zero hypersurface $\underset{\sim}{Z}(\boldsymbol{x}) = 0$. By (4.11) this may be written $\boldsymbol{Z} \cdot \boldsymbol{x} = 0$ in terms of the vector $\boldsymbol{Z}$ corresponding to $\underset{\sim}{Z}$. Using components derived from some inertial coordinates with origin at $O$, it satisfies

$$\underset{\sim}{Z}(\boldsymbol{x}) = \boldsymbol{Z} \cdot \boldsymbol{x} = Z_\mu x^\mu = 0; \tag{4.14}$$

because $x^\mu$, the components of the vector $\overrightarrow{OX}$, are also the coordinates of the point $X$, the linear equation (4.14), with constant coefficients $Z_\mu$, is recognised as the equation of a three-dimensional hyperplane. The hyperplane contains the point $O$ since $\boldsymbol{x} = \boldsymbol{0}$ satisfies the equation, as do the end points of all translation vectors $\overrightarrow{OX}$ orthogonal to $\boldsymbol{Z}$. The physical significance of these three-dimensional hyperplanes will be different depending on whether $\boldsymbol{Z}$ is timelike, spacelike or lightlike. If $\boldsymbol{Z}$ is timelike then all vectors $\boldsymbol{x}$ in the hyperplane are necessarily spacelike (by the lemma in Section 1.6); the hyperplane itself is called spacelike. On the other hand, if $\boldsymbol{Z}$ is spacelike or lightlike, $\boldsymbol{x}$ will have a timelike or lightlike solution, respectively, as well as two independent spacelike solutions. These hyperplanes are called timelike or lightlike.

The simplest example of a spacelike hyperplane may be written using the bases derived from some inertial frame:

$$\underset{\sim}{E}^0(\boldsymbol{x}) = 0 = \underset{\sim}{E}^0(x^0\boldsymbol{E}_0 + x^m\boldsymbol{E}_m) = x^0\underset{\sim}{E}^0(\boldsymbol{E}_0) = x^0.$$

This is just the $x^0 = 0$ spacelike hyperplane of simultaneity for the frame. As an example of a timelike hyperplane, there is $\underset{\sim}{E}_1(\boldsymbol{x}) = x^1 = 0$, which contains one timelike dimension and two spacelike dimensions. Finally, a simple lightlike hyperplane is given by

$$(\underset{\sim}{E}^0 + \underset{\sim}{E}^1)(\boldsymbol{x}) = x^0 + x^1 = 0.$$

It has a lightlike dimension and two spacelike ones.

The nonzero hyperplanes, satisfying $\underset{\sim}{Z}(\boldsymbol{x}) = C \neq 0$, are parallel to the zero hyperplane (Figure 4.1). (Certainly there can be no point in common.) Also, all solutions to $\underset{\sim}{Z}(\boldsymbol{x}) = C$ for any $C$ may be found from one solution to $\underset{\sim}{Z}(\boldsymbol{x}_1) = 1$ together with all points on the zero hypersurface ($\underset{\sim}{Z}(\boldsymbol{x}_0) = 0$) by $\boldsymbol{x} = C\boldsymbol{x}_1 + \boldsymbol{x}_0$. Therefore, given one point on $\underset{\sim}{Z}(\boldsymbol{x}) = C$, all others may be found by translating with the solutions $\boldsymbol{x}_0$ to $\underset{\sim}{Z}(\boldsymbol{x}_0) = 0$.

The stack of hyperplanes $\underset{\sim}{Z}(\boldsymbol{x}) = C$, for various constants $C$, provides the picture of covectors which is appropriate when vectors are pictured as arrows from the point $O$. The real number $\underset{\sim}{Z}(\boldsymbol{x})$ associated with a particular vector

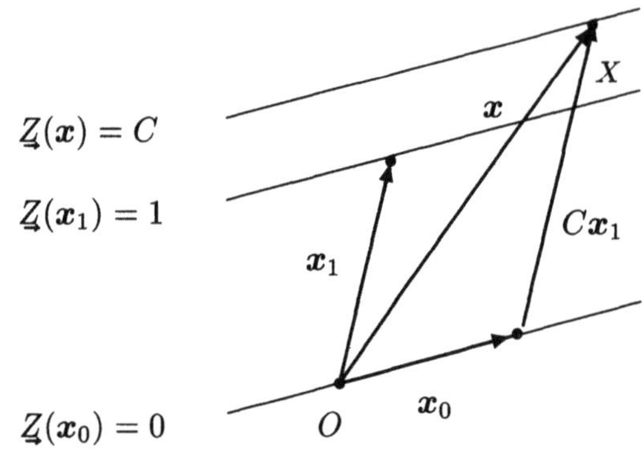


**Figure 4.1**

$\boldsymbol{x}$ is the constant $C$ labelling the particular hyperplane from the stack on which the end point of the arrow $\boldsymbol{x}$ lies. This is the picture which has been made popular by Misner, Thorne and Wheeler in their book.

## 4.2 Gradient of a Scalar Field

We consider a scalar field in spacetime, $\phi : \mathcal{M} \to \mathcal{R}$, which associates a real number $\phi(P)$ with each point $P$ of a region. A number of examples follow.

- The simplest example is a constant, $\phi(P) = C$ for all $P$.
- A covector may be used to construct a slightly more complicated example: $\phi(P) = \underset{\sim}{Z}(\boldsymbol{x})$ for $\boldsymbol{x} = \overrightarrow{OP}$ with fixed $O$. This is just the stack of hyperplanes of the previous section.
- If a particular frame is selected, $K$ say, and a particular one of its coordinates, $x^\mu$ say, a scalar field may be defined by $\phi(P) = x^\mu(P)$, where $x^\mu(P)$ is the $\mu$th coordinate of $P$ in the special frame. (If this field were expressed in different coordinates $x'^\lambda = a^\lambda{}_\rho x^\rho$, the particular coordinate would have to be expressed in terms of the new ones: $x^\mu(P) = a^{-1\mu}{}_\rho x'^\rho(P)$.) Actually, this example is just a special case of the covector example, with $\underset{\sim}{Z} = \underset{\sim}{E}^\mu$, where the cobasis element is the one associated with the particular frame.
- As a *physical* example of a scalar field, and one which suggests the wide possible variety, we might think of the temperature in a relativistic gas: at each $P$ the field $\phi(P)$ is the temperature in the local average rest frame of the gas at $P$.

### 4.2.1 Approximation of Scalar Fields and the Covector Gradient

To get an approximate picture of a field $\phi$ and a provisional understanding of $\phi$ near some point $P_0$, it is useful to think of the field's "level" hypersurfaces

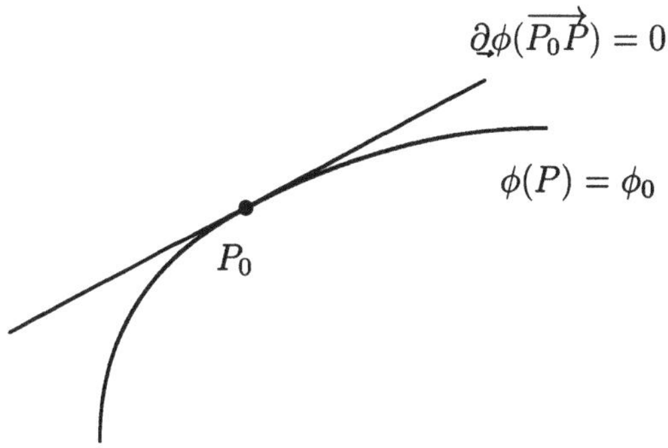


**Figure 4.2**

near $P_0$. The hypersurface containing $P_0$ consists of the points $P$ for which $\phi(P) = \phi(P_0) \equiv \phi_0$ (Figure 4.2). The nearby hypersurfaces, for values of $\phi$ near $\phi_0$, contain $P$ such that $\phi(P) = \phi_0 + \delta\phi_0$. The covector gradient of $\phi$, evaluated at $P_0$, is defined (in the next paragraph) to provide a linear approximation to $\phi$ near $P_0$ :

$$\phi(P) \approx \phi(P_0) + \underset{\rightarrow}{\partial}\phi(\overrightarrow{P_0P}). \tag{4.15}$$

In this approximation the level hypersurfaces of $\phi$ are replaced by the hyperplanes on which the covector gradient (evaluated at $P_0$) is constant, that is, $\underset{\rightarrow}{\partial}\phi(\overrightarrow{P_0P}) = \text{Const}$. This constant is zero for the hyperplane through $P_0$. The field $\phi$ is approximated near $P_0$ by the sum of $\phi$ at $P_0$ plus the value of the gradient evaluated at the displacement from $P_0$.

### 4.2.2 Components of the Covector Gradient

If $K$-coordinates are used to label points, $x_0^\mu$ for $P_0$, and $x^\mu = x_0^\mu + \delta x^\mu$ for the nearby $P$, and $\phi$ is represented by its $K$-coordinate form, then, from the linear term in the Taylor expansion, we get an explicit form[2] for $\underset{\rightarrow}{\partial}\phi$ :

$$\begin{aligned} \phi(P) = \phi(x_0^\mu + \delta x^\mu) &= \phi(x_0^\mu) + \delta x^\nu \partial_\nu \phi(x_0^\mu) + \cdots && (4.16) \\ &\equiv \phi(P_0) + \underset{\rightarrow}{\partial}\phi(\delta x) + \cdots . && (4.17) \end{aligned}$$

Hence,

[2] When fields in spacetime are expressed in terms of coordinates, as here, it is very common to use the notation $x^\mu$ to refer to the *set of 4 coordinates.* An alternative notation would be just $x$, but the former is preferred in this context because it produces correct chain-rule expressions automatically.

$$\underset{\sim}{\partial}\phi \equiv \underset{\sim}{E}^{\nu}\partial_{\nu}\phi. \tag{4.18}$$

The definition appears to depend on the use of $K$-coordinates, but in fact it does not; if $K'$-coordinates are used instead, with $\delta x'^{\mu} = a^{\mu}{}_{\nu}\delta x^{\nu}$, and $\phi(P) = \phi(x^{\mu}(P)) = \phi'(x'^{\mu}(P))$, then

$$\partial'_{\rho}\phi' = \frac{\partial x^{\tau}}{\partial x'^{\rho}}\partial_{\tau}\phi = a^{-1\tau}{}_{\rho}\partial_{\tau}\phi, \tag{4.19}$$

which, by (4.8), implies

$$\underset{\sim}{\partial}\phi \equiv (\partial_{\tau}\phi)\underset{\sim}{E}^{\tau} = (\partial'_{\tau}\phi')\underset{\sim}{E}'^{\tau}. \tag{4.20}$$

*Any* inertial frame can be used to express $\underset{\sim}{\partial}\phi$ in this way.

Notice that for the special case $\phi(P) = \underset{\sim}{Z}(\boldsymbol{x}) = Z_{\mu}x^{\mu}$,

$$\underset{\sim}{\partial}\phi = \partial_{\mu}(Z_{\nu}x^{\nu})\underset{\sim}{E}^{\mu} = (\delta^{\nu}_{\mu}Z_{\nu})\underset{\sim}{E}^{\mu} = Z_{\mu}\underset{\sim}{E}^{\mu} = \underset{\sim}{Z}, \tag{4.21}$$

as the construction was designed to produce. ($\underset{\sim}{\partial}\phi$ gives a linear approximation, but $\underset{\sim}{Z}$ is exactly linear.)

Consider a curve $\gamma(\lambda)$, parameterised with $\lambda$, and with $K$-coordinates $x^{\mu} = \gamma^{\mu}(\lambda)$. If we consider the values of $\phi$ along the curve, $\phi(\gamma(\lambda))$, we can differentiate with respect to $\lambda$ and use the chain rule to get

$$\frac{d\phi}{d\lambda} = \partial_{\mu}\phi\left(\frac{d\gamma^{\mu}}{d\lambda}\right) = \underset{\sim}{\partial}\phi\left(\frac{d\gamma}{d\lambda}\right),$$

where the tangent vector to the curve is given by

$$\frac{d\gamma}{d\lambda} = \frac{d\gamma^{\mu}}{d\lambda}\boldsymbol{E}_{\mu},$$

exactly as in eq. (2.39), although here we are not assuming that the curve is timelike. This form of the argument produces the covector gradient without the use of approximations in the analysis.

### 4.2.3 Vector Gradient

The *vector* gradient $\boldsymbol{\partial}\phi$ is related to the *covector* gradient $\underset{\sim}{\partial}\phi$ as a special case of the isomorphism $\underset{\sim}{Z}(\boldsymbol{A}) = \boldsymbol{Z}\cdot\boldsymbol{A}$ between vectors and covectors. By (4.13), $Z^{\mu} = \eta^{\mu\nu}Z_{\nu}$, the expansion in the inertial basis is

$$\boldsymbol{\partial}\phi \equiv \partial^{\mu}\phi\boldsymbol{E}_{\mu} = \eta^{\mu\nu}\partial_{\nu}\phi\boldsymbol{E}_{\mu}. \tag{4.22}$$

Notice the new definition of $\partial^{\mu}$ by $\partial^{\mu}\phi = \eta^{\mu\nu}\partial_{\nu}\phi$; it is used everywhere in the calculus of covectors, dyadics and tensors. The value of a covector gradient, applied to a certain displacement vector, for example, $\underset{\sim}{\partial}\phi(\delta\boldsymbol{x})$, may now be expressed as a scalar product $\boldsymbol{\partial}\phi\cdot\delta\boldsymbol{x}$.

### 4.2.4 Gradient Operators

If operators $\boldsymbol{\partial} = \boldsymbol{E}_\mu \partial^\mu$ and $\underset{\leftarrow}{\partial} = \underset{\leftarrow}{E}^\mu \partial_\mu$ are used in a vector calculus, it must be remembered that it is the *covector* components which are the simple partial derivatives, whereas, for the vector components, $\partial^\mu = \eta^{\mu\nu}\frac{\partial}{\partial x^\nu}$. Explicitly, the time component of $\boldsymbol{\partial}$ is $-\partial_0$. Expanding to show the difference, one has

$$\underset{\leftarrow}{\partial} = \underset{\leftarrow}{E}^0 \frac{\partial}{\partial x^0} + \underset{\leftarrow}{E}^m \frac{\partial}{\partial x^m}, \qquad \boldsymbol{\partial} = -\boldsymbol{E}_0 \frac{\partial}{\partial x^0} + \boldsymbol{E}_m \frac{\partial}{\partial x^m}. \tag{4.23}$$

The linear approximation to a scalar field $\phi$ near $P_0$ is, rewriting (4.15) in terms of the vector gradient,

$$\phi(P) \approx \phi(P_0) + (\overrightarrow{P_0 P}) \cdot \boldsymbol{\partial}\phi. \tag{4.24}$$

There is a generalisation of this formula for any sort of field, which serves to define the effect of the gradient operators on other geometrical fields, including vectors and dyadics.

### 4.2.5 Cobasis as the Covector Gradient of the Inertial Coordinates

It is worth noting that the elements of the cobasis $\underset{\leftarrow}{E}^\mu$ associated with the vector basis $\boldsymbol{E}_\nu$ are just the covector gradients of the coordinate functions for $K$. To see this, replace $\phi$ with $x^\mu$ (regarded here as a particular scalar field in spacetime) in (4.20):

$$\underset{\leftarrow}{\partial} x^\mu = \frac{\partial x^\mu}{\partial x^\tau} \underset{\leftarrow}{E}^\tau = \delta^\mu_\tau \underset{\leftarrow}{E}^\tau = \underset{\leftarrow}{E}^\mu \tag{4.25}$$

and

$$\underset{\leftarrow}{\partial} x^\mu (\boldsymbol{E}_\nu) = \underset{\leftarrow}{E}^\mu (\boldsymbol{E}_\nu) = \delta^\mu_\nu. \tag{4.26}$$

Then, for a completely general $\phi$, the formula (4.20) may be read as an instance of $\underset{\leftarrow}{\partial}\phi = \frac{\partial \phi}{\partial x^\mu} \underset{\leftarrow}{\partial} x^\mu$, itself a special case of $\underset{\leftarrow}{\partial}\phi(\psi) = \frac{\partial \phi}{\partial \psi} \underset{\leftarrow}{\partial}\psi$.

There is another special derivative, the *exterior derivative,* whose action on scalar fields coincides with that of $\underset{\leftarrow}{\partial}$. Consequently, the notation $\underset{\leftarrow}{d}\phi$ (and even more often without the underarrow, $d\phi$) is commonly used for $\underset{\leftarrow}{\partial}\phi$.

## 4.3 Dyadics in Spacetime

Dyadics form a space whose simplest elements, $\boldsymbol{A} \otimes \boldsymbol{B}$, may be thought of as a new type of "product" of vectors $\boldsymbol{A}$ and $\boldsymbol{B}$. These simple dyadics contain exactly the same information as the (ordered) pairs of vectors of which they are composed.

Dyadics are elements of a 16-dimensional vector space $\mathcal{V} \bigotimes \mathcal{V}$ (in the present context in which the vector space $\mathcal{V}$ is four-dimensional). A simple dyadic is written as a so-called *tensor product* of two vectors but linear combinations can be formed too and the most general dyadic may be represented (in many different ways) by an expression of the form $\Sigma \lambda_{ij} \boldsymbol{A}_i \otimes \boldsymbol{B}_j$, where $\boldsymbol{A}_i$ and $\boldsymbol{B}_j$ are vectors, and $\lambda_{ij}$ real constants.

Because the rules of operation are so elementary (see ahead: (4.28), (4.29), (4.30)), it is easiest to think of dyadics as formal sums of formal products. Two sums are equivalent, and so define the same dyadic, if one can be converted to the other by using these rules. At the present stage, the definition in terms of multilinear real-valued mappings is perhaps too abstract, but all the rules of operation may be well motivated in terms of linear transformations of vectors.[3]

In nonrelativistic physical theories in three-dimensional Euclidean space, many quantities are most appropriately regarded as dyadics: moment of inertia, angular momentum, stress, strain, dielectric tensor, and so on. In four-dimensional relativity two examples stand out.

- The electromagnetic field is a dyadic, although it cannot be a simple one because it must be antisymmetric. Just to illustrate how two vectors are used in its specification, consider a special field for which a frame exists in which it is a pure electric field. The dyadic is determined by the timelike basis vector of that frame, $\boldsymbol{E}_0$, and the electric field with respect to the frame, $\boldsymbol{E}$.
- The energy tensor for a flowing fluid is a symmetric dyadic. It is constructed from two vectors: one representing the spacetime velocity of the flow, the other giving the momentum density carried by the flow.

### 4.3.1 Linear Transformations as Dyadics

Dyadics may be put in one-one correspondence with linear mappings of $\mathcal{V}$ to itself. It is most convenient, in the first instance, to identify the dyadics and the linear mappings in order to explain the properties of the dyadics in terms of those of the mappings. The correspondence could be achieved in different ways, but the method we choose is determined by the scalar product of vectors (hence by the metric). The linear transformation corresponding to $\boldsymbol{A} \otimes \boldsymbol{B}$ is

$$\boldsymbol{A} \otimes \boldsymbol{B} : \boldsymbol{X} \mapsto (\boldsymbol{A} \otimes \boldsymbol{B}) \cdot \boldsymbol{X} \equiv \boldsymbol{A}(\boldsymbol{B} \cdot \boldsymbol{X}). \tag{4.27}$$

(We are using the same notation for the dyadic and the corresponding transformation.) The effect of the transformation $\boldsymbol{A} \otimes \boldsymbol{B}$ on $\boldsymbol{X}$ is to give a vector

[3] A complete discussion (in much greater generality) of the relation between different definitions is given in the first two sections of Chapter 2 of Volume 1 from Penrose and Rindler's book on spinors.

parallel to $\boldsymbol{A}$, namely, $\boldsymbol{A}$ times the scalar product of $\boldsymbol{X}$ with $\boldsymbol{B}$. Slightly more complicated linear transformations arise from linear combinations of the simplest ones.

### 4.3.2 Simplest Properties

The linear transformations (and so the dyadics) form a vector space in which multiplication by a scalar and addition are defined in an obvious way. The definition of multiplication by a scalar, rewritten for simple dyadics, is

$$\begin{aligned} [\lambda(\boldsymbol{A}\otimes\boldsymbol{B})]\cdot\boldsymbol{X} = \lambda(\boldsymbol{A}\otimes\boldsymbol{B}\cdot\boldsymbol{X}) &= \lambda(\boldsymbol{A}(\boldsymbol{B}\cdot\boldsymbol{X})) \\ &= (\lambda\boldsymbol{A})(\boldsymbol{B}\cdot\boldsymbol{X}) \\ &= \boldsymbol{A}(\lambda\boldsymbol{B}\cdot\boldsymbol{X}). \end{aligned}$$

Hence $\lambda(\boldsymbol{A}\otimes\boldsymbol{B})$ is defined and satisfies

$$\lambda(\boldsymbol{A}\otimes\boldsymbol{B}) = (\lambda\boldsymbol{A})\otimes\boldsymbol{B} = \boldsymbol{A}\otimes(\lambda\boldsymbol{B}). \tag{4.28}$$

Similarly, addition of linear transformations, rewritten in the dyadic notation, motivates the definition

$$[\boldsymbol{A}\otimes\boldsymbol{B}+\boldsymbol{C}\otimes\boldsymbol{D}]\cdot\boldsymbol{X} = \boldsymbol{A}(\boldsymbol{B}\cdot\boldsymbol{X})+\boldsymbol{C}(\boldsymbol{D}\cdot\boldsymbol{X}).$$

Hence, in particular, when $\boldsymbol{A}=\boldsymbol{C}$,

$$\boldsymbol{A}\otimes(\boldsymbol{B}+\boldsymbol{D}) = \boldsymbol{A}\otimes\boldsymbol{B}+\boldsymbol{A}\otimes\boldsymbol{D}, \tag{4.29}$$

and when $\boldsymbol{B}=\boldsymbol{D}$,

$$(\boldsymbol{A}+\boldsymbol{C})\otimes\boldsymbol{B} = \boldsymbol{A}\otimes\boldsymbol{B}+\boldsymbol{C}\otimes\boldsymbol{B}. \tag{4.30}$$

### 4.3.3 Bases for the Space of Dyadics

The rules for addition and multiplication by a scalar are as simple as those for real numbers. They account completely for the different expressions that exist for a particular dyadic. For any way that $\boldsymbol{A}$ may be written as a linear combination $\boldsymbol{A}=\sum_i \boldsymbol{A}_i$, and $\boldsymbol{B}=\sum_j \boldsymbol{B}_j$, there is an expansion $\boldsymbol{A}\otimes\boldsymbol{B} = \sum_{ij}\boldsymbol{A}_i\otimes\boldsymbol{B}_j$. In particular, both $\boldsymbol{A}$ and $\boldsymbol{B}$ can be decomposed in some inertial basis $\boldsymbol{E}_\mu$ (with the summation convention now expressing the sums),

$$\boldsymbol{A}\otimes\boldsymbol{B} = (A^\mu\boldsymbol{E}_\mu)\otimes(B^\nu\boldsymbol{E}_\nu) = A^\mu B^\nu(\boldsymbol{E}_\mu\otimes\boldsymbol{E}_\nu). \tag{4.31}$$

In fact, the sixteen dyadics $\boldsymbol{E}_\mu\otimes\boldsymbol{E}_\nu$, or the corresponding sixteen dyadics derived from any other inertial basis, form a basis for the space $\mathcal{V}\bigotimes\mathcal{V}$.

To show that $\boldsymbol{E}_\mu\otimes\boldsymbol{E}_\nu$ is a basis for all dyadics we consider an arbitrary linear transformation $T$, and the corresponding dyadic $\overset{\Rightarrow}{T}$. The transformation is determined, once a basis for $\mathcal{V}$ is specified, by sixteen numerical components $T^\nu{}_\mu$. We may write $T^\nu{}_\mu = T^{\nu\lambda}\eta_{\lambda\mu}$, where $T^{\nu\lambda} \equiv T^\nu{}_\mu\eta^{\mu\lambda}$. The transformation is

$$\begin{aligned} T(\boldsymbol{X}) &= T(X^\mu \boldsymbol{E}_\mu) = X^\mu T(\boldsymbol{E}_\mu) \\ &= X^\mu \boldsymbol{E}_\nu T^\nu{}_\mu = X^\mu \boldsymbol{E}_\nu T^{\nu\lambda}\eta_{\lambda\mu} \\ &= T^{\nu\lambda}(\boldsymbol{E}_\nu \otimes \boldsymbol{E}_\lambda) \cdot \boldsymbol{X} \\ &= \overset{\Rightarrow}{T} \cdot \boldsymbol{X}, \end{aligned} \tag{4.32}$$

where

$$\overset{\Rightarrow}{T} \equiv T^{\nu\lambda}\boldsymbol{E}_\nu \otimes \boldsymbol{E}_\lambda \tag{4.33}$$

is the corresponding dyadic. When a single symbol is used to denote a dyadic the two arrows appear one above the other. This is a natural extension of the arrow notation for vectors and covectors, $\boldsymbol{A}$ and $\underset{\leftarrow}{Z}$. Mathematically, there are further generalisations that would call for any number of arrows above and/or below. In practice this explicit notation is not often used for these more complicated objects. By the time such complicated objects are met, there is no longer a need for a notation explicitly providing the information that the arrows carry.

The formula (4.33) shows that the collection of simple dyadics $\boldsymbol{E}_\mu \otimes \boldsymbol{E}_\nu$ spans the space $\mathcal{V} \bigotimes \mathcal{V}$. It is elementary to see that the $\boldsymbol{E}_\mu \otimes \boldsymbol{E}_\nu$ are independent and so form a basis. Consider $\overset{\Rightarrow}{Z} = Z^{\nu\lambda}\boldsymbol{E}_\nu \otimes \boldsymbol{E}_\lambda = \overset{\Rightarrow}{0}$. This means $\overset{\Rightarrow}{Z} \cdot \boldsymbol{X} = \boldsymbol{0}$ for all $\boldsymbol{X}$, that is, $Z^{\nu\lambda}\eta_{\lambda\mu}X^\mu = 0$ for all $X^\mu$; which gives $Z^{\nu\rho} = 0$ by choosing $\eta^{\mu\rho}$ for $X^\mu$.

### 4.3.4 A Unit Dyadic: The Contravariant Metric

Because vectors and covectors can be related in a natural way using the metric (as discussed on page 128), it is possible to express transformations in several (metrically related) ways. When transformations are expressed in terms of dyadics and scalar products, as we have been doing, the dyadic

$$\overset{\Rightarrow}{\eta} = \eta^{\nu\lambda}\boldsymbol{E}_\nu \otimes \boldsymbol{E}_\lambda = -\boldsymbol{E}_0 \otimes \boldsymbol{E}_0 + \boldsymbol{E}_m \otimes \boldsymbol{E}_m \tag{4.34}$$

corresponds to the unit transformation and so may be called the unit dyadic. Note that $\overset{\Rightarrow}{\eta}$ has exactly the same form (and same components) when it is expanded in the basis derived from any other inertial frame. The equation

$$\overset{\Rightarrow}{\eta} \cdot \boldsymbol{X} = \overset{\Rightarrow}{\eta} \cdot \boldsymbol{E}_\mu X^\mu = \eta^{\nu\lambda}\boldsymbol{E}_\nu\eta_{\lambda\mu}X^\mu = \boldsymbol{E}_\mu X^\mu = \boldsymbol{X}$$

is just the expression of (4.32) for the transformation whose matrix is the unit matrix $T^{\nu}{}_{\mu} = \delta^{\nu}{}_{\mu}$. An alternative name for $\overset{\Rightarrow}{\eta}$ is the contravariant metric tensor[4].

### 4.3.5 A Geometrical Example: Reflection Dyadics

The reflection dyadics are both extremely simple and amazingly helpful for geometrical constructions in spacetime.

A spacelike hyperplane has a unit, timelike normal vector $\boldsymbol{v}$. Any vector $\boldsymbol{X}$ can be decomposed into a sum of two vectors, $-\boldsymbol{v}(\boldsymbol{v}\cdot\boldsymbol{X})$, proportional to $\boldsymbol{v}$, and the other, $\boldsymbol{X}_{\perp} = \boldsymbol{X} + \boldsymbol{v}(\boldsymbol{v}\cdot\boldsymbol{X})$, orthogonal to $\boldsymbol{v}$ :

$$\boldsymbol{X} = -\boldsymbol{v}(\boldsymbol{v}\cdot\boldsymbol{X}) + \boldsymbol{X}_{\perp}. \tag{4.35}$$

The orthogonal reflection of $\boldsymbol{X}$ in the hyperplane is then

$$\boldsymbol{X}_R \equiv +\boldsymbol{v}(\boldsymbol{v}\cdot\boldsymbol{X}) + \boldsymbol{X}_{\perp} = +2\boldsymbol{v}(\boldsymbol{v}\cdot\boldsymbol{X}) + \boldsymbol{X} \tag{4.36}$$

(Figure 4.3). The dyadic that performs the transformation is the reflection dyadic:

$$\boldsymbol{X}_R = (\overset{\Rightarrow}{\eta} + 2\boldsymbol{v}\otimes\boldsymbol{v})\cdot\boldsymbol{X}. \tag{4.37}$$

If the hyperplane is *timelike,* it has a unit *spacelike* normal vector, $\boldsymbol{n}$, say. The similar construction has a sign difference:

$$\boldsymbol{Y}_R = -\boldsymbol{n}(\boldsymbol{n}\cdot\boldsymbol{Y}) + \boldsymbol{Y}_{\perp} = -2\boldsymbol{n}(\boldsymbol{n}\cdot\boldsymbol{Y}) + \boldsymbol{Y} = (\overset{\Rightarrow}{\eta} - 2\boldsymbol{n}\otimes\boldsymbol{n})\cdot\boldsymbol{Y}. \tag{4.38}$$

Products of reflection dyadics will be used in the next section to construct active rotation and boost transformations.

### 4.3.6 Transformation Law for Components

Expanding some arbitrary dyadic $\overset{\Rightarrow}{T}$ in the distinct bases derived from two different inertial frames involves two versions of (4.33). If the coordinates of the two frames are connected by the relation $x'^{\mu} = a^{\mu}{}_{\nu}x^{\nu} + a^{\mu}$, then the relation between the sets of basis vectors may be introduced:

$$\begin{aligned}\overset{\Rightarrow}{T} &= T^{\mu\nu}\boldsymbol{E}_{\mu}\otimes\boldsymbol{E}_{\nu} = T'^{\mu\nu}\boldsymbol{E}'_{\mu}\otimes\boldsymbol{E}'_{\nu}\\ &= T^{\mu\nu}(a^{\rho}{}_{\mu}\boldsymbol{E}'_{\rho})\otimes(a^{\tau}{}_{\nu}\boldsymbol{E}'_{\tau}).\end{aligned}$$

[4] It is only in general relativity, where the metric becomes the fundamental physical field in the theory, that different names are used for related components. Then $g_{\mu\nu}$, which replaces $\eta_{\mu\nu}$, is called the metric, or covariant metric, and $g^{\mu\nu}$, the matrix inverse, is called the inverse metric, or contravariant metric.

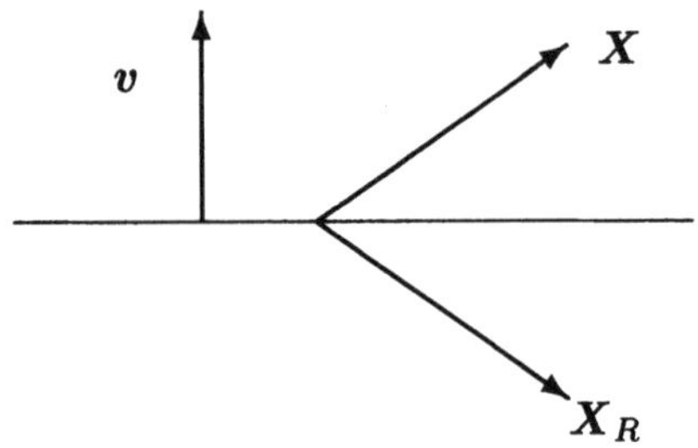


**Figure 4.3**

One obtains

$$T'^{\rho\tau} = a^{\rho}{}_{\mu} a^{\tau}{}_{\nu} T^{\mu\nu}, \tag{4.39}$$

the transformation law for the components of the dyadic (or second rank contravariant tensor, as it is otherwise called). The law is a sort of second power of the transformation law for the components of vectors.

### 4.3.7 Transposed Dyadics and Symmetries

The transpose of a simple dyadic $\boldsymbol{A} \otimes \boldsymbol{B}$ is defined by

$$(\boldsymbol{A} \otimes \boldsymbol{B})^T \equiv \boldsymbol{B} \otimes \boldsymbol{A}, \tag{4.40}$$

with the obvious linear extension to linear combinations, including

$$(\overset{\Rightarrow}{T})^T = T^{\mu\nu}(\boldsymbol{E}_\mu \otimes \boldsymbol{E}_\nu)^T = T^{\mu\nu} \boldsymbol{E}_\nu \otimes \boldsymbol{E}_\mu = T^{\nu\mu}(\boldsymbol{E}_\mu \otimes \boldsymbol{E}_\nu), \tag{4.41}$$

hence $T^{T\mu\nu} = T^{\nu\mu}$. (If this equation holds for the decomposition with respect to one inertial basis, then the same relation is true for the components with respect to any other inertial basis, $T'^{T\mu\nu} = T'^{\nu\mu}$.) A dyadic is called *symmetric* (a physical example is the energy tensor) if

$$(\overset{\Rightarrow}{S})^T = \overset{\Rightarrow}{S} \Longleftrightarrow S^{T\nu\mu} = S^{\mu\nu} = S^{\nu\mu}, \tag{4.42}$$

or *antisymmetric* (physical examples are angular momentum tensors and the electromagnetic field tensor) if

$$\overset{\Rightarrow}{A}^T = -\overset{\Rightarrow}{A} \Longleftrightarrow A^{T\nu\mu} = A^{\mu\nu} = -A^{\nu\mu}. \tag{4.43}$$

(These relations between the components hold with respect to all inertial frames.) For antisymmetric dyadics, which are especially important in physical applications, the special notation

$$\boldsymbol{A} \wedge \boldsymbol{B} = \boldsymbol{A} \otimes \boldsymbol{B} - \boldsymbol{B} \otimes \boldsymbol{A} \tag{4.44}$$

is often used. There is a special algebra, exterior algebra, based on and generalising the $\wedge$ product between vectors. In terms of this, an antisymmetric dyadic has a second expansion

$$\overset{\Rightarrow}{A} = A^{\mu\nu} \boldsymbol{E}_\mu \otimes \boldsymbol{E}_\nu = \frac{1}{2} A^{\mu\nu} \boldsymbol{E}_\mu \wedge \boldsymbol{E}_\nu. \tag{4.45}$$

### 4.3.8 Scalar Products

There are several "dot" products involving dyadics, all derived from the scalar product between two vectors. In addition to the product with which we began, of a dyadic $\boldsymbol{A} \otimes \boldsymbol{B}$ with a vector on the right $(\boldsymbol{A} \otimes \boldsymbol{B}) \cdot \boldsymbol{X}$, we can define a product with the vector on the left by using the transpose $(\boldsymbol{A} \otimes \boldsymbol{B})^T$ :

$$(\boldsymbol{A} \otimes \boldsymbol{B})^T \cdot \boldsymbol{X} = (\boldsymbol{B} \otimes \boldsymbol{A}) \cdot \boldsymbol{X} = \boldsymbol{B}(\boldsymbol{A} \cdot \boldsymbol{X}) \equiv \boldsymbol{X} \cdot (\boldsymbol{A} \otimes \boldsymbol{B}). \tag{4.46}$$

Also, there is a dot product between two dyadics $\boldsymbol{A} \otimes \boldsymbol{B}$ and $\boldsymbol{C} \otimes \boldsymbol{D}$ :

$$(\boldsymbol{A} \otimes \boldsymbol{B}) \cdot (\boldsymbol{C} \otimes \boldsymbol{D}) = (\boldsymbol{A} \otimes \boldsymbol{D})(\boldsymbol{B} \cdot \boldsymbol{C}), \tag{4.47}$$

and a similar product with the dyadics the other way round. All these products are generalised in an obvious way to dyadics which are linear combinations of simple ones.

### 4.3.9 Trace

A last contraction, or dot product, is the *trace* of a dyadic. It is defined by

$$\mathrm{tr}(\boldsymbol{A} \otimes \boldsymbol{B}) = \boldsymbol{A} \cdot \boldsymbol{B}, \tag{4.48}$$

or, for a linear combination,

$$\mathrm{tr}\Sigma\lambda_i(\boldsymbol{A}_i \otimes \boldsymbol{B}_i) = \Sigma\lambda_i(\boldsymbol{A}_i \cdot \boldsymbol{B}_i). \tag{4.49}$$

If the dyadic is expanded in terms of an inertial basis, one has

$$\mathrm{tr}(\overset{\Rightarrow}{T}) = \mathrm{tr}(T^{\mu\nu} \boldsymbol{E}_\mu \otimes \boldsymbol{E}_\nu) = T^{\mu\nu} \boldsymbol{E}_\mu \cdot \boldsymbol{E}_\nu = T^{\mu\nu} \eta_{\mu\nu}. \tag{4.50}$$

In the notation used in (4.32), the final form may simply be written $T^\mu{}_\mu$, which accounts for the name "trace". It is easy to check that one gets the same answer no matter which inertial basis is used for the computation.

## 4.4 Rotation and Boost Dyadics

Reflection dyadics such as (4.38) can be used to construct *rotations* with respect to an inertial frame $K$, whose time basis vector is $\boldsymbol{E}_0$. If $\boldsymbol{n}$ and $\boldsymbol{Y}$ are both spatial in $K$, then so will $\boldsymbol{Y}_R$ be. Furthermore, $\boldsymbol{Y}$ and $\boldsymbol{Y}_R$ have the same magnitude, since, from (4.38), $\boldsymbol{Y}\cdot\boldsymbol{Y} = \boldsymbol{Y}_R\cdot\boldsymbol{Y}_R$. The transformation $\boldsymbol{Y}\mapsto\boldsymbol{Y}_R$ is not, however, a rotation because its determinant is $-1$. But if a second reflection is performed, one *does* get a rotation $\boldsymbol{Y}\mapsto\boldsymbol{Y}_{Rot}$,

$$\boldsymbol{Y}_{Rot} = (\overset{\Rightarrow}{\eta} - 2\boldsymbol{m}\otimes\boldsymbol{m})\cdot(\overset{\Rightarrow}{\eta} - 2\boldsymbol{n}\otimes\boldsymbol{n})\cdot\boldsymbol{Y}. \tag{4.51}$$

The second vector $\boldsymbol{m}$ must also be spatial in $K$, and the rotation is around the spatial axis in $K$ which is perpendicular to both $\boldsymbol{n}$ and $\boldsymbol{m}$.

Very interestingly, two reflections along timelike axes, as in (4.37), generate an active version of the *boost* transformation. Such a transformation is a special case of a general active Lorentz transformation. Corresponding to a Lorentz transformation $x'^{\mu} = a^{\mu}{}_{\nu}x^{\nu}$, there is the associated transformation from the original basis $\boldsymbol{E}_\mu$ of $\mathcal{V}$ to the new one $\boldsymbol{E}'_\mu = \eta_{\mu\beta}a^{\beta}{}_{\lambda}\eta^{\lambda\alpha}\boldsymbol{E}_\alpha$. This transformation may be simply generated by an appropriate dyadic:

$$\overset{\Rightarrow}{L} \equiv \eta^{\mu\nu}\boldsymbol{E}'_\mu\otimes\boldsymbol{E}_\nu, \qquad \overset{\Rightarrow}{L}\cdot\boldsymbol{E}_\rho = \boldsymbol{E}'_\rho. \tag{4.52}$$

The condition that the $\boldsymbol{E}'_\mu$ form an inertial basis may be expressed by

$$\overset{\Rightarrow}{L}{}^T\cdot\overset{\Rightarrow}{L} = \eta^{\mu\nu}\boldsymbol{E}_\nu\otimes\boldsymbol{E}'_\mu\cdot\eta^{\alpha\beta}\boldsymbol{E}'_\alpha\otimes\boldsymbol{E}_\beta = \eta^{\mu\nu}\eta_{\mu\alpha}\eta^{\alpha\beta}\boldsymbol{E}_\nu\otimes\boldsymbol{E}_\beta = \overset{\Rightarrow}{\eta}. \tag{4.53}$$

The general dyadic $\overset{\Rightarrow}{L}$ may be written

$$\overset{\Rightarrow}{L} = -\boldsymbol{E}'_0\otimes\boldsymbol{E}_0 + \boldsymbol{E}'_1\otimes\boldsymbol{E}_1 + \boldsymbol{E}'_2\otimes\boldsymbol{E}_2 + \boldsymbol{E}'_3\otimes\boldsymbol{E}_3. \tag{4.54}$$

A simple rotation, adapted to the $K$-basis, will change only two vectors, say $\boldsymbol{E}_2$ and $\boldsymbol{E}_3$, for a rotation about the $x^1$-axis. Correspondingly, a boost, adapted to the frame, will change only two vectors, say $\boldsymbol{E}_0$ and $\boldsymbol{E}_1$. The characterisation of the rotation and the boost can be expressed more geometrically, without reference to the frame $K$, by saying that a rotation dyadic $\overset{\Rightarrow}{L}$ leaves invariant a timelike direction and a spacelike direction (orthogonal to the former), whereas a boost dyadic leaves invariant two orthogonal spacelike directions. The boost is therefore fully determined by $\boldsymbol{E}_0$ and $\boldsymbol{E}'_0$. In the case of the standard boost in eq. (2.12), we can solve for $\boldsymbol{E}_1$ and $\boldsymbol{E}'_1$,

$$\boldsymbol{E}_1 = \frac{\boldsymbol{E}'_0 - \gamma\boldsymbol{E}_0}{V\gamma}, \qquad \boldsymbol{E}'_1 = \frac{\gamma\boldsymbol{E}'_0 - \boldsymbol{E}_0}{V\gamma}, \tag{4.55}$$

and substitute into the appropriate version of (4.54) (writing $\overset{\Rightarrow}{B}$ for the boost dyadic instead of $\overset{\Rightarrow}{L}$)

$$\begin{aligned}\vec{\vec{B}} &= -2\boldsymbol{E}'_0 \otimes \boldsymbol{E}_0 + \boldsymbol{E}'_0 \otimes \boldsymbol{E}_0 + \boldsymbol{E}'_1 \otimes \boldsymbol{E}_1 + \boldsymbol{E}_2 \otimes \boldsymbol{E}_2 + \boldsymbol{E}_3 \otimes \boldsymbol{E}_3 \\ &= -2\boldsymbol{E}'_0 \otimes \boldsymbol{E}_0 + (\boldsymbol{E}'_0 + \boldsymbol{E}_0) \otimes \boldsymbol{E}_0 + (\boldsymbol{E}'_1 - \boldsymbol{E}_1) \otimes \boldsymbol{E}_1 + \vec{\vec{\eta}}.\end{aligned}$$

After simplification (writing $V$ in terms of $\gamma$), one gets

$$\vec{\vec{B}}(\boldsymbol{E}_0 \mapsto \boldsymbol{E}'_0) = \vec{\vec{\eta}} - 2\boldsymbol{E}'_0 \otimes \boldsymbol{E}_0 + \frac{(\boldsymbol{E}_0 + \boldsymbol{E}'_0) \otimes (\boldsymbol{E}_0 + \boldsymbol{E}'_0)}{\gamma + 1}. \tag{4.56}$$

Although the equation for $B(\boldsymbol{E}_0 \mapsto \boldsymbol{E}'_0)$ was derived for the standard boost, it holds for any boost. To see this we have only to consider that

$$\begin{aligned}B(\boldsymbol{E}_0 \mapsto \boldsymbol{E}'_0) \cdot \boldsymbol{E}_0 &= \boldsymbol{E}_0 - 2\boldsymbol{E}'_0(-1) + \frac{(\boldsymbol{E}_0 + \boldsymbol{E}'_0)(-1 + \boldsymbol{E}_0 \cdot \boldsymbol{E}'_0)}{\gamma + 1} \\ &= \boldsymbol{E}'_0,\end{aligned}$$

and that any direction orthogonal to the plane of $\boldsymbol{E}_0$ and $\boldsymbol{E}'_0$ is obviously invariant.

Extensive use of the rotation and boost dyadics will be made in the discussion of spin in Chapter 7.

## 4.5 Gradient of a Vector Field

The gradient of a vector field, $\boldsymbol{\partial} \otimes \boldsymbol{A}$, may be used near a point to provide a linear approximation to the field. It can be constructed by finding the gradient of the components when the field is expanded in an inertial basis. In special relativity the result is independent of which basis is used.

Consider a (smooth) vector field, a vector, $\boldsymbol{A}(P)$, say, defined at each point $P$ of a region in spacetime. As an example, one might think of the field of spacetime velocity vectors $\boldsymbol{v}(P)$ for an accelerating fluid; each particle in the continuum, at each point of its worldline, has a spacetime velocity.

If we expand $\boldsymbol{A}(P)$ in the bases generated by typical inertial frames $K$ and $K'$,

$$\boldsymbol{A}(P) = A^{\mu}(P)\boldsymbol{E}_{\mu} = A'^{\mu}(P)\boldsymbol{E}'_{\mu}, \tag{4.57}$$

the components will be (smooth) functions of the points in spacetime. If point-coordinates with respect to $K$ and $K'$ are related by $x'^{\mu} = a^{\mu}{}_{\nu}x^{\nu} + a^{\mu}$, then the different components of the vector will be related by

$$A'^{\mu}(P) = a^{\mu}{}_{\nu}A^{\nu}(P). \tag{4.58}$$

The same notation for the components is used when they are regarded as functions of the coordinates of $P$.

Unless there is some special reason not to, one regards $A'^{\mu}$, without change of notation, as a function of the $K'$-coordinates $x'^{\lambda}$, and $A^{\nu}$, similarly, as a function of the $K$-coordinates $x^{\rho}$. Instead of the proliferation of functions that would arise if we used a different notation for the components regarded as functions of each different set of coordinates, as in

$$A'^{\mu}(P) = A'^{\mu}_{K'}(x'^{\lambda}(P)) = a^{\mu}{}_{\nu}A^{\nu}(P) = a^{\mu}{}_{\nu}A^{\nu}_{K}(x^{\rho}(P)),$$

we put up with the small amount of ambiguity in interpreting the relation between the components as

$$A'^{\mu}(x'^{\lambda}(P)) = a^{\mu}{}_{\nu}A^{\nu}(x^{\rho}(P)),$$

or simply

$$A'^{\mu}(x'^{\lambda}) = a^{\mu}{}_{\nu}A^{\nu}(x^{\rho}), \tag{4.59}$$

with the relation $x^{\rho}(x'^{\lambda})$ understood.

Differentiating (4.59) partially with respect to $x'^{\lambda}$ gives

$$\partial'_{\lambda}A'^{\mu} = a^{\mu}{}_{\nu}\frac{\partial x^{\rho}}{\partial x'^{\lambda}}\partial_{\rho}A^{\nu}.$$

From the inverse Poincaré transformation, $x^{\rho} = a^{-1\rho}{}_{\lambda}(x'^{\lambda} - a^{\lambda})$, and (1.19),

$$\frac{\partial x^{\rho}}{\partial x'^{\lambda}} = a^{-1\rho}{}_{\lambda} = \eta^{\rho\tau}a^{\omega}{}_{\tau}\eta_{\lambda\omega}, \tag{4.60}$$

so that

$$\partial'_{\lambda}A'^{\mu} = a^{-1\rho}{}_{\lambda}a^{\mu}{}_{\nu}\partial_{\rho}A^{\nu}. \tag{4.61}$$

Hence, multiplying by $\eta^{\alpha\lambda}$,

$$\partial'^{\alpha}A'^{\mu} \equiv \eta^{\alpha\lambda}\partial'_{\lambda}A'^{\mu} = a^{\alpha}{}_{\tau}a^{\mu}{}_{\nu}\eta^{\tau\rho}\partial_{\rho}A^{\nu} \equiv a^{\alpha}{}_{\tau}a^{\mu}{}_{\nu}\partial^{\tau}A^{\nu}. \tag{4.62}$$

Comparing this with (4.39), the transformation formula for dyadic components, we see that the derivatives of the vector components are the components of a dyadic. More precisely, there is a dyadic $\boldsymbol{\partial}\otimes\boldsymbol{A}$ whose components with respect to each inertial basis have the same form (derivatives of the relevant vector components) relative to the inertial frame:

$$\boldsymbol{\partial}\otimes\boldsymbol{A} = \partial^{\mu}A^{\nu}\boldsymbol{E}_{\mu}\otimes\boldsymbol{E}_{\nu} = \partial'^{\mu}A'^{\nu}\boldsymbol{E}'_{\mu}\otimes\boldsymbol{E}'_{\nu}. \tag{4.63}$$

This is the vector gradient of $\boldsymbol{A}$, independent of the inertial frame used to express it.

Note that the scalar gradient may be used to write the vector gradient: by expanding $\boldsymbol{A} = A^{\mu}\boldsymbol{E}_{\mu}$, and applying $\boldsymbol{\partial}$ to the components,

$$\boldsymbol{\partial}\otimes\boldsymbol{A} = (\boldsymbol{\partial}A^{\mu})\otimes\boldsymbol{E}_{\mu}. \tag{4.64}$$

The dyadic $\boldsymbol{\partial}\otimes\boldsymbol{A}$ at $P_0$ has an analogous significance in approximating the vector field $\boldsymbol{A}$ near $P_0$ as the gradient $\boldsymbol{\partial}\phi$ has in approximating $\phi$. Suppose $P$ is a point near $P_0$, with $\overrightarrow{P_0P} = \delta x^{\mu}\boldsymbol{E}_{\mu}$. If the Taylor expansion is used to approximate the $K$-component functions $A^{\mu}(x_0^{\rho} + \delta x^{\rho})$ one obtains

$$\begin{aligned}\boldsymbol{A}(P) &= A^\mu(x_0^\rho + \delta x^\rho)\boldsymbol{E}_\mu \\ &= A^\mu(x_0^\rho)\boldsymbol{E}_\mu + \delta x^\rho \partial_\rho A^\mu \boldsymbol{E}_\mu + \cdots \\ &= \boldsymbol{A}(P_0) + \delta x^\rho \eta_{\rho\tau}\partial^\tau A^\mu \boldsymbol{E}_\mu + \cdots \\ &= \boldsymbol{A}(P_0) + \overrightarrow{P_0P}\cdot \boldsymbol{\partial}\otimes \boldsymbol{A} + \cdots .\end{aligned} \tag{4.65}$$

A coordinate system was used to get this result, but only in order to exploit the conventional Taylor expansion; the approximation itself is quite independent of the coordinates used to obtain it.

The trace of the gradient of $\boldsymbol{A}$ is called the divergence of $\boldsymbol{A}$ :

$$\mathrm{tr}\boldsymbol{\partial}\otimes\boldsymbol{A} \equiv \boldsymbol{\partial}\cdot\boldsymbol{A} = \partial^\mu A^\nu \eta_{\mu\nu} = \partial_\nu A^\nu, \tag{4.66}$$

a result which is independent of the inertial frame (the expression has the same form with respect to any inertial frame and the numerical value is the same).

The integral of a divergence over a four-dimensional volume may be expressed as an integral over the three-dimensional boundary hypersurface. This relation, the divergence theorem, will be discussed in Section 4.9. It provides one of the most useful ways of expressing the physical content of the conservation laws in physical theories.

## 4.6 Extensions

In addition to the space of linear combinations of simple dyadics $\boldsymbol{A}\otimes\boldsymbol{B}$, the tensor product space $\mathcal{V}\bigotimes\mathcal{V}$, there are three other spaces in which covectors replace one or both of the vectors, for example

$$\underset{\sim}{Z}\otimes\boldsymbol{B}\in\mathcal{V}^*\bigotimes\mathcal{V}, \quad \underset{\sim}{Z}\otimes\underset{\sim}{Y}\in\mathcal{V}^*\bigotimes\mathcal{V}^*.$$

In view of the natural isomorphism between $\mathcal{V}$ and $\mathcal{V}^*$ given by $\underset{\sim}{Z}\leftrightarrow\boldsymbol{Z}$, where $\underset{\sim}{Z}(\boldsymbol{X}) = \boldsymbol{Z}\cdot\boldsymbol{X}$, there are natural isomorphisms between all of these spaces.

The dot notation is extended to give the same result, under the natural isomorphism, in all four pairs of spaces

$$\underset{\sim}{Z}\cdot\boldsymbol{X}\equiv\underset{\sim}{Z}(\boldsymbol{X}) = \boldsymbol{Z}\cdot\boldsymbol{X}\equiv\underset{\sim}{Z}\cdot\underset{\sim}{X} = \underset{\sim}{X}(\boldsymbol{Z}). \tag{4.67}$$

It may also be extended for linear transformations

$$(\boldsymbol{A}\otimes\underset{\sim}{Z})\cdot\boldsymbol{X}\equiv\boldsymbol{A}(\underset{\sim}{Z}(\boldsymbol{X})) = \boldsymbol{A}(\boldsymbol{Z}\cdot\boldsymbol{X}) = (\boldsymbol{A}\otimes\boldsymbol{Z})\cdot\boldsymbol{X} \tag{4.68}$$

$$\boldsymbol{X}\cdot(\underset{\sim}{Z}\otimes\boldsymbol{B}) = \underset{\sim}{Z}(\boldsymbol{X})\boldsymbol{B} = (\boldsymbol{X}\cdot\boldsymbol{Z})\boldsymbol{B} = \boldsymbol{X}\cdot(\boldsymbol{Z}\otimes\boldsymbol{B}). \tag{4.69}$$

In particular, a definite linear transformation of vectors, $\mathcal{V} \to \mathcal{V}$, may be represented either by an element of $\mathcal{V} \bigotimes \mathcal{V}$, or by an element of $\mathcal{V} \bigotimes \mathcal{V}^*$.

The approximation (4.65) for $\boldsymbol{A}(P)$ near $P_0$ can be rewritten

$$\boldsymbol{A}(P) = \boldsymbol{A}(P_0) + (\overrightarrow{P_0 P}) \cdot \underset{\rightarrow}{\partial} \otimes \boldsymbol{A} + \cdots , \tag{4.70}$$

where $\underset{\rightarrow}{\partial} \otimes \boldsymbol{A} = \partial_\mu A^\nu \underset{\rightarrow}{E}^\mu \otimes \boldsymbol{E}_\nu \in \mathcal{V}^* \bigotimes \mathcal{V}$ is evaluated at $P_0$, and

$$\delta \boldsymbol{x} \cdot \underset{\rightarrow}{\partial} \otimes \boldsymbol{A} = \partial_\mu A^\nu \underset{\rightarrow}{E}^\mu (\delta \boldsymbol{x}) \boldsymbol{E}_\nu = \partial_\mu A^\nu (\delta x^\mu) \boldsymbol{E}_\nu .$$

The two-factor tensor products $\mathcal{V} \bigotimes \mathcal{V}$, $\mathcal{V} \bigotimes \mathcal{V}^*, \cdots$ are only the beginning. One can go on to three factors, four factors, and so on. For the simplest applications in electromagnetism these appear only implicitly. From a dyadic $\overset{\Rightarrow}{T} = T^{\mu\nu} \boldsymbol{E}_\mu \otimes \boldsymbol{E}_\nu$ one can construct a gradient which is a triadic (an element of $\mathcal{V} \bigotimes \mathcal{V} \bigotimes \mathcal{V}$)

$$\boldsymbol{\partial} \otimes \overset{\Rightarrow}{T} = \partial^\lambda T^{\mu\nu} \boldsymbol{E}_\lambda \otimes \boldsymbol{E}_\mu \otimes \boldsymbol{E}_\nu . \tag{4.71}$$

This gradient is used in the approximation of the dyadic field

$$\overset{\Rightarrow}{T}(P) = \overset{\Rightarrow}{T}(P_0) + (\overrightarrow{P_0 P}) \cdot \boldsymbol{\partial} \otimes \overset{\Rightarrow}{T} + \cdots . \tag{4.72}$$

Taking the trace of the first two factors in the triadic gradient (4.71) gives the *divergence* of the dyadic $\overset{\Rightarrow}{T}$,

$$\boldsymbol{\partial} \cdot \overset{\Rightarrow}{T} = \partial_\mu T^{\mu\nu} \boldsymbol{E}_\nu , \tag{4.73}$$

which is a vector. The dyadic divergence appears in the mathematical expression of very many physical laws (most notably Maxwell's equations and the local conservation of four-momentum).

## 4.7 Dual of an Antisymmetric Dyadic

The operation of forming the dual (this is an entirely new use of the word, not connected with the earlier meaning), or $*$ operation, is applied here to any antisymmetric dyadic $\overset{\Rightarrow}{F} = -(\overset{\Rightarrow}{F})^T$. It appears in electromagnetism, where it generates an interchange of electric and magnetic fields. The operation may be simply defined in terms of the components of $\overset{\Rightarrow}{F}$ with respect to an inertial basis

$$\overset{\Rightarrow}{F} = F^{\mu\nu} \boldsymbol{E}_\mu \otimes \boldsymbol{E}_\nu .$$

It is, however, important that the $\boldsymbol{E}_\mu$ be the basis corresponding to one of the right-handed inertial frames (and not, for example, such a basis after spatial reflection $\{\boldsymbol{E}_0, -\boldsymbol{E}_m\}$, which would be left-handed). The definition of the dual $\overset{\Rightarrow}{F}^*$ uses the orientation of the basis; it would take a different form in a left-handed basis.

Assuming we have a right-handed frame, the dual $\overset{\Rightarrow}{F}^*$ is defined by

$$\overset{\Rightarrow}{F}^* = F^{*\mu\nu}\boldsymbol{E}_\mu \otimes \boldsymbol{E}_\nu, \quad F^{*\mu\nu} \equiv \frac{1}{2}\epsilon^{\mu\nu\lambda\rho}F_{\lambda\rho}, \tag{4.74}$$

where

$$F_{\lambda\rho} \equiv \eta_{\lambda\mu}\eta_{\rho\nu}F^{\mu\nu} \tag{4.75}$$

and

$$\epsilon^{\mu\nu\lambda\rho} = \{+1, -1, 0\} \tag{4.76}$$

according to whether $\mu\nu\lambda\rho$ is {an even, an odd, or no} permutation of 0123. In particular, $\epsilon^{0123} = +1$.[5] When the scripts are lowered, each nonzero term in the sum has one factor $(-1)$ and three factors $(+1)$, hence

$$\epsilon_{\alpha\beta\gamma\delta} \equiv \eta_{\alpha\mu}\eta_{\beta\nu}\eta_{\gamma\lambda}\eta_{\delta\rho}\epsilon^{\mu\nu\lambda\rho} \tag{4.77}$$

satisfies $\epsilon_{0123} = -\epsilon^{0123}$. $\epsilon^{\mu\nu\lambda\rho}$ is completely antisymmetric; it changes sign if any two scripts are interchanged (note that it is *not* cyclic, as $\epsilon_{mnl} (= \epsilon^{0mnl})$ is).

Both eq. (4.75), in which the scripts of $F^{\mu\nu}$ are lowered, and eq. (4.77), in which the same happens to $\epsilon^{\mu\nu\lambda\rho}$, are generalisations of (4.12), in which the components of the covector corresponding naturally to a vector are computed.

### 4.7.1 The Definition is Basis-independent

We would like eq. (4.74) to hold using the components of any frame, but we can only define $\overset{\Rightarrow}{F}^*$ once! The definition above, in terms of components with respect to some arbitrary basis $\boldsymbol{E}_\mu$, will only be sensible if the transformed components with respect to another right-handed basis $\boldsymbol{E}'_\mu$ satisfy a similar equation

$$F'^{*\mu\nu} = \frac{1}{2}\epsilon^{\mu\nu\lambda\rho}F'_{\lambda\rho}, \qquad F'_{\lambda\rho} = \eta_{\lambda\mu}\eta_{\rho\nu}F'^{\mu\nu}. \tag{4.78}$$

In this case the dual *could* have been defined in the same way with respect to the new frame. The definition would then be independent of the frame used.

If the coordinates of the dashed and undashed frames are related by $x'^\mu = a^\mu{}_\nu x^\nu + a^\mu$, then the components of the dual transform according to (4.39),

$$F'^{*\mu\nu} = a^\mu{}_\alpha a^\nu{}_\beta F^{*\alpha\beta}, \tag{4.79}$$

[5] The symbol $\epsilon^{\mu\nu\lambda\rho}$ is sometimes defined slightly differently, for example, with the different basic ordering 1230, or with the opposite convention for $\eta_{\mu\nu}$, or when the definition is made for the covariant components, and can result in the opposite sign.

and the covariant components of the original dyadic transform as

$$F_{\gamma\delta} = a^{\lambda}{}_{\gamma} a^{\rho}{}_{\delta} F'_{\lambda\rho}. \tag{4.80}$$

This "backward" form of the transformation, which is the standard transformation law for elements of $\mathcal{V}^* \bigotimes \mathcal{V}^*$, is the most useful version here. It may be derived from the formula for $F'^{\mu\nu}$, analogous to (4.79), and two uses of the inverse, (4.60).

The determinant of the matrix $(a^{\mu}{}_{\nu})$, written $a \equiv \det(a^{\mu}{}_{\nu})$, may be expressed by

$$a\epsilon^{\mu\nu\lambda\rho} = \epsilon^{\alpha\beta\gamma\delta} a^{\mu}{}_{\alpha} a^{\nu}{}_{\beta} a^{\lambda}{}_{\gamma} a^{\rho}{}_{\delta}. \tag{4.81}$$

It satisfies $a = 1$ for the transformation between any of the right-handed orthochronous inertial frames. Consequently, multiplying each side of (4.81) by $F'_{\lambda\rho}$ (and summing automatically as the scripts imply) we can write

$$\frac{1}{2}\epsilon^{\mu\nu\lambda\rho} F'_{\lambda\rho} = a^{\mu}{}_{\alpha} a^{\nu}{}_{\beta} \frac{1}{2}\epsilon^{\alpha\beta\gamma\delta} a^{\lambda}{}_{\gamma} a^{\rho}{}_{\delta} F'_{\lambda\rho}.$$

Using (4.80) and (4.79) the equation becomes

$$\frac{1}{2}\epsilon^{\mu\nu\lambda\rho} F'_{\lambda\rho} = a^{\mu}{}_{\alpha} a^{\nu}{}_{\beta} \frac{1}{2}\epsilon^{\alpha\beta\gamma\delta} F_{\gamma\delta} = a^{\mu}{}_{\alpha} a^{\nu}{}_{\beta} F^{*\alpha\beta} = F'^{*\mu\nu}, \tag{4.82}$$

which is what we wanted to verify. The calculation shows that the dual can be defined in the same way using components with respect to any (right-handed) inertial frame and the same dyadic will result.

### 4.7.2 Explicit Components of the Dual

It is often efficient to work with the explicit components of the dual. Components with one superscript equal to zero satisfy

$$F^{*0n} = \frac{1}{2}\epsilon^{0nrl} F_{rl} = \frac{1}{2}\epsilon^{nrl} F^{rl} = -F^{*n0}, \tag{4.83}$$

hence

$$F^{*01} = F^{23}, \quad F^{*02} = F^{31}, \quad F^{*03} = F^{12}. \tag{4.84}$$

For the "spatial" components, using an explicit instance of $F^{\mu\nu} = \eta^{\mu\alpha}\eta^{\nu\beta}F_{\alpha\beta}$,

$$F^{*mn} = \frac{1}{2}\epsilon^{mn\rho\lambda} F_{\rho\lambda} = \epsilon^{mn0l} F_{0l} = -\epsilon^{mnl} F^{0l}, \tag{4.85}$$

hence

$$F^{*12} = -F^{03}, \quad F^{*31} = -F^{02}, \quad F^{*23} = -F^{01}. \tag{4.86}$$

As a particular consequence, when the $*$ operation is applied twice, we find $F^{**12} = -F^{*03} = -F^{12}$, etc., and $F^{**01} = F^{*23} = -F^{01}$, etc. These relations may be summarised by $** = -1$; when the dual is applied twice, the dyadic changes sign.

### 4.7.3 An Antisymmetric Basis

For an antisymmetric dyadic, the expansion $\overset{\Rightarrow}{F} = F^{\mu\nu}\boldsymbol{E}_\mu \otimes \boldsymbol{E}_\nu$ is often inconvenient because the basis is not antisymmetric; if one uses $\overset{\Rightarrow}{F} = \frac{1}{2}F^{\mu\nu}\boldsymbol{E}_\mu \wedge \boldsymbol{E}_\nu$, the summation convention runs over the antisymmetric basis twice. As a temporary expedient one can write out the independent terms explicitly:

$$\begin{aligned}\overset{\Rightarrow}{F} &= F^{01}\boldsymbol{E}_0 \wedge \boldsymbol{E}_1 + F^{02}\boldsymbol{E}_0 \wedge \boldsymbol{E}_2 + F^{03}\boldsymbol{E}_0 \wedge \boldsymbol{E}_3 + \\ &+ F^{23}\boldsymbol{E}_2 \wedge \boldsymbol{E}_3 + F^{31}\boldsymbol{E}_3 \wedge \boldsymbol{E}_1 + F^{12}\boldsymbol{E}_1 \wedge \boldsymbol{E}_2. \end{aligned} \tag{4.87}$$

The dual can be written similarly,

$$\begin{aligned}\overset{\Rightarrow}{F}{}^* &= F^{*01}\boldsymbol{E}_0 \wedge \boldsymbol{E}_1 + F^{*02}\boldsymbol{E}_0 \wedge \boldsymbol{E}_2 + F^{*03}\boldsymbol{E}_0 \wedge \boldsymbol{E}_3 + \\ &+ F^{*23}\boldsymbol{E}_2 \wedge \boldsymbol{E}_3 + F^{*31}\boldsymbol{E}_3 \wedge \boldsymbol{E}_1 + F^{*12}\boldsymbol{E}_1 \wedge \boldsymbol{E}_2, \end{aligned} \tag{4.88}$$

or as a transformation of the decomposition of $\overset{\Rightarrow}{F}$ in the antisymmetric basis

$$\begin{aligned}\overset{\Rightarrow}{F}{}^* &= F^{01}(\boldsymbol{E}_0 \wedge \boldsymbol{E}_1)^* + F^{02}(\boldsymbol{E}_0 \wedge \boldsymbol{E}_2)^* + F^{03}(\boldsymbol{E}_0 \wedge \boldsymbol{E}_3)^* + \\ &+ F^{23}(\boldsymbol{E}_2 \wedge \boldsymbol{E}_3)^* + F^{31}(\boldsymbol{E}_3 \wedge \boldsymbol{E}_1)^* + F^{12}(\boldsymbol{E}_1 \wedge \boldsymbol{E}_2)^*. \end{aligned} \tag{4.89}$$

The expressions for the components of the dual, (4.84) and (4.86), are equivalent to

$$\left.\begin{aligned}(\boldsymbol{E}_0 \wedge \boldsymbol{E}_l)^* &= -(\boldsymbol{E}_m \wedge \boldsymbol{E}_n) \\ (\boldsymbol{E}_m \wedge \boldsymbol{E}_n)^* &= +(\boldsymbol{E}_0 \wedge \boldsymbol{E}_l)\end{aligned}\right\} \tag{4.90}$$

where $lmn$ is a cyclic permutation of 123. By taking the dual of either equation, and using the other, one verifies $** = -1$.

An attractive and compact expression for the original dyadic can now be written by introducing the shorthand $F^{0n} \equiv E^n$, $F^{mn} \equiv B^l$ ($lmn$ cyclic) to form the spatial (but *frame-dependent*) vectors $\boldsymbol{E} = E^n\boldsymbol{E}_n$, $\boldsymbol{B} = B^l\boldsymbol{E}_l$. From (4.87) and (4.90) one has

$$\overset{\Rightarrow}{F} = \boldsymbol{E}_0 \wedge \boldsymbol{E} - (\boldsymbol{E}_0 \wedge \boldsymbol{B})^*. \tag{4.91}$$

The dual is then obtained by using $** = -1$ (or from (4.89))

$$\overset{\Rightarrow}{F}{}^* = (\boldsymbol{E}_0 \wedge \boldsymbol{E})^* + (\boldsymbol{E}_0 \wedge \boldsymbol{B}). \tag{4.92}$$

One obtains $\overset{\Rightarrow}{F}{}^*$ from $\overset{\Rightarrow}{F}$ by replacing $\boldsymbol{E} \mapsto \boldsymbol{B}$, $\boldsymbol{B} \mapsto -\boldsymbol{E}$ in the latter. The notation has been chosen to be useful in applications to electromagnetism, where $\boldsymbol{E}$ and $\boldsymbol{B}$ are the frame-dependent electric and magnetic fields.

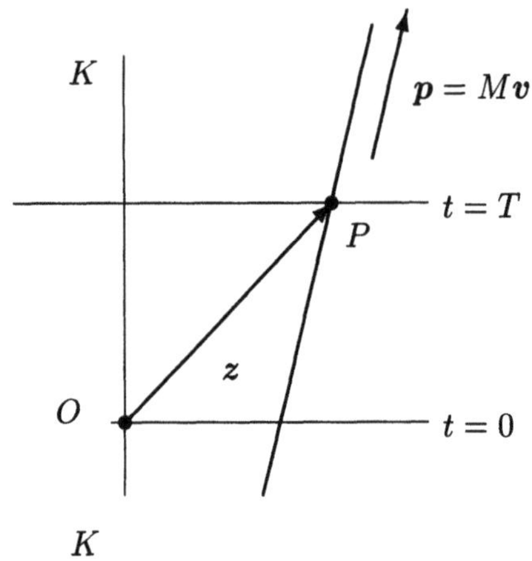


**Figure 4.4**

### 4.7.4 Angular Momentum Dyadic for a Freely Moving Particle

To give a satisfactory discussion of angular momentum requires the energy tensor (Chapter 6) but for the case of a single particle with mass our familiarity with the nonrelativistic case leads us to the correct expression. The dyadic representing the angular momentum about $O$ of a particle with momentum $\boldsymbol{p}$ is given by

$$\vec{\vec{L}}_O = \boldsymbol{z} \wedge \boldsymbol{p} = \boldsymbol{z} \otimes \boldsymbol{p} - \boldsymbol{p} \otimes \boldsymbol{z}, \tag{4.93}$$

where $\boldsymbol{z} = \overrightarrow{OP}$ locates the particle with respect to the point $O$ (Figure 4.4). Any point $P$ on the worldline of a free particle will do for this construction.

If $\boldsymbol{z}$ and $\boldsymbol{p}$ are decomposed with respect to a $K$-basis, then

$$\begin{aligned}\vec{\vec{L}}_O &= \boldsymbol{z} \wedge \boldsymbol{p} = (t\boldsymbol{E}_0 + \boldsymbol{Z}(t)) \wedge (E\boldsymbol{E}_0 + \boldsymbol{P}) \\ &= \boldsymbol{E}_0 \wedge (t\boldsymbol{P} - E\boldsymbol{Z}(t)) + \boldsymbol{Z}(t) \wedge \boldsymbol{P} \\ &= -E\boldsymbol{E}_0 \wedge \left(\boldsymbol{Z}(t) - \frac{\boldsymbol{P}}{E}t\right) + \boldsymbol{Z}(t) \wedge \boldsymbol{P}. \end{aligned} \tag{4.94}$$

We recognise the expression in round brackets as the particle's $K$-position at $K$-time $t = 0$, in the hyperplane containing $O$. Using (4.90) in the form $Z^m P^n (\boldsymbol{E}_m \wedge \boldsymbol{E}_n)^* = Z^m P^n \epsilon^{mnl} \boldsymbol{E}_0 \wedge \boldsymbol{E}_l$, $\vec{\vec{L}}_O$ can be rewritten in the same form as (4.91):

$$\vec{\vec{L}}_O = \boldsymbol{E}_0 \wedge \boldsymbol{K} - (\boldsymbol{E}_0 \wedge \boldsymbol{J})^*, \tag{4.95}$$

where $\boldsymbol{K} = -E\boldsymbol{Z}(0)$ and $\boldsymbol{J} = \boldsymbol{Z}(t) \times \boldsymbol{P} = \boldsymbol{Z}(0) \times \boldsymbol{P}$. The vectors $\boldsymbol{K}$ and $\boldsymbol{J}$ determine the position of the particle $\boldsymbol{Z}(0)$ on the hyperplane through $O$, and its three-dimensional orbital angular momentum.

## 4.8 Concept of Volume in Spacetime

The volume element that is used for integrals over four-dimensional regions in spacetime is defined in terms of *inertial* coordinates:

$$\int f(x)\, d^4x = \int f(x)\, dx^0 dx^1 dx^2 dx^3 = \int f'(x')\, dx'^0 dx'^1 dx'^2 dx'^3, \quad (4.96)$$

where the expressions using different coordinate systems are equal because the Jacobian for the transformation is unity (by (1.14))

$$\frac{\partial(x'^0, x'^1, x'^2, x'^3)}{\partial(x^0, x^1, x^2, x^3)} = \det(a^\mu{}_\nu) = 1. \quad (4.97)$$

The spacetime metric (which in its dyadic form is a coordinate-independent, geometrical idea) enters the definition of volume only *implicitly* when the inertial coordinates are given a privileged status. However, the actual concept of volume, as it is used in integrals, is not really a metric one but it is a *coordinate-dependent* measure of how a coordinate system divides up spacetime. A coordinate system divides spacetime into cells, and then the volume of a region is just the (limit of the) number of cells in it. The idea is easiest to understand on a two-dimensional plane.

### 4.8.1 Dimension Two

A flat two-dimensional affine space will have sets of special coordinates $x^\mu (\mu = 1, 2)$ corresponding to inertial coordinates, which serve to define straight lines and parallels.

For a general coordinate system $y^\mu = Y^\mu(x^\nu)$, the lines $y^1 = y_0^1 + n$ and $y^2 = y_0^2 + m$, for integer $n, m$, determine a coordinate grid on a portion of the plane near the point with $y$-coordinates $(y_0^1, y_0^2)$. The plane is thus divided into unit $y$-cells, each consisting of all the points whose coordinates satisfy $n < y^1 - y_0^1 < n+1,\ \ m < y^2 - y_0^2 < m+1$ (Figure 4.5). If the functions $Y^\mu$ vary rapidly, it may be necessary to work with a denser grid and subdivided cells in order to get coordinate curves which differ negligibly from straight lines.

### 4.8.2 Unoriented Region

The unoriented $y$-measure of an unoriented region (a set of points, nothing more) may be defined just as the number of elementary unit $y$-cells in the region (Figure 4.6). Such a measure is always positive and, in a special case, with special coordinates including an appropriate unit, corresponds to Euclidean area. If the region is awkward, the grid can be refined and the measure counted in fractions of a unit cell. Of course, if the region is *very* awkward, serious measure theory, for which this is not the place, might be needed.

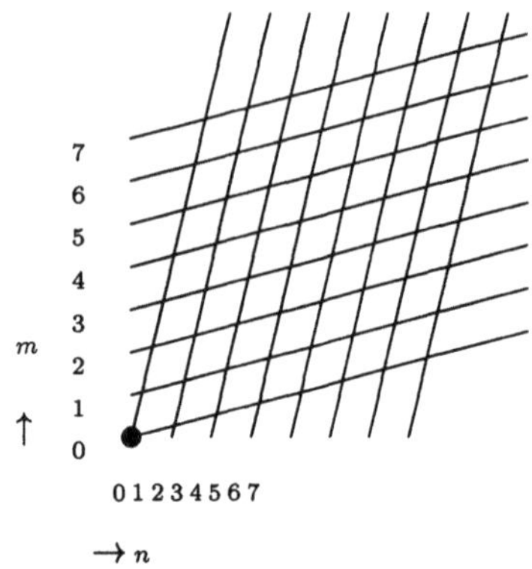


Figure 4.5

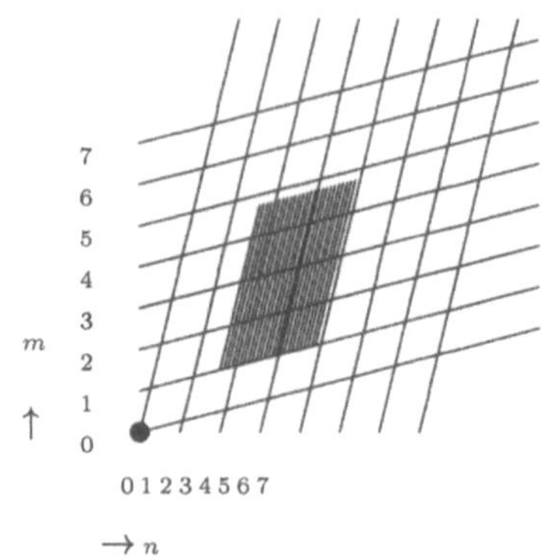


Figure 4.6

The "volume" of a large, general unoriented region is obtained by dividing it up into subregions in each ofwhich simple cell counting is possible, then adding the results (a limiting procedure may of course be necessary).

### 4.8.3 Oriented Region

If the region is given an orientation, done here by specifying an *order* for the boundary vectors $\boldsymbol{A}$, $\boldsymbol{B}$, an *oriented* $y$-measure can be defined. The order of the vectors determines the sense of a path round the boundary. The opposite order reverses it. The order $\boldsymbol{A}$, $\boldsymbol{B}$ defines the same orientation as

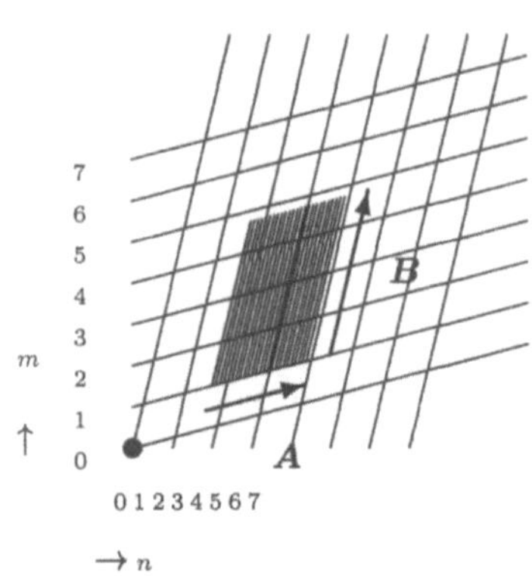


Figure 4.7

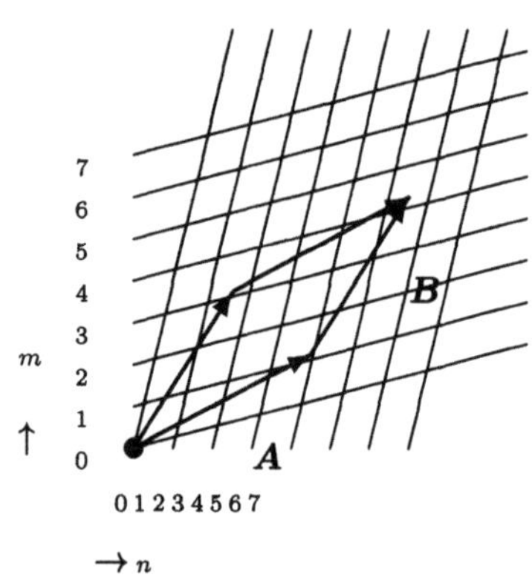


Figure 4.8

the unit basis vectors $\boldsymbol{\partial}_1, \boldsymbol{\partial}_2$ for the coordinates $y^1, y^2$. For the simple case (Figure 4.7), in which the boundary vectors are parallel to the coordinate axes, $\boldsymbol{A} = A\boldsymbol{\partial}_1$, $\boldsymbol{B} = B\boldsymbol{\partial}_2$, with $A > 0$, and $B > 0$. The oriented $y$-measure in the present case is $AB$, the number of unit $y$-cells in the region. For the opposite orientation the oriented $y$-measure is $-AB$.

In the general case the oriented parallelogram is bounded by vectors which are not parallel to the coordinate axes (which serve to define the cells). But there is a simple formula for the oriented $y$-measure of a region bounded by the (ordered) vectors $\boldsymbol{A} = A^1\boldsymbol{\partial}_1 + A^2\boldsymbol{\partial}_2$, $\boldsymbol{B} = B^1\boldsymbol{\partial}_1 + B^2\boldsymbol{\partial}_2$. For the case in Figure 4.8, $\boldsymbol{A}, \boldsymbol{B}$ define the same orientation as the basis vectors $\boldsymbol{\partial}_1, \boldsymbol{\partial}_2$. One can check that the number of unit $y$-cells in the enclosed region is $A^1B^2 - A^2B^1$ (it is easy to find a region with the same number of $y$-cells but with boundary vectors parallel to the coordinate vectors). The expression gives the oriented measure in all cases. Using the definitions $\underset{\sim}{d}y^\mu(\boldsymbol{\partial}_\nu) = \delta^\mu_\nu$ of the cobasis[6] and an antisymmetric wedge product for covectors corresponding to (4.44), one may put the expression in the form

$$\det\begin{pmatrix} A^1 & A^2 \\ B^1 & B^2 \end{pmatrix} = (\underset{\sim}{d}y^1 \otimes \underset{\sim}{d}y^2 - \underset{\sim}{d}y^2 \otimes \underset{\sim}{d}y^1)(\boldsymbol{A}, \boldsymbol{B}) = (\underset{\sim}{d}y^1 \wedge \underset{\sim}{d}y^2)(\boldsymbol{A}, \boldsymbol{B}). \tag{4.98}$$

Equation (4.98) is the count of $y$-cells when the orientation of $\boldsymbol{A}, \boldsymbol{B}$ agrees with that of $\boldsymbol{\partial}_1, \boldsymbol{\partial}_2$, and changes sign if the boundary vectors are interchanged. ***The formula gives the $y$-measure of volume for the region bounded by the ordered vectors $\boldsymbol{A}$ and $\boldsymbol{B}$.***

The expression for the oriented volume in two dimensions may be generalised easily to dimensions three and four.

### 4.8.4 Dimensions Three and Four

In a three dimensional affine space there are affine coordinates $x^\mu$ ($\mu = 1, 2, 3$) that define parallels. For a general coordinate system $y^\mu = Y^\mu(x^\nu)$, the two-dimensional surfaces $y^1 = y^1_0 + n$, $y^2 = y^2_0 + m$, $y^3 = y^3_0 + l$ for integer $n, m, l$ divide up space into three-dimensional unit $y$-cells. In a sufficiently small region the cells are very nearly parallelepipeds (or, if necessary, subdivided cells will be). Thus each coordinate system gives rise to an unoriented measure which is just the count of unit cells. For an *oriented* parallelepiped bounded by the (*ordered*) vectors $\boldsymbol{A}, \boldsymbol{B}, \boldsymbol{C}$, the oriented $y$-measure is

$$\det\begin{pmatrix} A^1 & A^2 & A^3 \\ B^1 & B^2 & B^3 \\ C^1 & C^2 & C^3 \end{pmatrix} = (\underset{\sim}{d}y^1 \wedge \underset{\sim}{d}y^2 \wedge \underset{\sim}{d}y^3)(\boldsymbol{A}, \boldsymbol{B}, \boldsymbol{C}), \tag{4.99}$$

[6] There are three notational schemes for the cobasis of an inertial basis, $\underset{\sim}{E}^\mu = \underset{\sim}{\partial}x^\mu = \underset{\sim}{d}x^\mu$; the latter two are also used for arbitrary coordinates.

where the triple wedge product is the antisymmetric extension of the double wedge product in (4.98),

$$\underset{\sim}{d}y^1 \wedge \underset{\sim}{d}y^2 \wedge \underset{\sim}{d}y^3 = \Sigma_\pi \mathrm{sign}(\pi) \underset{\sim}{d}y^{\pi(1)} \otimes \underset{\sim}{d}y^{\pi(2)} \otimes \underset{\sim}{d}y^{\pi(3)}. \tag{4.100}$$

The six terms in the sum are the six permutations $\pi$ of 123 with a minus sign when the permutation is odd. For the case $\boldsymbol{A} = A\boldsymbol{\partial}_1,\ \boldsymbol{B} = B\boldsymbol{\partial}_2,\ \boldsymbol{C} = C\boldsymbol{\partial}_3$ equation (4.99) gives $ABC$; the orientation of the vectors is the same as that of the coordinates when $A, B, C$ are all positive, and then $ABC$ is equal to the number of unit cells.

Finally, in four dimensions, with physically meaningful inertial coordinates $x^\mu (\mu = 0, 1, 2, 3)$ defining straight lines and parallels, any coordinates $y^\mu$ define a grid of unit $y$-cells (or subdivided cells, if necessary, to get "parallelepipeds") as in three dimensions but with a fourth hypersurface $y^0 = y^0_0 + k$ to complete the definition of the unit $y$-cell. The unoriented $y$-measure of a four-dimensional region is given by counting the unit $y$-cells within it. The oriented $y$-measure of the "parallelepiped" determined by the (ordered) vectors $\boldsymbol{A}, \boldsymbol{B}, \boldsymbol{C}, \boldsymbol{D}$ is the determinant of components which may be written

$$(\underset{\sim}{d}y^0 \wedge \underset{\sim}{d}y^1 \wedge \underset{\sim}{d}y^2 \wedge \underset{\sim}{d}y^3)(\boldsymbol{A}, \boldsymbol{B}, \boldsymbol{C}, \boldsymbol{D}), \tag{4.101}$$

where

$$\underset{\sim}{d}y^0 \wedge \underset{\sim}{d}y^1 \wedge \underset{\sim}{d}y^2 \wedge \underset{\sim}{d}y^3 = \Sigma_\pi \mathrm{sign}(\pi) \underset{\sim}{d}y^{\pi(0)} \otimes \underset{\sim}{d}y^{\pi(1)} \otimes \underset{\sim}{d}y^{\pi(2)} \otimes \underset{\sim}{d}y^{\pi(3)}, \tag{4.102}$$

and $\pi$ is now a permutation of 0123. Despite the unfamiliar notation these equations express a very simple geometrical idea: the number of $y$-cells in a "parallelepiped" with vector edges, and a sign depending on the relative orientation of the edges and the coordinate basis vectors.

### 4.8.5 The Common Measure in Minkowski Spacetime

The inertial frames, as a class of coordinate systems, determine a single measure, common to them all. Thanks to the metric $\eta_{\mu\nu}$, the unit cells for the inertial coordinates may be regarded as *orthogonal* parallelepipeds with edges of a unit, *physical* length.

To understand the common measure, consider any particular inertial coordinates $x^\mu$, and calculate the $x$-measure of the "parallelepiped" determined by the basis vectors $\boldsymbol{\partial}_\mu$ for some arbitrary, possibly curvilinear, coordinates $y^\mu$. Using $\boldsymbol{\partial}_\mu \equiv \partial\overrightarrow{OP}/\partial y^\mu = \partial\boldsymbol{x}/\partial y^\mu = \boldsymbol{E}_\nu \partial x^\nu/\partial y^\mu$ and the linearity of the covectors,

$$
\begin{aligned}
(\underset{\sim}{d}x^0 \wedge \underset{\sim}{d}x^1 \wedge \underset{\sim}{d}x^2 \wedge \underset{\sim}{d}x^3)(\boldsymbol{\partial}_0, \boldsymbol{\partial}_1, \boldsymbol{\partial}_2, \boldsymbol{\partial}_3)& \\
&= \frac{\partial x^\mu}{\partial y^0}\frac{\partial x^\nu}{\partial y^1}\frac{\partial x^\lambda}{\partial y^2}\frac{\partial x^\rho}{\partial y^3}(\underset{\sim}{d}x^0 \wedge \underset{\sim}{d}x^1 \wedge \underset{\sim}{d}x^2 \wedge \underset{\sim}{d}x^3) \\
&\qquad\qquad \times (\boldsymbol{E}_\mu, \boldsymbol{E}_\nu, \boldsymbol{E}_\lambda, \boldsymbol{E}_\rho) \\
&= \frac{\partial x^\mu}{\partial y^0}\frac{\partial x^\nu}{\partial y^1}\frac{\partial x^\lambda}{\partial y^2}\frac{\partial x^\rho}{\partial y^3}(-\epsilon_{\mu\nu\lambda\rho}) \\
&= \det\left(\frac{\partial x^\mu}{\partial y^\nu}\right). \qquad (4.103)
\end{aligned}
$$

The second equality uses (4.102) with $x$, and (4.76) and (4.77). The third equality uses (4.81). If the $y$-coordinates are *inertial* then the final determinant is unity, by (1.14), and eq. (4.103) expresses the fact that the $x$-measure of a unit $y$-cell is one, as is the $y$-measure. Since the inertial frames all have the same orientation too, this shows that their oriented measures are all equivalent.

Since the $y$-measure of a cell with edges $\boldsymbol{\partial}_0, \boldsymbol{\partial}_1, \boldsymbol{\partial}_2, \boldsymbol{\partial}_3$, is one, or, equivalently,

$$(\underset{\sim}{d}y^0 \wedge \underset{\sim}{d}y^1 \wedge \underset{\sim}{d}y^2 \wedge \underset{\sim}{d}y^3)(\boldsymbol{\partial}_0, \boldsymbol{\partial}_1, \boldsymbol{\partial}_2, \boldsymbol{\partial}_3) = 1,$$

eq. (4.103) directly gives the relation between $x$-measure and $y$-measure,

$$\underset{\sim}{d}x^0 \wedge \underset{\sim}{d}x^1 \wedge \underset{\sim}{d}x^2 \wedge \underset{\sim}{d}x^3 = \det\left(\frac{\partial x^\mu}{\partial y^\nu}\right)(\underset{\sim}{d}y^0 \wedge \underset{\sim}{d}y^1 \wedge \underset{\sim}{d}y^2 \wedge \underset{\sim}{d}y^3). \qquad (4.104)$$

(If both $x$- and $y$-coordinates are inertial, the determinant factor is one.) This relation is the basis for the "change of variables" formula in spacetime integrals.

### 4.8.6 Change of Variables Formula

Let $\mathcal{D}$ be a four-dimensional region of spacetime, regarded as being oriented consistently with the inertial frames, and $f(P)$ a numerical valued function in spacetime. The integral $I$ of $f$ over $\mathcal{D}$, defined as the limit of a sum of the common inertial measure for sufficiently small cells of spacetime volume multiplied by the value of $f$ at a typical point of the cell, may be written as a numerical integral

$$I = \int_{\mathcal{D}} f(P)(\underset{\sim}{d}x^0 \wedge \underset{\sim}{d}x^1 \wedge \underset{\sim}{d}x^2 \wedge \underset{\sim}{d}x^3) = \int_{K(\mathcal{D})} f(x(P))dx^0dx^1dx^2dx^3, \qquad (4.105)$$

where $K(\mathcal{D})$ is the $K$-coordinate region of $\mathcal{R}^4$ corresponding to $\mathcal{D}$, and $x(P) \in \mathcal{R}^4$ is the $K$-coordinate for a typical point in $\mathcal{D}$. The integral on the right is a standard integral over a region of $\mathcal{R}^4$ which can be evaluated as a repeated integral over the four coordinates of the (arbitrary) frame $K$. For any (not necessarily inertial) coordinates $y^\mu = Y^\mu(x^\nu)$ and corresponding function $F(y)$ which expresses $f(P)$,

$$f(P) = f(x(P)) = f(x) = F(y(P)) = F(Y(x)) = F(y),$$

the integral may be written, using (4.104), in the new variables

$$\begin{aligned} I &= \int_{\mathcal{D}} f(P) \det\left(\frac{\partial x^\mu}{\partial y^\nu}\right) (\underset{\sim}{d}y^0 \wedge \underset{\sim}{d}y^1 \wedge \underset{\sim}{d}y^2 \wedge \underset{\sim}{d}y^3) \\ &= \int_{Y(K(\mathcal{D}))} F(y) \det\left(\frac{\partial x^\mu}{\partial y^\nu}\right) dy^0 dy^1 dy^2 dy^3. \end{aligned} \tag{4.106}$$

The $y$-integral is over the $y$-coordinate region of $\mathcal{R}^4$ corresponding to $\mathcal{D}$, and the determinant is assumed to be positive (so a simple repeated integral is correct for the $y$-integrals too).

## 4.9 Divergence Theorem in Spacetime

As will be discussed in the next chapter, the differential expression, with respect to a frame $K$, of conservation of electric charge has the form

$$\frac{\partial \rho}{\partial t} + \partial_m J^m = 0, \tag{4.107}$$

where $\rho$ is the $K$-charge density, and $\boldsymbol{J} = J^m \boldsymbol{E}_m$ is the corresponding electric current density in the same frame (the concept will be discussed more fully in the next chapter). In fact, $\rho$ and $\boldsymbol{J}$ can be combined together to form a spacetime vector current

$$\boldsymbol{j} = \rho \boldsymbol{E}_0 + \boldsymbol{J} = \rho \boldsymbol{E}_0 + J^m \boldsymbol{E}_m = \rho' \boldsymbol{E}'_0 + \boldsymbol{J}' = \rho' \boldsymbol{E}'_0 + J'^m \boldsymbol{E}'_m, \tag{4.108}$$

in which $\rho'$ is the charge density in a new frame $K'$, and $\boldsymbol{J}'$ is the current density in $K'$. (Both $\rho$ and $\boldsymbol{J}$ are *frame-dependent.*) Equation (4.107) is seen to involve the divergence, (4.66), of the spacetime vector current

$$\frac{\partial \rho}{\partial t} + \partial_m J^m = \boldsymbol{\partial} \cdot \boldsymbol{j} = \partial_\mu j^\mu = 0. \tag{4.109}$$

### 4.9.1 Integral Form of the Conservation Law

The simplest integral consequence of the conservation law (4.107) may now be obtained by integrating the expression over a three dimensional $K$-volume $V$ for a $K$-time interval $[t_1, t_2]$ (Figure 4.9):

$$\begin{aligned} 0 &= \int_{t_1}^{t_2} \int_V [\frac{\partial \rho}{\partial t} + \partial_m J^m] d^3x \, dt \\ &= \int_V [\rho]_{t_1}^{t_2} \, d^3x + \int_{t_1}^{t_2} \int_S J^m dS_m dt, \end{aligned} \tag{4.110}$$

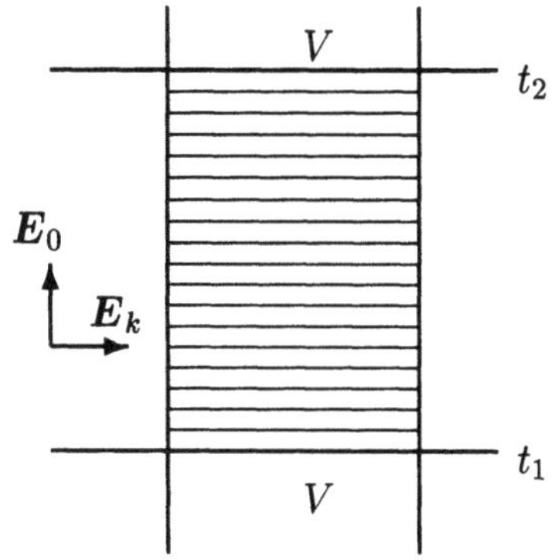


**Figure 4.9**

where $S$ is the two-dimensional boundary of $V$.

Equation (4.110) can be rewritten in a form which explains the way charge conservation is achieved:

$$\int_V [\rho](t_1)\, d^3x = \int_V [\rho](t_2)\, d^3x + \int_{t_1}^{t_2} \int_S J^m dS_m dt. \tag{4.111}$$

On the left is the total charge in the volume $V$ at time $t_1$. It equals the charge in $V$ at time $t_2$ plus the net outflow of charge over the boundary $S$ during the time interval $[t_1, t_2]$. This physical interpretation justifies calling $\rho$ the $K$-charge density, and $\boldsymbol{J}$ the $K$-current density. (Each is a density over a three-dimensional hypersurface, but the language is different for a spacelike hypersurface and a timelike hypersurface. Recall the discussion of different kinds of hyperplane at the end of Section 4.1.)

If the volume $V$ is so large that the current $\boldsymbol{J}$ vanishes on its boundary $S$, then the last term in (4.111) drops out and the equation reduces to

$$\int_V [\rho](t_1)\, d^3x = \int_V [\rho](t_2)\, d^3x. \tag{4.112}$$

The total charge, as measured in $K$, is the same for all times. However, as one expects physically, an even more general statement is true: the total charge, in no matter what frame it is measured, and at no matter what time, is a constant. The general argument has the same structure as for (4.110), but the boundary hyperplanes at constant time refer to different inertial frames so they are no longer parallel. (The proof appears in the Problems.)

Equation (4.110) is the simplest instance of the divergence theorem in spacetime: a four-dimensional integral of a divergence is rewritten as a three-dimensional integral over the boundary.

In physical applications, integrals over arbitrary four-dimensional regions $V_4 \subset \mathcal{M}$, of the form

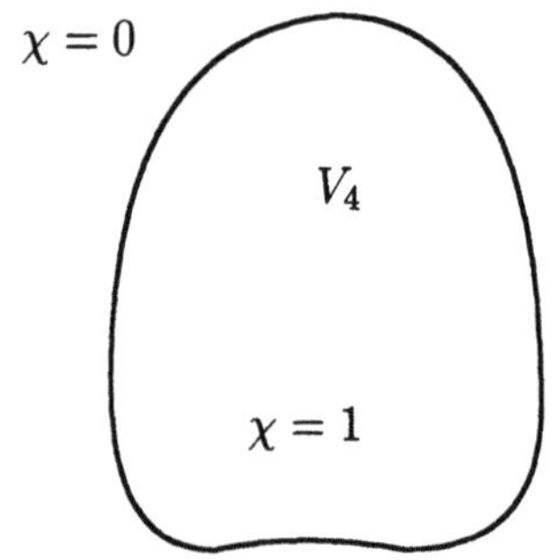


**Figure 4.10**

$$I = \int_{V_4} \boldsymbol{\partial} \cdot \boldsymbol{A}\, d^4x = \int_{V_4} \partial_\mu A^\mu \, d^4x = \int_{V_4} \underset{\sim}{\partial} \cdot \boldsymbol{A}\, d^4x, \tag{4.113}$$

can appear. The measure is the common inertial measure, denoted by $d^4x = dx^0 dx^1 dx^2 dx^3$. The integrand is a frame-independent real-valued function of position, (4.66), as the geometrical notation $\boldsymbol{\partial} \cdot \boldsymbol{A}$ indicates. The alternative notation, $\partial_\mu A^\mu$, does use the components and coordinates of a frame, but *any* inertial frame may be used. The Jacobian is unity when one changes from one set of inertial coordinates to another set (because, for any proper Lorentz transformation, $\det(a^\mu{}_\nu) = 1$). Finally, $V_4$ is a fixed subset of spacetime $\mathcal{M}$; its specification in terms of inertial coordinates *will* be frame-dependent.

The integral $I$ may be considered to be a special case of an integral over the whole of spacetime, whose integrand is a product, of the form

$$\begin{aligned} \int \chi \boldsymbol{\partial} \cdot \boldsymbol{A}\, d^4x = \int \chi \partial_\mu A^\mu \, d^4x &= -\int (\partial_\mu \chi) A^\mu \, d^4x \\ &= -\int \underset{\sim}{\partial}\chi(\boldsymbol{A})\, d^4x. \end{aligned} \tag{4.114}$$

The integration by parts is valid if $\chi$ vanishes for the infinite reaches of the coordinates, or, if not, $\boldsymbol{A}$ vanishes in these regions instead. Most instances arising in physical applications involve integrals of this type.

### 4.9.2 Divergence Theorem in Spacetime: Geometrical Expression

If the function $\chi$ is the characteristic function for the four-dimensional region $V_4$, that is,

$$\chi(P) = 1 \text{ if } P \in V_4, \quad 0 \text{ otherwise (Figure 4.10)}, \tag{4.115}$$

the integral (4.114) reduces to the previous case (4.113)

$$I = \int_{V_4} \boldsymbol{\partial} \cdot \boldsymbol{A}\, d^4x = -\int \underset{\sim}{\partial}\chi(\boldsymbol{A})\, d^4x. \tag{4.116}$$

Formula (4.116) is the most generally useful representation of the divergence theorem in spacetime. Since $\underset{\sim}{\partial}\chi$ vanishes except on the boundary of $V_4$, where $\chi$ changes from 1 to 0, the integral on the right-hand side is effectively a three-dimensional one. It may happen that the hypersurface integral can be handled analytically (as is often the case with integrals in electromagnetism). Otherwise methods must be developed to compute it.

Note the use of the covector gradient $\underset{\sim}{\partial}\chi$ in (4.116). Although one could equally well write $\boldsymbol{\partial}\chi \cdot \boldsymbol{A}$ for $\underset{\sim}{\partial}\chi(\boldsymbol{A})$, the latter expresses the result most simply. The Minkowski scalar product is quite irrelevant in this situation. If the boundary of $V_4$ consists, piecewise, of either spacelike or timelike hyperplanes, then near each piece of boundary there will be inertial coordinates such that the boundary is given by $x^\mu = 0$ for one particular $\mu$. If $x^\mu > 0$ in the interior of $V_4$, then $\underset{\sim}{\partial}\chi = \delta(x^\mu)\underset{\sim}{\partial}x^\mu$(no sum!) and so $\underset{\sim}{\partial}\chi(\boldsymbol{A}) = \delta(x^\mu)A^\mu$ (no sum!). The total integral $I$ in (4.116) is a sum of contributions from different parts of the boundary of $V_4$, each of the form $-\int \delta(x^\mu)A^\mu\, dx^0 dx^1 dx^2 dx^3$. (Recall, however, that the oriented inertial coordinates all have future-pointing time axes, so there may be hypersurfaces for which $x^\mu < 0$ in the interior.)

### 4.9.3 Divergence Theorem: Analytical Expression

In the general case we suppose that the region containing the boundary of $V_4$ can be divided into subregions, in each of which a coordinate change is possible to coordinates $\xi^\mu$ so that the boundary of $V_4$ is given by $\xi^0 = 0$, with $\xi^0 > 0$ outside $V_4$ and $\xi^0 < 0$ inside. Then, using the fact that the derivative of the step-function $\theta$ is the Dirac delta function,

$$\chi = \theta(-\xi^0), \qquad \underset{\sim}{\partial}\chi = -\delta(\xi^0)\underset{\sim}{\partial}\xi^0. \tag{4.117}$$

The new coordinates $\xi^\mu$ are, of course, in general not inertial (and in particular $\xi^0$ has no necessary connection with timelike directions), but we can arrange that the orientation is right-handed so the Jacobian is positive

$$J = \det\left(\frac{\partial x^\mu}{\partial \xi^\nu}\right) = -\epsilon_{\mu\nu\lambda\rho}\frac{\partial x^\mu}{\partial \xi^0}\frac{\partial x^\nu}{\partial \xi^1}\frac{\partial x^\lambda}{\partial \xi^2}\frac{\partial x^\rho}{\partial \xi^3} > 0. \tag{4.118}$$

(Recall formula (4.81) for the determinant, and $\epsilon^{0123} = +1 = -\epsilon_{0123}$.)

Using (4.117) the integral over $V_4$ can now be written according to the change of variables formula,

$$\begin{aligned} I &= -\int d\xi^0 d\xi^1 d\xi^2 d\xi^3 J(-\delta(\xi^0))\underset{\sim}{\partial}\xi^0(\boldsymbol{A}) \\ &= \int d\xi^1 d\xi^2 d\xi^3 J\underset{\sim}{\partial}\xi^0(\boldsymbol{A}) \qquad (4.119) \\ &= \int J(\boldsymbol{\partial}\xi^0 \cdot \boldsymbol{A}) d\xi^1 d\xi^2 d\xi^3, \qquad (4.120) \end{aligned}$$

an expression which looks a little simpler than it really is since the division into subregions has been suppressed and the limits on the integration variables have been omitted.

It is possible to get a relatively simple expression for $J\underset{\sim}{\partial}\xi^0$. The collection of vectors $\boldsymbol{\partial}_\mu \equiv \frac{\partial \boldsymbol{x}}{\partial \xi^\mu} = \frac{\partial x^\rho}{\partial \xi^\mu}\boldsymbol{E}_\rho$ is a basis for $\mathcal{V}$ (since the Jacobian is nonzero the four vectors are independent) and its cobasis is $\underset{\sim}{\partial}\xi^\mu = \frac{\partial \xi^\mu}{\partial x^\nu}\underset{\sim}{\partial}x^\nu$, because

$$\underset{\sim}{\partial}\xi^\mu\left(\frac{\partial \boldsymbol{x}}{\partial \xi^\nu}\right) = \frac{\partial \xi^\mu}{\partial x^\tau}\underset{\sim}{\partial}x^\tau(\boldsymbol{E}_\rho)\frac{\partial x^\rho}{\partial \xi^\nu} = \frac{\partial \xi^\mu}{\partial x^\rho}\frac{\partial x^\rho}{\partial \xi^\nu} = \frac{\partial \xi^\mu}{\partial \xi^\nu} = \delta^\mu_\nu.$$

We may now verify the relation between covectors,

$$J\underset{\sim}{\partial}\xi^0 = -\underset{\sim}{\partial}x^\mu \epsilon_{\mu\nu\lambda\rho}\frac{\partial x^\nu}{\partial \xi^1}\frac{\partial x^\lambda}{\partial \xi^2}\frac{\partial x^\rho}{\partial \xi^3}, \qquad (4.121)$$

by evaluating the right-hand side on the basis vectors $\frac{\partial \boldsymbol{x}}{\partial \xi^\tau}$. Since $\underset{\sim}{\partial}x^\mu(\frac{\partial \boldsymbol{x}}{\partial \xi^\tau}) = \frac{\partial x^\mu}{\partial \xi^\tau}$, we see that this gives $J\delta^0_\tau$, as does the left-hand side. Since the set of $\frac{\partial \boldsymbol{x}}{\partial \xi^\mu}$ form a basis, these values determine the covector $J\underset{\sim}{\partial}\xi^0$, as required.

The integral over $V_4$ may now be written, using $\underset{\sim}{\partial}x^\mu(\boldsymbol{A}) = A^\mu$, as

$$I = \int d\xi^1 d\xi^2 d\xi^3 A^\mu(-\epsilon_{\mu\nu\lambda\rho})\frac{\partial x^\nu}{\partial \xi^1}\frac{\partial x^\lambda}{\partial \xi^2}\frac{\partial x^\rho}{\partial \xi^3}, \qquad (4.122)$$

a computable form.

## 4.10 Problems

1. If two vectors satisfy $\boldsymbol{R}\otimes\boldsymbol{N} = \overset{\Rightarrow}{0}$, show that $\boldsymbol{R} = \boldsymbol{0}$ or $\boldsymbol{N} = \boldsymbol{0}$. Show that the same alternative results if $\boldsymbol{R}\otimes\boldsymbol{N} + \boldsymbol{N}\otimes\boldsymbol{R} = \overset{\Rightarrow}{0}$.

2. The covector gradient of a spacetime scalar field $\phi$ is given by

$$\underset{\sim}{\partial}\phi = (\partial_\mu\phi)\underset{\sim}{E}^\mu = \frac{\partial \phi}{\partial x^\mu}\underset{\sim}{E}^\mu = \partial_0\phi\underset{\sim}{E}^0 + \partial_m\phi\underset{\sim}{E}^m.$$

Show that this relation is frame-independent in the sense that if $\underset{\rightarrow}{\partial}\phi$ is defined by it using the $K$-frame cobasis $\underset{\rightarrow}{E}^{\mu}$, then also $\underset{\rightarrow}{\partial}\phi = \partial'_{\mu}\phi \underset{\rightarrow}{E}'^{\mu}$ in terms of the $K'$-frame cobasis $\underset{\rightarrow}{E}'^{\mu}$ and the $K'$-coordinate partial derivatives. The corresponding *vector* gradient is

$$\boldsymbol{\partial}\phi = (\eta^{\mu\nu}\partial_{\nu}\phi)\boldsymbol{E}_{\mu} = -\partial_0\phi\boldsymbol{E}_0 + \partial_m\phi\boldsymbol{E}_m.$$

Show that this relation too is frame-independent in a similar sense.

3. Verify that the unit dyadic has the same form with respect to any inertial frame, that is,

$$\overset{\Rightarrow}{\eta} = \eta^{\mu\nu}\boldsymbol{E}_{\mu}\otimes\boldsymbol{E}_{\nu} = \eta^{\mu\nu}\boldsymbol{E}'_{\mu}\otimes\boldsymbol{E}'_{\nu}.$$

4. a) Show that the definition of the trace of a dyadic,

$$\mathrm{tr}\overset{\Rightarrow}{T} = T^{\mu\nu}\eta_{\mu\nu} = T^{\mu\nu}\boldsymbol{E}_{\mu}\cdot\boldsymbol{E}_{\nu},$$

gives a result independent of the inertial frame used, i.e., that $T'^{\mu\nu}\eta_{\mu\nu} = T^{\mu\nu}\eta_{\mu\nu}$ (use the transformation law for dyadic components).

   b) If $\overset{\Rightarrow}{T} = \Sigma\boldsymbol{\alpha}\otimes\boldsymbol{\beta}$, then $\mathrm{tr}\overset{\Rightarrow}{T} = \Sigma\boldsymbol{\alpha}\cdot\boldsymbol{\beta}$.

   c) $\mathrm{tr}\overset{\Rightarrow}{\eta} = 4$.

5. An antisymmetric dyadic $\overset{\Rightarrow}{A} = A^{\mu\nu}\boldsymbol{E}_{\mu}\otimes\boldsymbol{E}_{\nu}$ satisfies $A^{\mu\nu} = -A^{\nu\mu}$. Show that the antisymmetry relation holds for the components with respect to any inertial basis if it holds for one. The antisymmetry is thus frame-independent. For a symmetric dyadic $\overset{\Rightarrow}{S}$ the components satisfy $S^{\mu\nu} = S^{\nu\mu}$ whatever the basis, again consistent with the transformation law.

6. a) If two covectors, $\underset{\rightarrow}{J}$ and $\underset{\rightarrow}{K}$, satisfy $\underset{\rightarrow}{J}\cdot\boldsymbol{v} = \underset{\rightarrow}{K}\cdot\boldsymbol{v}$, for every future-pointing unit timelike vector $\boldsymbol{v}$, then they are the same, $\underset{\rightarrow}{J} = \underset{\rightarrow}{K}$.

   b) If two symmetric codyadics, $\underset{\Rightarrow}{T}$ and $\underset{\Rightarrow}{U}$, satisfy $\underset{\Rightarrow}{T}(\boldsymbol{v},\boldsymbol{v}) = \underset{\Rightarrow}{U}(\boldsymbol{v},\boldsymbol{v})$, for every future-pointing unit timelike vector $\boldsymbol{v}$, then they are the same, that is, $\underset{\Rightarrow}{T}(\boldsymbol{A},\boldsymbol{B}) = \underset{\Rightarrow}{U}(\boldsymbol{A},\boldsymbol{B})$, for all vectors $\boldsymbol{A}$, $\boldsymbol{B}$.

7. The (vector) gradient of a vector field $\boldsymbol{A}$ is the dyadic defined by

$$\boldsymbol{\partial}\otimes\boldsymbol{A} = \boldsymbol{E}_{\mu}\otimes\boldsymbol{E}_{\nu}\eta^{\mu\rho}\partial_{\rho}A^{\nu} \equiv (\partial^{\mu}A^{\nu})\boldsymbol{E}_{\mu}\otimes\boldsymbol{E}_{\nu}.$$

Verify that the $\boldsymbol{E}'_{\mu}\otimes\boldsymbol{E}'_{\nu}$-components of $\boldsymbol{\partial}\otimes\boldsymbol{A}$, defined in exactly the same way but with respect to $K'$-components and $K'$-vectors, are related to the displayed components by the transformation law of dyadic components.

The divergence of $\boldsymbol{A}$ is $\boldsymbol{\partial}\cdot\boldsymbol{A} = \mathrm{tr}\boldsymbol{\partial}\otimes\boldsymbol{A}$. Verify that this may also be written $\partial_{\mu}A^{\mu} = \partial'_{\mu}A'^{\mu}$, i.e., using any inertial basis.

8. The (vector) gradient of a dyadic $\overset{\Rightarrow}{T}$ is a triadic

$$\boldsymbol{\partial} \otimes \overset{\Rightarrow}{T} = \boldsymbol{E}_\mu \otimes \boldsymbol{E}_\alpha \otimes \boldsymbol{E}_\beta \eta^{\mu\nu} \partial_\nu T^{\alpha\beta}.$$

The transformation law for triadic components is $T'^{\alpha\beta\gamma} = a^\alpha{}_\mu a^\beta{}_\nu a^\gamma{}_\lambda T^{\mu\nu\lambda}$.
Show that it is satisfied here using the $K'$-components of $\boldsymbol{\partial} \otimes \overset{\Rightarrow}{T}$.
The (left) divergence of $\overset{\Rightarrow}{T}$ is the vector which may be written

$$\boldsymbol{\partial} \cdot \overset{\Rightarrow}{T} = (\partial_\mu T^{\mu\nu})\boldsymbol{E}_\nu = (\partial'_\mu T'^{\mu\nu})\boldsymbol{E}'_\nu,$$

the same form with respect to any inertial frame. Show that the components with respect to different frames transform properly as vector components.

9. The alternating symbol $\epsilon^{\mu\nu\lambda\rho}$ is defined to equal $(+1, -1, 0)$ according to whether $\mu\nu\lambda\rho$ is (an even, an odd, no) permutation of 0123. It produces the determinant:

$$\det(\mathrm{a})\epsilon^{\mu\nu\lambda\rho} = a^\mu{}_\alpha a^\nu{}_\beta a^\lambda{}_\gamma a^\rho{}_\delta \epsilon^{\alpha\beta\gamma\delta}.$$

Verify similar formulae for $2 \times 2$ and $3 \times 3$ matrices.
From the components $g_{\mu\nu}$ define the determinant

$$\det(\mathrm{g}) = \epsilon^{\alpha\beta\gamma\delta} g_{0\alpha} g_{1\beta} g_{2\gamma} g_{3\delta},$$

and show from it that, for a small variation $\delta g_{\mu\nu}$ of the components, $\delta\det(\mathrm{g}) = \delta g_{\mu\nu}\mathrm{Cofactor}^{\mu\nu}$, where the cofactor of the element $g_{\mu\nu}$ in the determinant is meant, and a full sum over $\mu$ and $\nu$.

10. Verify the useful formulae involving sums of products of $\epsilon$-symbols:
   a) $\epsilon^{\mu\nu\lambda\rho}\epsilon_{\mu\nu\lambda\rho} = -4!$
   b) $\epsilon^{\mu\nu\lambda\rho}\epsilon_{\mu\nu\lambda\tau} = -\delta^\rho_\tau$
   c) $\epsilon^{\mu\nu\lambda\rho}\epsilon_{\mu\nu\omega\tau} = -2(\delta^\lambda_\omega\delta^\rho_\tau - \delta^\lambda_\tau\delta^\rho_\omega) \equiv -2\delta^{\lambda\rho}_{\omega\tau}$
   d) $\epsilon^{\mu\nu\lambda\rho}\epsilon_{\mu\pi\omega\tau} \equiv -\delta^{\nu\lambda\rho}_{\pi\omega\tau}$
   e) $\epsilon^{\mu\nu\lambda\rho}\epsilon_{\alpha\pi\omega\tau} \equiv -\delta^{\mu\nu\lambda\rho}_{\alpha\pi\omega\tau}$

11. There are many identities, similar to those of three-dimensional vector analysis, that are very useful.
   a) $\boldsymbol{\partial}\phi(\psi) = \frac{\partial\phi}{\partial\psi}\boldsymbol{\partial}\psi$ (The field $\phi$ is a function of the field $\psi$.)
   b) $\boldsymbol{\partial}(\boldsymbol{A}(P) \cdot \boldsymbol{B}(P)) = (\boldsymbol{\partial} \otimes \boldsymbol{A}) \cdot \boldsymbol{B} + (\boldsymbol{\partial} \otimes \boldsymbol{B}) \cdot \boldsymbol{A}$
   c) $\boldsymbol{\partial} \otimes (\psi\boldsymbol{A}) = \boldsymbol{\partial}\psi \otimes \boldsymbol{A} + \psi\boldsymbol{\partial} \otimes \boldsymbol{A}$
   d) $\boldsymbol{\partial} \otimes \boldsymbol{A}(\psi) = \boldsymbol{\partial}\psi \otimes \frac{\partial\boldsymbol{A}}{\partial\psi}$ (The field $\boldsymbol{A}$ is a function of the field $\psi$.)
   e) $\boldsymbol{\partial} \otimes (\psi\overset{\Rightarrow}{T}) = \boldsymbol{\partial}\psi \otimes \overset{\Rightarrow}{T} + \psi\boldsymbol{\partial} \otimes \overset{\Rightarrow}{T}$
   f) $\boldsymbol{\partial} \otimes (\boldsymbol{A} \otimes (\boldsymbol{B}(\psi)) = (\boldsymbol{\partial} \otimes \boldsymbol{A}) \otimes \boldsymbol{B} + \boldsymbol{\partial}\psi \otimes \boldsymbol{A} \otimes \frac{\partial\boldsymbol{B}}{\partial\psi}$
   g) $\boldsymbol{\partial} \otimes (\boldsymbol{A} \otimes \boldsymbol{B}) = (\boldsymbol{\partial} \otimes \boldsymbol{A}) \otimes \boldsymbol{B} + \boldsymbol{\partial}B^\mu \otimes \boldsymbol{A} \otimes \boldsymbol{E}_\mu$

12. a) $\boldsymbol{\partial} \wedge \boldsymbol{\partial}\phi = \vec{\vec{0}}$
b) $\boldsymbol{\partial} \cdot (\boldsymbol{\partial} \wedge \boldsymbol{A})^* = \boldsymbol{0}$

13. Consider the integral of a smooth covector field $\underset{\sim}{A}$

$$\oint_\Gamma \underset{\sim}{A}(d\boldsymbol{x}) = \oint_\Gamma \underset{\sim}{A}\left(\frac{d\boldsymbol{x}}{d\lambda}\right) d\lambda = \oint_\Gamma A_\mu dx^\mu$$

round a closed loop $\Gamma$ which starts from a point $P$ then translates successively by $\epsilon\boldsymbol{\Delta}$, $\epsilon\boldsymbol{\delta}$, $-\epsilon\boldsymbol{\Delta}$, $-\epsilon\boldsymbol{\delta}$, for fixed linearly independent vectors $\boldsymbol{\Delta}$ and $\boldsymbol{\delta}$. By using appropriate expansions of $\underset{\sim}{A}$ in Taylor series, show that in the limit $\epsilon \to 0$,

$$\lim_{\epsilon\to 0} \frac{1}{\epsilon^2} \oint_\Gamma \underset{\sim}{A}(d\boldsymbol{x}) = \boldsymbol{\Delta} \cdot (\underset{\sim}{\partial} \wedge \underset{\sim}{A}) \cdot \boldsymbol{\delta},$$

evaluated at $P$.

14. Show that the dyadic

$$\vec{\vec{P}}(\boldsymbol{v}) \equiv \vec{\vec{\eta}} + \boldsymbol{v} \otimes \boldsymbol{v}$$

is a projection dyadic into the space orthogonal to the unit, timelike vector $\boldsymbol{v}$. (This requires $\vec{\vec{P}} \cdot \vec{\vec{P}} = \vec{\vec{P}}$ and $\boldsymbol{v} \cdot \vec{\vec{P}} \cdot \boldsymbol{X} = 0$ for all $\boldsymbol{X}$.) What is the projection dyadic into the space orthogonal to a unit, spacelike vector $\boldsymbol{n}$?

15. The vectors $\boldsymbol{E}_\mu$ are basis vectors derived from an inertial frame $K$ and $\boldsymbol{F}_0$ is a normalised timelike vector. A dyadic $\vec{\vec{B}}$ is defined by

$$\vec{\vec{B}} = (\vec{\vec{\eta}} + 2\boldsymbol{F}_0 \otimes \boldsymbol{F}_0) \cdot (\vec{\vec{\eta}} + 2\boldsymbol{E}_0 \otimes \boldsymbol{E}_0).$$

Show that each factor of $\vec{\vec{B}}$ has the unit dyadic $\vec{\vec{\eta}}$ as its square; each factor determines a reflection in a hyperplane with timelike normal. Show that $\vec{\vec{B}}$ itself generates a boosted basis by $\boldsymbol{E}'_\mu = \vec{\vec{B}} \cdot \boldsymbol{E}_\mu$. (That is, the vectors $\boldsymbol{E}'_\mu$ are the basis vectors of a frame $K'$ reached from $K$ by a boost.) If $\boldsymbol{F}_0 = \Gamma(\boldsymbol{E}_0 + \boldsymbol{U})$, then the relative speed of the implied frame $K'$ with respect to $K$ is $V = 2U/(1+U^2)$.

16. In spacetime $\mathcal{M}$ points have coordinates $x^\mu$ and $x'^\mu$ with respect to two inertial frames $K$ and $K'$ which are related by a Poincaré transformation $x'^\mu = a^\mu{}_\nu x^\nu + a^\mu$. Recall the relations satisfied by the Lorentz matrix $(a^\mu{}_\nu)$, and write down the relation between the orthonormal basis vectors $\boldsymbol{E}_\mu$, $\boldsymbol{E}'_\mu$.
a) If the dyadic $\vec{\vec{L}}$ is defined by $\vec{\vec{L}} = \eta^{\mu\nu} \boldsymbol{E}_\mu \otimes \boldsymbol{E}'_\nu$, show that $\vec{\vec{L}} \cdot \vec{\vec{L}}^T = \vec{\vec{\eta}}$.

b) If $\overset{\Rightarrow}{J} = J^{\mu\nu}\boldsymbol{E}_\mu \otimes \boldsymbol{E}_\nu$, and $\overset{\Rightarrow}{J} \cdot \overset{\Rightarrow}{J}^T = \overset{\Rightarrow}{\eta}$, show that $J^{\mu\lambda}\eta_{\lambda\nu} \equiv J^\mu{}_\nu$ defines a Lorentz matrix.

c) Consider the mapping of $\mathcal{M}$ to itself in which points $P$ are mapped to points $Q$ determined by the translation vector

$$\overrightarrow{PQ} = \boldsymbol{a} + (\overset{\Rightarrow}{L} - \overset{\Rightarrow}{\eta}) \cdot \overrightarrow{OP},$$

where $\boldsymbol{a} = a^\mu \boldsymbol{E}_\mu$, $\overset{\Rightarrow}{L}$ is as above, and point $O$ is the fixed spacetime origin for $K$. For given $Q$, what $P$ is mapped to $Q$? If $Q_1 = Q_2$, then $P_1 = P_2$. Show that $\overrightarrow{P_1P_2} \cdot \overrightarrow{P_1P_2} = \overrightarrow{Q_1Q_2} \cdot \overrightarrow{Q_1Q_2}$. The $K$-coordinates of $Q$ equal the $K'$-coordinates of $P$. The mapping $\mathcal{M} \to \mathcal{M}$ is an isometry.

17. Instead of using the inertial basis $\boldsymbol{E}_\mu$ derived from frame $K$, use a basis developed from it with two lightlike vectors: $\boldsymbol{E}_+,\ \boldsymbol{E}_-,\ \boldsymbol{E}_2,\ \boldsymbol{E}_3$, where $\boldsymbol{E}_\pm = 2^{-\frac{1}{2}}(\boldsymbol{E}_0 \pm \boldsymbol{E}_1)$. Express the components of a general vector

$$\boldsymbol{A} = A^+\boldsymbol{E}_+ + A^-\boldsymbol{E}_- + A^2\boldsymbol{E}_2 + A^3\boldsymbol{E}_3$$

in terms of scalar products of $\boldsymbol{A}$ with the new basis vectors. Express the unit dyadic $\overset{\Rightarrow}{\eta}$ in terms of the new basis. Write the scalar product of two vectors $\boldsymbol{A}$ and $\boldsymbol{B}$ in terms of the new components.

If $K'$ is a frame travelling with relative velocity $V\boldsymbol{E}_1$ with respect to $K$, $\boldsymbol{E}'_\pm$ are defined analogously to $\boldsymbol{E}_\pm$, and $\boldsymbol{E}'_2 = \boldsymbol{E}_2,\ \boldsymbol{E}'_3 = \boldsymbol{E}_3$, then express the components of $\boldsymbol{A}$ with respect to the basis $\boldsymbol{E}'_+,\ \boldsymbol{E}'_-,\ \boldsymbol{E}'_2,\ \boldsymbol{E}'_3$ in terms of $A^\pm,\ A^2,\ A^3$.

For the basis $\boldsymbol{E}_\pm,\ \boldsymbol{E}_2,\ \boldsymbol{E}_3$ find a cobasis $\underset{\sim}{E}^\pm,\ \underset{\sim}{E}^2,\ \underset{\sim}{E}^3$ expressed in terms of $\underset{\sim}{\partial} x^\mu$. Define $x^\pm$ so that

$$\underset{\sim}{\partial}\phi = \partial_+\phi\underset{\sim}{E}^+ + \partial_-\phi\underset{\sim}{E}^- + \partial_2\phi\underset{\sim}{E}^2 + \partial_3\phi\underset{\sim}{E}^3.$$

Notice that $\delta\boldsymbol{x} = \delta x^\mu \boldsymbol{E}_\mu = \delta x^+\boldsymbol{E}_+ + \delta x^-\boldsymbol{E}_- + \delta x^2\boldsymbol{E}_2 + \delta x^3\boldsymbol{E}_3$ with this choice of variables.

18. For any two future-pointing unit timelike vectors $\boldsymbol{E}_0,\ \boldsymbol{E}'_0$ define the dyadic

$$\overset{\Rightarrow}{B} = \overset{\Rightarrow}{\eta} - 2\boldsymbol{E}'_0 \otimes \boldsymbol{E}_0 + \frac{(\boldsymbol{E}_0 + \boldsymbol{E}'_0) \otimes (\boldsymbol{E}_0 + \boldsymbol{E}'_0)}{\gamma + 1},$$

where $\gamma \equiv -\boldsymbol{E}_0 \cdot \boldsymbol{E}'_0$.

a) Show that if $\boldsymbol{E}_\mu$ is an orthonormal inertial basis, then so is $\boldsymbol{E}'_\mu = \overset{\Rightarrow}{B} \cdot \boldsymbol{E}_\mu$. This is a particularly clear way of characterising the boost between bases with timelike vectors $\boldsymbol{E}_0$ and $\boldsymbol{E}'_0$.

b) Factor $\overset{\Rightarrow}{B}$ into reflections by finding a suitable vector $\boldsymbol{F}$ so that $\overset{\Rightarrow}{B} = (\overset{\Rightarrow}{\eta} + 2\boldsymbol{F} \otimes \boldsymbol{F}) \cdot (\overset{\Rightarrow}{\eta} + 2\boldsymbol{E}_0 \otimes \boldsymbol{E}_0)$.

c) If $\boldsymbol{E}'_0 = \gamma(\boldsymbol{E}_0 - V^n \boldsymbol{E}_n)$, find an explicit expression for $\boldsymbol{E}'_m$ in terms of $\boldsymbol{E}_\mu$. Then, from $x^\mu \boldsymbol{E}_\mu = x'^\mu \boldsymbol{E}'_\mu$, find expressions for $x'^0$, $x'^m$ in terms of $x^0$, $x^m$.

d) Show that the boost taking $\boldsymbol{E}_0$ to $\boldsymbol{E}'_0$ may be written

$$\overset{\Rightarrow}{B} = \overset{\Rightarrow}{\eta} + \boldsymbol{E}_0 \wedge \boldsymbol{E}'_0 + \frac{(\boldsymbol{E}_0 \wedge \boldsymbol{E}'_0) \cdot (\boldsymbol{E}_0 \wedge \boldsymbol{E}'_0)}{1 - \boldsymbol{E}_0 \cdot \boldsymbol{E}'_0},$$

where $\boldsymbol{a} \wedge \boldsymbol{b} \equiv \boldsymbol{a} \otimes \boldsymbol{b} - \boldsymbol{b} \otimes \boldsymbol{a}$.

19. The inertial bases for two frames are related by

$$\begin{aligned}
\boldsymbol{E}'_0 &= \gamma(\boldsymbol{E}_0 + V\cos\vartheta\, \boldsymbol{E}_1 + V\sin\vartheta\, \boldsymbol{E}_2), \quad \gamma^{-2} = 1 - V^2, \\
\boldsymbol{E}'_1 &= \frac{\cos\vartheta - V}{1 - V\cos\vartheta}\gamma(V\boldsymbol{E}_0 + \cos\vartheta\, \boldsymbol{E}_1 + \sin\vartheta\, \boldsymbol{E}_2) + \\
&\qquad + \frac{\sin\vartheta}{\gamma(1 - V\cos\vartheta)}(\sin\vartheta\, \boldsymbol{E}_1 - \cos\vartheta\, \boldsymbol{E}_2), \\
\boldsymbol{E}'_2 &= \boldsymbol{E}_2 + \frac{V\sin\vartheta}{1 - V\cos\vartheta}(\boldsymbol{E}_0 + \boldsymbol{E}_1), \quad \boldsymbol{E}'_3 = \boldsymbol{E}_3.
\end{aligned}$$

Show that the $K'$-basis can be reached from the $K$-basis by a boost followed by a clockwise rotation about $\boldsymbol{E}_3$ by an angle $\alpha \geq 0$ for which

$$\cos\alpha = 1 - \frac{\gamma V^2 \sin^2\vartheta}{(\gamma + 1)(1 - V\cos\vartheta)}.$$

20. Suppose a scalar field $\Phi$ and an antisymmetric dyadic $G^{\lambda\rho}$ are related by

$$\partial_\mu \Phi = \frac{1}{2}\epsilon_{\mu\nu\lambda\rho}\partial^\nu G^{\lambda\rho}.$$

Observe that $\partial^\mu \partial_\mu \Phi = 0$ and $\epsilon^{\mu\nu\lambda\rho}\partial_\lambda \partial_\rho \Phi = 0$. Verify that

$$\epsilon^{\alpha\beta\tau\mu}\partial_\mu \Phi = \partial^\alpha G^{\beta\tau} + \partial^\tau G^{\alpha\beta} + \partial^\beta G^{\tau\alpha} \equiv H^{\alpha\beta\tau}.$$

Check that $H^{\alpha\beta\tau}$ is antisymmetric under interchange of any two superscripts, and that $\partial_\alpha H^{\alpha\beta\tau} = 0$. Verify that

$$\partial_\mu \Phi \partial^\nu \Phi - \frac{1}{2}\delta^\nu_\mu \partial_\alpha \Phi \partial^\alpha \Phi = \frac{1}{2}\left[H^{\nu\beta\tau}H_{\mu\beta\tau} - \frac{1}{6}\delta^\nu_\mu H^{\alpha\beta\tau}H_{\alpha\beta\tau}\right].$$

21. In this problem the theta function is used:

$$\theta(x) = \begin{cases} 1, & \text{if } x > 0 \\ 0, & \text{if } x < 0. \end{cases}$$

It is to be treated as continuous and differentiable, with derivative $\theta'(x) = \delta(x)$. Let $K$ and $K'$ be the standard boost-related inertial frames. Define a scalar function in spacetime by

$$\chi(P) = \theta(-x'^0(P)) - \theta(-x^0(P)),$$

and draw a diagram showing its values. Integrate by parts to show

$$\int \chi \partial_\mu j^\mu \, d^4x = \int_{(x'^0=0)} j'^0 \, d^3x' - \int_{(x^0=0)} j^0 \, d^3x,$$

provided that spatial components of $\boldsymbol{j}$ vanish in sufficiently distant spatial directions. Note that if $\boldsymbol{j}$ is conserved, $\boldsymbol{\partial} \cdot \boldsymbol{j} = 0$ or $\partial_\mu j^\mu = 0$, that this is equivalent to $\int \rho \, dV = \int \rho' \, dV'$ and the charge as measured in $K$ at a $K$-instant equals the charge measured in $K'$ at a $K'$-instant.

## References

B. Felsager, Geometry,Particles, and Fields (Springer, 1998)
C.W. Misner, K.S. Thorne and J.A. Wheeler, Gravitation (Freeman, 1973)
R. Penrose and W. Rindler, Spinors and space-time, Volume 1 (CUP, 1984)

# 5. Electromagnetism

*Action on the Lightcone*

In earlier chapters a model of light has been used based simply on light rays, sometimes plain, sometimes with added frequency (Chapter 2) and sometimes as a carrier of four-momentum (Chapter 3). Here the formulation of the classical field theory of electromagnetism (the classical theory of light) is undertaken, omitting the aspects concerned with energy and momentum, which are left to the next chapter.

The sections of the present chapter which deal with the current density and fields of charged point particles are at times more technical and mathematically demanding, since they require four-dimensional delta-functions and solutions of partial differential equations with such sources. These sections may be skipped on a first reading.

## 5.1 Maxwell's Equations

It was Einstein's inspired idea that the equations of electromagnetism should take the same form in all inertial reference frames. Since there was no longer a rôle for a special frame, this idea ended the history of the æther as a physical continuum and it created special relativity. In their most fundamental form Maxwell's equations govern the generation of electric and magnetic fields $\mathbf{E}$ and $\mathbf{B}$ from their sources, the charge density $\rho$ and current density $\mathbf{J}$ :

$$\nabla \cdot \mathbf{E} = 4\pi\rho \tag{5.1}$$

$$\nabla \times \mathbf{E} = -\frac{\partial \mathbf{B}}{\partial t} \tag{5.2}$$

$$\nabla \cdot \mathbf{B} = 0 \tag{5.3}$$

$$\nabla \times \mathbf{B} = 4\pi\mathbf{J} + \frac{\partial \mathbf{E}}{\partial t}. \tag{5.4}$$

These are the so-called *in vacuo* equations, with $c = 1$, not the phenomenological equations for electromagnetism in a material medium; the sources $\rho$ and $\mathbf{J}$ are a model for the density and flux of *all* the charge that is present, including, if appropriate, microscopic polarisation charge and magnetisation current.

One can ensure that Maxwell's equations hold in every inertial frame if it is possible to formulate electromagnetism as a set of *geometrical* (frame-independent) equations whose decomposition with respect to any particular inertial frame produces Maxwell's equations with respect to that frame. The required geometrical formulation proves to be possible and the geometrical equations are [1]

$$\boldsymbol{\partial} \cdot \overset{\rightrightarrows}{F} = -4\pi \boldsymbol{j} \tag{5.5}$$

$$\boldsymbol{\partial} \cdot \overset{\rightrightarrows}{F}{}^{*} = \boldsymbol{0} \tag{5.6}$$

for an *antisymmetric* electromagnetic field tensor (dyadic) $\overset{\rightrightarrows}{F}$ and its dual $\overset{\rightrightarrows}{F}{}^{*}$. The dual was defined in Section 4.7 and it is completely determined by $\overset{\rightrightarrows}{F}$, so both equations must be regarded as restricting $\overset{\rightrightarrows}{F}$. Although these tensors are geometrical objects, once an inertial frame is given, they determine (or are determined by) two vectors $\boldsymbol{E}$ and $\boldsymbol{B}$, spatial in the given frame, according to (4.91). The four-current $\boldsymbol{j}$, when it is decomposed with respect to the inertial frame, leads to both the charge density and the three-current with respect to that frame,

$$\boldsymbol{j} = j^{\mu} \boldsymbol{E}_{\mu} = \rho \boldsymbol{E}_0 + \boldsymbol{J}, \qquad \boldsymbol{J} \cdot \boldsymbol{E}_0 = 0. \tag{5.7}$$

If the electromagnetic field tensor is decomposed with respect to the $K$-basis, $\overset{\rightrightarrows}{F} = F^{\mu\nu} \boldsymbol{E}_{\mu} \otimes \boldsymbol{E}_{\nu}$, then the geometrical equations (5.5) and (5.6) give rise to two sets of equations for $F^{\mu\nu} = -F^{\nu\mu}$,

$$\partial_{\mu} F^{\mu\nu} = -4\pi j^{\nu} \tag{5.8}$$

$$\partial_{\mu} F^{*\mu\nu} = 0 \tag{5.9}$$

for the components. Decomposing $\overset{\rightrightarrows}{F}$ and $\boldsymbol{j}$ with respect to the $K'$-basis instead leads to equations of *exactly the same form* for the $K'$-components of $\overset{\rightrightarrows}{F}$, involving $\partial'_{\mu}$ derivatives, and the $K'$-densities for the sources. On the other hand, exactly these equations for the $K'$-components *also arise* by transforming (5.8) and (5.9) with the help of the transformation equations (4.39). One gets the same $K'$-equations by transforming the relative $K$-equations as by decomposing the geometrical equations. ***The invariant form of the equations under change of inertial coordinates is the clue that an absolute, geometrical formulation exists.***

---

[1] In special relativity, with a fixed metric tensor $\overset{\rightrightarrows}{\eta}$, the spaces $\mathcal{V} \otimes \mathcal{V}$ and $\mathcal{V}^* \otimes \mathcal{V}^*$ are isomorphic (as are their antisymmetric subspaces) and largely interchangeable for constructing the electromagnetic tensors. The former space has been chosen here simply because it may be more familiar, being composed of products of vectors. However, in curved space, in general relativity with a dynamical metric, the latter space has significant advantages.

### 5.1.1 Verification of Maxwell's Equations

To verify that Maxwell's equations can indeed be generated from (5.5) and (5.6), one looks at each equation for the cases $\nu = 0$ (the timelike component) and $\nu = n$ (the spatial components) separately.

- Consider first the $\nu = 0$ case of (5.8), with $F^{0n} = -F^{n0} \equiv E^n$ (the notation has already been used in (4.91)) and $F^{00} = 0$,

$$\partial_\mu F^{\mu 0} = \partial_m F^{m0} = -\partial_m F^{0m} = -\partial_m E^m = -4\pi j^0.$$

  This is the same as $\nabla \cdot \mathbf{E} = 4\pi\rho$ if $j^0 = \rho$ and $E^n$ is the component of the $K$-electric field in the $K$-spatial direction along $\boldsymbol{E}_n$.
- For the case $\nu = n$ of (5.8), with $F^{mn} = \epsilon^{mnk} B^k$ (again compare with (4.91)),

$$\partial_\mu F^{\mu n} = \partial_0 F^{0n} + \partial_m F^{mn} = \partial_0 E^n - \epsilon^{nmk} \partial_m B^k = -4\pi j^n.$$

  If $j^n = J^n$ and $B^k$ is the $k^{th}$ component of the magnetic field, we have reproduced the component version of the second inhomogeneous Maxwell equation, (5.4), in the form $\partial_0 \mathbf{E} - \nabla \times \mathbf{B} = -4\pi \mathbf{J}$.
- To get the components of the dual, $F^{*\mu\nu}$, from $F^{\mu\nu}$ we recall the explicit equations (4.83) and (4.85). Then $F^{*0n} = \frac{1}{2}\epsilon^{nlm} F^{lm} = B^n$ and $F^{*mn} = -\epsilon^{mnl} F^{0l} = -\epsilon^{mnl} E^l$; in short, we change $F^{\mu\nu}$ to $F^{*\mu\nu}$ by replacing $E^n$ with $B^n$ and $B^n$ with $-E^n$. It is then clear, by comparison with the calculations for the inhomogeneous equations, that $\partial_\mu F^{*\mu\nu} = 0$ gives rise to both homogeneous Maxwell equations, $\nabla \cdot \mathbf{B} = 0$ and $\partial_0 \mathbf{B} + \nabla \times \mathbf{E} = \mathbf{0}$.

We have verified that equations (5.5) and (5.6) produce Maxwell's equations in each inertial frame of reference provided that the relativistic $\overset{\Rightarrow}{F}$ is related to the Euclidean $\mathbf{E}$, $\mathbf{B}$ by

$$\overset{\Rightarrow}{F} = \boldsymbol{E}_0 \wedge \boldsymbol{E} - (\boldsymbol{E}_0 \wedge \boldsymbol{B})^*,$$

and the relativistic four-current $\boldsymbol{j}$ is related to $\rho$ and $\mathbf{J}$ by (5.7).

However, both the electric and magnetic fields and the source densities are *frame-dependent.* One aspect of this is that the four-vectors $\boldsymbol{E}, \boldsymbol{B}, \boldsymbol{J}$ are orthogonal to the time basis vector of the frame used to specify them. (The situation is quite different from what it is (say) with the nonrelativistic Newton's equation, $m\ddot{\mathbf{X}} = \mathbf{F}$. The *components* of the acceleration and force vectors change when the Galilean inertial frame used to decompose them is changed, but the vectors themselves remain the same.) The relation of the electric and magnetic fields with respect to $K$ (components $E^n = F^{0n}$ and $B^l = \frac{1}{2}\epsilon^{lmn} F^{mn}$) and those with respect to $K'$ (similar components in terms of $F'^{0n}$ and $F'^{mn}$) are determined by the geometrical nature of $\overset{\Rightarrow}{F}$ and will be discussed in the next section. The transformation of the Euclidean components of the four-current $\boldsymbol{j} = \rho \boldsymbol{E}_0 + J^n \boldsymbol{E}_n$ will be considered afterwards. It is essential for the viability of this formulation of electromagnetism that the mathematical transformations of electric and magnetic fields, and of charge and current densities, can be verified experimentally.

## 5.2 Transformation of Electric and Magnetic Fields

We have formulated electromagnetism in terms of the geometrical, frame-independent dyadic $\overset{\Rightarrow}{F}$ and its dual. The electric and magnetic fields are also geometrical but the geometrical relations *involve the frame*. In developing Maxwell's equations we have assumed that the components of the electric and magnetic fields with respect to $K$ are given by $F^{0n} = E^n$ and $F^{mn} = \epsilon^{mnk} B^k$. With these definitions the $K$-electric and $K$-magnetic fields, both spatial with respect to $K$, are given by $\boldsymbol{E} = E^n \boldsymbol{E}_n$ and $\boldsymbol{B} = B^n \boldsymbol{E}_n$. In terms of them, as in Section 4.7, a variety of decompositions are available,

$$\begin{aligned}\overset{\Rightarrow}{F} &= F^{\mu\nu} \boldsymbol{E}_\mu \otimes \boldsymbol{E}_\nu &(5.10)\\ &= \boldsymbol{E}_0 \wedge \boldsymbol{E} - (\boldsymbol{E}_0 \wedge \boldsymbol{B})^* &(5.11)\\ &= E^1(\boldsymbol{E}_0 \wedge \boldsymbol{E}_1) + \cdots + B^1(\boldsymbol{E}_2 \wedge \boldsymbol{E}_3) + \cdots . &(5.12)\end{aligned}$$

The electric and magnetic fields with respect to a second inertial frame $K'$ are given by $\boldsymbol{E}' = E'^n \boldsymbol{E}'_n$ and $\boldsymbol{B}' = B'^n \boldsymbol{E}'_n$, where $F'^{0n} = E'^n$ and $F'^{mn} = \epsilon^{mnk} B'^k$. They are spatial with respect to $K'$, i.e. orthogonal to $\boldsymbol{E}'_0$. Using *them* to decompose $\overset{\Rightarrow}{F}$ gives

$$\begin{aligned}\overset{\Rightarrow}{F} &= F'^{\mu\nu} \boldsymbol{E}'_\mu \otimes \boldsymbol{E}'_\nu &(5.13)\\ &= \boldsymbol{E}'_0 \wedge \boldsymbol{E}' - (\boldsymbol{E}'_0 \wedge \boldsymbol{B}')^* &(5.14)\\ &= E'^1(\boldsymbol{E}'_0 \wedge \boldsymbol{E}'_1) + \cdots + B'^1(\boldsymbol{E}'_2 \wedge \boldsymbol{E}'_3) + \cdots . &(5.15)\end{aligned}$$

- As a special case of (4.39), when the frames $K$ and $K'$ are related by $x'^\mu = a^\mu{}_\nu x^\nu + a^\mu$, the electromagnetic dyadic components are related by

$$F'^{\mu\nu} = a^\mu{}_\alpha a^\nu{}_\beta F^{\alpha\beta}. \tag{5.16}$$

If the Poincaré transformation is the standard boost, $x'^0 = \gamma(x^0 - Vx^1)$, $x'^1 = \gamma(x^1 - Vx^0)$, $x'^2 = x^2$, $x'^3 = x^3$, (5.16) gives (for example)

$$\begin{aligned}E'^2 = F'^{02} = a^0{}_\alpha a^2{}_\beta F^{\alpha\beta} = a^0{}_\alpha a^2{}_2 F^{\alpha 2} &= a^0{}_0 F^{02} + a^0{}_1 F^{12}\\ &= \gamma(E^2 - VB^3),\end{aligned}$$

and hence, in complete form,

$$\boldsymbol{E}' = E'^n \boldsymbol{E}'_n = E^1 \boldsymbol{E}'_1 + \gamma(E^2 - VB^3)\boldsymbol{E}'_2 + \gamma(E^3 + VB^2)\boldsymbol{E}'_3 \tag{5.17}$$

$$\boldsymbol{B}' = B'^n \boldsymbol{E}'_n = B^1 \boldsymbol{E}'_1 + \gamma(B^2 + VE^3)\boldsymbol{E}'_2 + \gamma(B^3 - VE^2)\boldsymbol{E}'_3. \tag{5.18}$$

The new electric and magnetic fields, in terms of the old, can be worked out for any Poincaré transformation in exactly the same way, although the result will be more complicated.

- Alternatively, (5.17) and (5.18) may be derived by replacing in (5.12) the $K$-basis vectors with their $K'$-basis expansions, $\boldsymbol{E}_0 = \gamma(\boldsymbol{E}'_0 - V\boldsymbol{E}'_1)$, $\boldsymbol{E}_1 = \gamma(\boldsymbol{E}'_1 - V\boldsymbol{E}'_0)$, $\boldsymbol{E}_2 = \boldsymbol{E}'_2$, $\boldsymbol{E}_3 = \boldsymbol{E}'_3$, and then comparing the new decomposition with (5.15). As an illustration, if the relative fields in $K$ are $\boldsymbol{E} = E\boldsymbol{E}_2$, $\boldsymbol{B} = B\boldsymbol{E}_3$, so the geometrical dyadic is

$$\overset{\rightrightarrows}{F} = E(\boldsymbol{E}_0 \wedge \boldsymbol{E}_2) + B(\boldsymbol{E}_1 \wedge \boldsymbol{E}_2),$$

then the relative fields in $K'$ are found by expressing $\overset{\rightrightarrows}{F}$ in the new basis,

$$\begin{aligned}\overset{\rightrightarrows}{F} &= \gamma E(\boldsymbol{E}'_0 \wedge \boldsymbol{E}'_2) - \gamma VE(\boldsymbol{E}'_1 \wedge \boldsymbol{E}'_2) + \gamma B(\boldsymbol{E}'_1 \wedge \boldsymbol{E}'_2) \\ &\qquad - \gamma VB(\boldsymbol{E}'_0 \wedge \boldsymbol{E}'_2) \\ &= \gamma(E - VB)(\boldsymbol{E}'_0 \wedge \boldsymbol{E}'_2) + \gamma(B - VE)(\boldsymbol{E}'_1 \wedge \boldsymbol{E}'_2).\end{aligned}$$

Hence the relative fields in $K'$ are $\boldsymbol{E}' = \gamma(E - VB)\boldsymbol{E}'_2$ and $\boldsymbol{B}' = \gamma(B - VE)\boldsymbol{E}'_3$.

- Yet another way to get the relation between the $K'$-components of the $K'$-electric and magnetic fields and $K$-components of the $K$-electric and magnetic fields is to reinterpret (5.16) as a matrix equation, relating the matrix $(F'^{\mu\nu}) = F'$ to the matrix $(F^{\mu\nu}) = F$ by

$$F' = aFa^T, \tag{5.19}$$

where $a = (a^\mu{}_\nu)$ and $a^T$ is the transpose (which is the same as $a$ for the standard boost, but not for a general Lorentz transformation).

## 5.3 Example: An Infinite Line of Charge

There is a simple example which allows us easily to compare the transformation law for electric and magnetic fields from two points of view: by solving Maxwell's equations in two different frames with the appropriate sources, and by mathematically transforming the components of $\overset{\rightrightarrows}{F}$ as in the last section. In this example, which has such high symmetry that everything may be calculated easily, there is an infinite line of charge, of uniform charge density

$\lambda_0$ per unit length in $K$, a frame in which the charge is at rest. The current density is zero and, with respect to $K$, only an electric field exists. Observed from a suitable frame $K'$, the same line of charge moves with relative speed $V$ in the direction of the line. With respect to $K'$ there is both a nonzero current density and nonzero charge density, hence both electric and magnetic fields are present in the $K'$-description.

### 5.3.1 Fields and Sources with Respect to the Rest Frame $K$

Assume that the $z$-axis for the frame $K$ is aligned with the static charge (Figure 5.1). Then $\mathbf{B} = \mathbf{0}$, and the high symmetry implies that the electric field has the form, using cylindrical coordinates, $\mathbf{E}(\rho, \varphi, z) = E(\rho)\mathbf{E}_\rho(\varphi)$. The unit basis vector $\mathbf{E}_\rho(\varphi)$ depends on the azimuthal angle $\varphi$, but the component $E(\rho)$, which is the only unknown, does not. Using the divergence theorem in three-space to integrate the Maxwell equation (5.1) over a cylindrical pill box of unit height and radius $\rho$, with axis along the line of charge,

$$\int \mathbf{E} \cdot \mathbf{n}\, dS = E(\rho) 2\pi\rho = 4\pi\lambda_0.$$

(The top and bottom of the pill box do not contribute because the outward unit normal vector $\mathbf{n}$ is orthogonal to $\mathbf{E}$ for those surfaces.) The static electric field in this case is now known: $\mathbf{E}(\rho, \varphi, z) = (2\lambda_0/\rho)\mathbf{E}_\rho(\varphi)$.

For some spatial point $\mathbf{P}$ with respect to $K$, spatially rotated if necessary, we may use coordinates such that $\mathbf{P}$ is on the $x$-axis, as in Figure 5.2. Then the electric field that has just been calculated is $\boldsymbol{E}(P) = (2\lambda_0/x)\boldsymbol{E}_1$, and the electromagnetic field tensor, in the form (5.11), is

$$\vec{\vec{F}}(P) = \frac{2\lambda_0}{x}(\boldsymbol{E}_0 \wedge \boldsymbol{E}_1) \tag{5.20}$$

at any event $P$ in the $K$-history of $\mathbf{P}$.

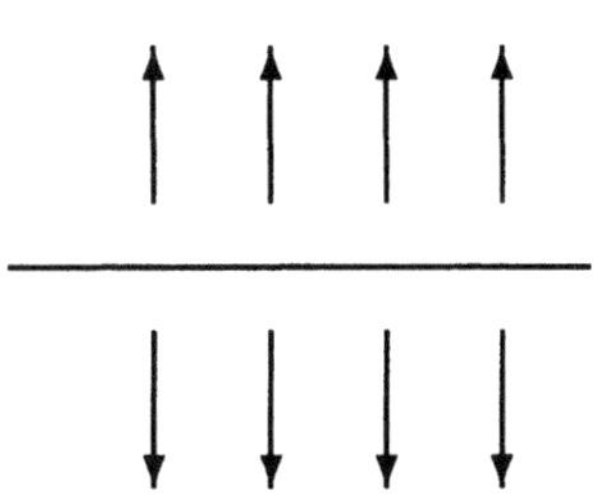


**Figure 5.1**

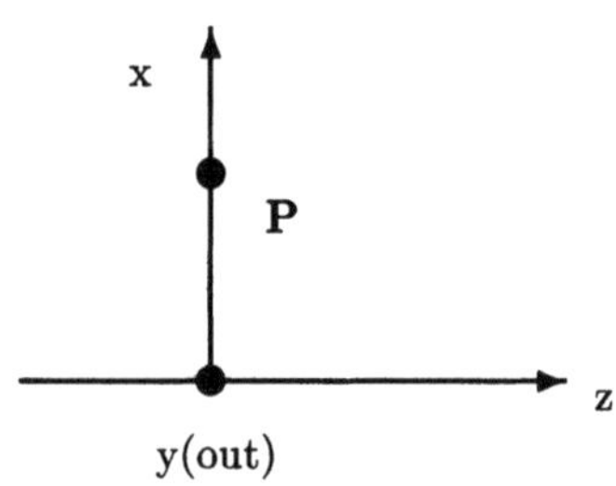


**Figure 5.2**

### 5.3.2 Fields and Sources with Respect to $K'$

With respect to $K'$ every bit of charge travels along the $z'$-axis with relative velocity $\mathbf{V}' = V\mathbf{E}'_{z'}$ (Figure 5.3). A unit *rest length,* which contains an absolute amount of charge $\lambda_0$, has in $K'$ a contracted length $1/\gamma$. Therefore, the linear density of charge with respect to $K'$ is $\lambda_0/(1/\gamma) = \gamma\lambda_0$. Since the sources of charge and current are *steady,* they and the electric and magnetic fields in $K'$ are independent of time. Maxwell's equations split into two equations for the electric field in terms of the charge density, and two equations for the magnetic field in terms of the current. The equations for the electric field are exactly like those of the previous case except for an extra factor $\gamma$ in the charge density. This field is therefore

$$\mathbf{E}'(\rho', \varphi', z') = \frac{\gamma\lambda_0}{2\rho'}\mathbf{E}'_{\rho'}(\varphi'). \tag{5.21}$$

The current density $\mathbf{J}'$ is determined by the amount of charge carried past a fixed point on the $z'$-axis per unit $K'$-time. It is the volume charge density multiplied by the relative velocity:

$$\mathbf{J}' = \delta(x')\delta(y')(\gamma\lambda_0)V\mathbf{E}'_{z'}.$$

The symmetry of the physical situation tells us this time that the $K'$-magnetic field has the form $\mathbf{B}'(\rho', \varphi', z') = B'(\rho')\mathbf{E}'_{\varphi'}(\varphi')$. If the Maxwell equation $\nabla' \times \mathbf{B}' = 4\pi\mathbf{J}'$ is integrated over a disk of radius $\rho'$ with axis in the direction $\mathbf{n}' = \mathbf{E}'_{z'}$, then Stokes' theorem and the explicit $\mathbf{J}'$ give

$$4\pi \int dS'\mathbf{n}' \cdot \mathbf{J}' = 4\pi\gamma\lambda_0 V = \int \mathbf{B}' \cdot \mathbf{t}'dl' = B'(\rho')2\pi\rho',$$

where $\mathbf{t}' = \mathbf{E}'_{\varphi'}(\varphi')$ is a unit tangent vector to the rim of the disk. Hence, $B'(\rho')$ is known and so

$$\mathbf{B}' = \frac{2\gamma\lambda_0 V}{\rho'}\mathbf{E}'_{\varphi'}(\varphi'). \tag{5.22}$$

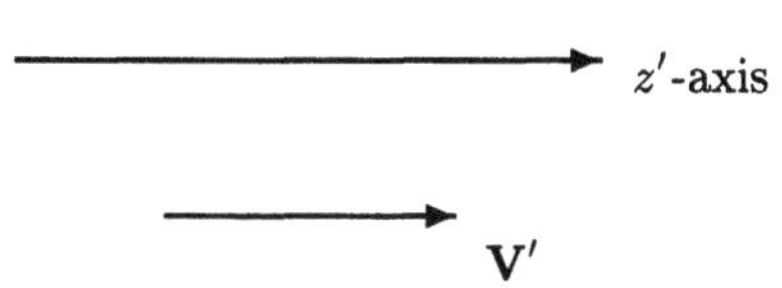


Figure 5.3

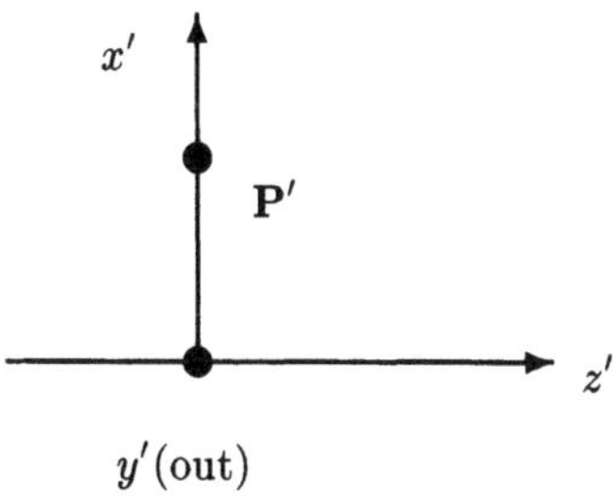


Figure 5.4

For some spatial point $\mathbf{P}'$ with respect to $K'$, spatially rotated if necessary, we may use coordinates such that $\mathbf{P}'$ is on the $x'$-axis, as in Figure 5.4. Then the electric field (5.21) is $\boldsymbol{E}' = (2\gamma\lambda_0/x')\boldsymbol{E}'_1$, and the magnetic field (5.22) is $\boldsymbol{B}' = (2\gamma\lambda_0 V/x')\boldsymbol{E}'_2$. Constructing the field tensor according to (5.15), we have

$$\overrightarrow{\overrightarrow{F}} = \frac{2\gamma\lambda_0}{x'}\left(\boldsymbol{E}'_0 \wedge \boldsymbol{E}'_1 + V\boldsymbol{E}'_3 \wedge \boldsymbol{E}'_1\right). \tag{5.23}$$

The rest frame for the line of charge, $K$, moves at relative velocity $V\boldsymbol{E}'_3$ with respect to $K'$, so the boost relation between the bases gives

$$\boldsymbol{E}_0 = \gamma(\boldsymbol{E}'_0 + V\boldsymbol{E}'_3), \quad \boldsymbol{E}_3 = \gamma(\boldsymbol{E}'_3 + V\boldsymbol{E}'_0), \quad \boldsymbol{E}_1 = \boldsymbol{E}'_1, \quad \boldsymbol{E}_2 = \boldsymbol{E}'_2.$$

At a spacetime event $P$ for which the spatial $\mathbf{P}$ and $\mathbf{P}'$ correspond, $x = x'$. Consequently, $\overrightarrow{\overrightarrow{F}}(P)$ in (5.23) may be rewritten with the $K$-basis vectors

$$\overrightarrow{\overrightarrow{F}} = \frac{2\lambda_0}{x}\boldsymbol{E}_0 \wedge \boldsymbol{E}_1,$$

which is precisely the form (5.20) previously found. This is exactly what is expected. There is really only one physical situation, which is being analysed from two points of view. The geometrical, or absolute, expression of both relative views will be the same if the geometrical theory properly reflects the physical reality.

## 5.4 Vector Potential

It is an interesting and useful consequence of the homogeneous Maxwell equation, $\boldsymbol{\partial}\cdot\overrightarrow{\overrightarrow{F}}^* = \mathbf{0}$ (so long as it really *is* always homogeneous, that is, so long as magnetic poles, which would generate a magnetic counterpart to the electric current $\boldsymbol{j}$, do not exist), that it implies the existence of a so-called vector potential, $\boldsymbol{A}$. A vector potential allows the electromagnetic field to be written in derivative form. In terms of the covariant components of the electromagnetic field (the components $F_{\lambda\rho} \equiv \eta_{\lambda\mu}\eta_{\rho\nu}F^{\mu\nu}$ of eq. (4.75)), the definition of the (covector) components of the vector potential requires it to generate the field by

$$F_{\lambda\rho} = \partial_\lambda A_\rho - \partial_\rho A_\lambda. \tag{5.24}$$

Of course the contravariant version of the equation,

$$F^{\lambda\rho} = \partial^\lambda A^\rho - \partial^\rho A^\lambda, \tag{5.25}$$

is equally true, but it is less useful for the argument below because it is not a simple antisymmetrisation of a derivative $\partial/\partial x^\lambda$ applied to four functions $A_\rho$ (because the implied $\eta^{\mu\nu}$ introduce sign differences).

For a given field dyadic $\overrightarrow{\overrightarrow{F}}$ there are many different vector potentials (see below for so-called changes of gauge), but it is relatively easy to write down one of them explicitly in an analytic form. This is done using a new form for the homogeneous Maxwell equation.

### 5.4.1 Alternative Form for the Homogeneous Maxwell Equation

One begins by writing the component form of the Maxwell equation, (5.9), in terms of the covariant components

$$\partial_\mu F^{*\mu\nu} = 0 = \frac{1}{2}\epsilon^{\mu\nu\lambda\rho}\partial_\mu F_{\lambda\rho}. \tag{5.26}$$

Note that if $F_{\lambda\rho}$ has the form (5.24) with any $A_\rho$, then (5.26) is trivially satisfied by the symmetry of $\partial_\mu\partial_\lambda$ and the antisymmetry of $\epsilon^{\mu\nu\lambda\rho}$. The deeper result is that the converse (called Poincaré's lemma) is true: if (5.26) is satisfied, then a vector potential exists so that $F_{\lambda\rho}$ may be given the form (5.24). Bearing in mind the antisymmetry of both $\epsilon^{\mu\nu\lambda\rho}$ and $F_{\lambda\rho}$ it is easy to see that (5.26) is equivalent to the relation

$$\partial_\mu F_{\lambda\rho} + \partial_\rho F_{\mu\lambda} + \partial_\lambda F_{\rho\mu} = 0 \tag{5.27}$$

for all values of $\lambda, \mu, \rho$. The argument goes: for each $\nu$, take $\lambda, \rho, \mu$ to be different integers, different from $\nu$. The sums in (5.26) produce the terms in (5.27). If $\lambda, \rho, \mu$ are not all different the equation is trivially true because of the antisymmetry of $F_{\mu\nu}$.

### 5.4.2 An Explicit Vector Potential

The Poincaré lemma is proved just by exhibiting a possible vector potential. With the help of (5.27) one can differentiate

$$\psi_\nu(x^0, x^1, x^2, x^3) = x^\mu \int_0^1 tF_{\mu\nu}(tx^0, tx^1, tx^2, tx^3)\, dt \tag{5.28}$$

to verify

$$F_{\lambda\rho} = \partial_\lambda\psi_\rho - \partial_\rho\psi_\lambda.$$

A different symbol, $\psi_\nu$, has been used here because the relation is solely mathematical; in the physical context, the vector potential is often subject to extra conditions, such as gauge conditions.

If $\boldsymbol{A}$ is a geometrical vector potential, determined by the components in (5.24), or in (5.25), the defining relation may be written in geometrical form

$$\overset{\Rightarrow}{F} = \boldsymbol{\partial} \otimes \boldsymbol{A} - (\boldsymbol{\partial} \otimes \boldsymbol{A})^T \equiv \boldsymbol{\partial} \wedge \boldsymbol{A}. \tag{5.29}$$

This is equivalent to (5.24) being satisfied, with the appropriate components, with respect to every inertial frame.

### 5.4.3 Lorentz Condition

A convenient way to make the vector potential more nearly unique is to subject it to the so-called Lorentz condition. Because $\boldsymbol{\partial}\wedge\boldsymbol{\partial}\phi = \vec{\vec{0}}$ for any scalar field $\phi$, when (5.29) is satisfied with $\boldsymbol{A}$, it is also satisfied with $\boldsymbol{A}' = \boldsymbol{A} + \boldsymbol{\partial}\phi$. Then $\vec{\vec{F}} = \boldsymbol{\partial} \wedge \boldsymbol{A}'$, and so $\boldsymbol{A}'$ is also a vector potential. This construction works for any scalar $\phi$. By choosing $\phi$ to satisfy $0 = \boldsymbol{\partial} \cdot \boldsymbol{A} + \boldsymbol{\partial} \cdot \boldsymbol{\partial}\phi$ one has arranged that $\boldsymbol{\partial} \cdot \boldsymbol{A}' = 0$. When this has been done and the prime has been dropped, Maxwell's equations are all satisfied if

$$\boldsymbol{\partial} \cdot \vec{\vec{F}} = \boldsymbol{\partial} \cdot (\boldsymbol{\partial} \otimes \boldsymbol{A}) \equiv \partial^2 \boldsymbol{A} = -4\pi \boldsymbol{j}, \qquad \boldsymbol{\partial} \cdot \boldsymbol{A} = 0. \tag{5.30}$$

The condition $\boldsymbol{\partial} \cdot \boldsymbol{A} = 0$, imposed to simplify the equations, and not for a physical reason, is called the Lorentz condition.

## 5.5 Electric Current Density

The current $\boldsymbol{j}$ appearing in $\boldsymbol{\partial} \cdot \vec{\vec{F}} = -4\pi\boldsymbol{j}$ is a geometrical vector field whose decomposition with respect to an arbitrary inertial basis,

$$\boldsymbol{j} = j^\mu \boldsymbol{E}_\mu = \rho \boldsymbol{E}_0 + \boldsymbol{J}, \qquad \boldsymbol{J} \cdot \boldsymbol{E}_0 = 0, \tag{5.31}$$

gives the (relative) electric charge density $\rho$ and the (relative) electric three-current density $\boldsymbol{J}$. The above decomposition properties are needed in order to deduce the inhomogeneous Maxwell's equations, (5.1) and (5.4), from their relativistic version, (5.5). It is of interest to find a physical explanation for the transformation law relating charge and current in one frame to those in another.

Because $\vec{\vec{F}}$ is antisymmetric, as a matter of consistency with Maxwell's equations, the divergence of $\boldsymbol{j}$ must vanish:

$$\boldsymbol{\partial} \cdot \boldsymbol{j} = \partial_\mu j^\mu = -\frac{1}{4\pi}\partial_\nu \partial_\mu F^{\mu\nu} = 0. \tag{5.32}$$

This formula expresses the conservation of charge. The discussion of the integral formulation of conservation was begun in Section 4.9.

Different models of the conserved current $\boldsymbol{j}$ are used in different situations. For a single point charge $q$, associated with a particle of mass, the current $\boldsymbol{j}$ is nonzero only on the worldline of the particle. This is a singular case which requires the use of delta-functions and so will be considered separately in the next section. For matter viewed microscopically, but classically, charge is carried by particles in fixed, characteristic amounts on each type of particle ($-e$ for an electron, $+e$ for a proton, $+2e$ for a Helium nucleus, $+\frac{2}{3}e$ for an up quark, and so on). Classical models of electric current for matter involve some sort of smooth average instead of an assembly of point particles.

### 5.5.1 Charged Dust

The simplest model of electric charge, from the mathematical point of view, is a continuous version of charged dust. Charge conservation is clear in this model, as are the transformation laws for the components of $\boldsymbol{j}$.

The history of each individual particle is its worldline, in general curved to represent accelerated motion. If thermal motion can be disregarded, one envisages the worldlines of the particles as field lines, or integral curves, of a smooth, interpolating, future-pointing, unit timelike vector field $\boldsymbol{v}$ ($\boldsymbol{v} \cdot \boldsymbol{v} = -1$). Near any point $P$, $\boldsymbol{v}(P)$ is the spacetime velocity of the particles in the immediate vicinity (Figure 5.5). Any *inertial* frame $\tilde{K}_P$ whose time basis vector is $\tilde{\boldsymbol{E}}_0 = \boldsymbol{v}(P)$ is called *a local rest frame at P.* The particles in the vicinity of $P$ are instantaneously at rest in $\tilde{K}_P$. In any such local rest frame at $P$ the averaged number density of particles (the number of particles, by construction at rest, per unit local rest frame three-volume) is a smooth positive function $n_{(0)}(P) > 0$. If each particle in the present case is supposed to have a charge $q$, an appropriate smoothed current density is given by

$$\boldsymbol{j} = q n_{(0)} \boldsymbol{v}. \tag{5.33}$$

This is the correct form because it equals, at each event $P$, $q n_{(0)} \tilde{\boldsymbol{E}}_0$, which by comparison with (5.31) applied to the local rest frame gives the correct rest frame charge density and the correct rest frame three-current ($\boldsymbol{0}$).

The charge density in a local rest frame is $\rho_{(0)} \equiv q n_{(0)}$, which may be written $\rho_{(0)} = q n_{(0)} = -\boldsymbol{v} \cdot \boldsymbol{j}$. To construct a continuous model for the assembly of particles, the vector field $\boldsymbol{v}$ is assumed to interpolate smoothly between the spacetime velocity vectors of the individual particles, and the

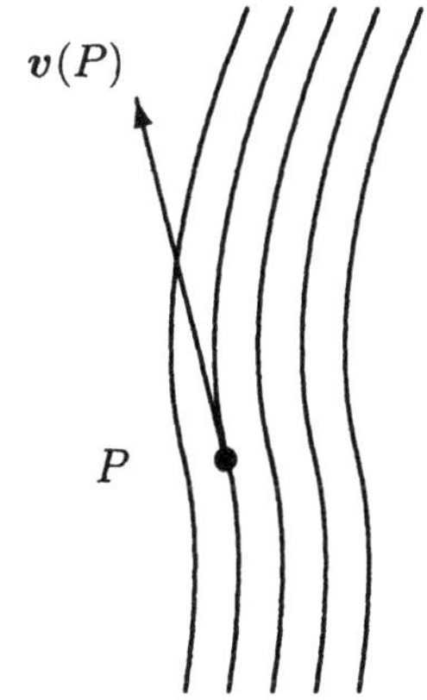


**Figure 5.5**

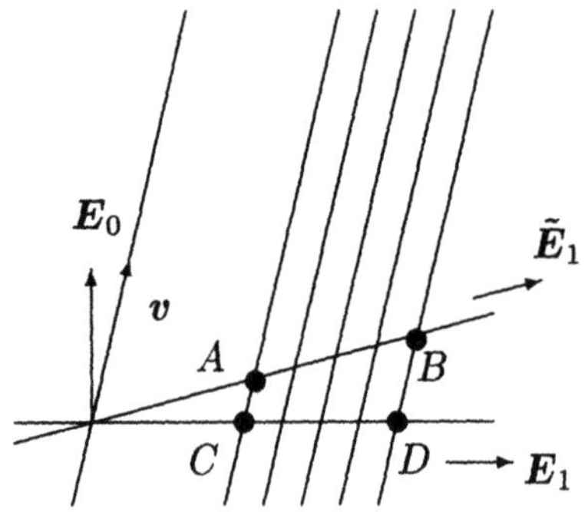


**Figure 5.6**

density functions, which for point particles have a delta-function character, are assumed to be smoothly averaged.

In order to appreciate the simplicity and elegance of the frame-independent form of the current (5.33) for a single type of particle, we verify that it decomposes in an arbitrary frame as expected in (5.31). If the spacetime velocity vector $\boldsymbol{v}$ in (5.33) is decomposed by (2.52) to show the relative velocity with respect to an inertial frame $K$, the corresponding decomposition (5.31) is obtained:

$$\boldsymbol{j} = qn_{(0)}\gamma(\boldsymbol{E}_0 + \boldsymbol{V}) = \rho\boldsymbol{E}_0 + \boldsymbol{J}. \tag{5.34}$$

Hence, the charge density with respect to $K$ is $\rho = qn_{(0)}\gamma = \gamma\rho_{(0)}$, and the three-current with respect to $K$ is $\boldsymbol{J} = qn_{(0)}\gamma\boldsymbol{V} = \rho\boldsymbol{V}$.

### 5.5.2 Density of Charge

Figure 5.6 illustrates a portion of the plane in spacetime determined by the time basis vectors for $K$ and $\tilde{K}_P$. The scale is adjusted so the worldlines appear straight in the region shown. The spatial basis vectors, $\boldsymbol{E}_1$ and $\tilde{\boldsymbol{E}}_1$, are adapted to the situation and lie in the plane. Orthogonally, $\tilde{\boldsymbol{E}}_2 = \boldsymbol{E}_2$, and $\tilde{\boldsymbol{E}}_3 = \boldsymbol{E}_3$. The density of particles with respect to $K$ is based on the number of particles in a $K$-volume at a $K$-instant, say $t = 0$. The projection on the plane of the $K$-volume at $t = 0$ is $CD$. Exactly the same particles (and so evidently the same number of particles) are in a corresponding $\tilde{K}$-volume, at a $\tilde{K}$-instant $\tilde{t} = 0$, whose projection on the plane is $AB$.

Write $\overrightarrow{CD} = L\boldsymbol{E}_1$ and $\overrightarrow{AB} = \tilde{L}\tilde{\boldsymbol{E}}_1$. These are corresponding edges of the $K$-volume and $\tilde{K}$-volume, whose orthogonal edges are the same, $M\boldsymbol{E}_2$ and $N\boldsymbol{E}_3$, say. Since $L\boldsymbol{E}_1 = \tilde{L}\tilde{\boldsymbol{E}}_1 + \lambda\boldsymbol{v}$, for some $\lambda$, taking the scalar product with $\tilde{\boldsymbol{E}}_1$, orthogonal to $\boldsymbol{v}$, gives $\gamma L = \tilde{L}$. The volume in the local rest frame and the corresponding volume in $K$ are related by

$$\mathrm{Vol}(\tilde{K}) = \tilde{L}MN = \gamma LMN = \gamma\mathrm{Vol}(K). \tag{5.35}$$

(Of course, this argument is exactly the same as the one for length contraction, except that here we are not making a definition of an abstract concept of length. The motivation for making the two volumes correspond is so that they contain the same particles.) Expressing the fact that the number of particles is the same in the two volumes gives

$$n_{(0)}\mathrm{Vol}(\tilde{K}) = n\mathrm{Vol}(K) = n\mathrm{Vol}(\tilde{K})/\gamma, \tag{5.36}$$

and hence $n = \gamma n_{(0)}$. Finally, since the charge per particle is independent of the frame (our presumption in building the model), the charge densities are related by $\rho = qn = qn_{(0)}\gamma = \gamma\rho_{(0)}$. Therefore the geometrical argument confirms one of the relations in (5.34) that the proposed current (5.33) gives rise to. The relation for the three-current may similarly be verified by studying the flow of particles.

### 5.5.3 Flux of Charge

In the local rest frame $\tilde{K}_P$ the particles near $P$ are instantaneously at rest. There is no flux of particles relative to any spatial surface at rest with respect to $\tilde{K}_P$. But from the $K$-point of view the charged particles have some velocity $\boldsymbol{V}$, their relative velocity with respect to the inertial frame $K$. The flux of particles, like the density, is a frame-dependent concept. The flux ($K$-flux here) of particles through a surface at rest in $K$ is the net number of particles passing through the surface per unit $K$-time. The surface has an orientation (a choice of one side as the "out" side) and a unit spatial normal $\boldsymbol{S}$, which we may suppose points towards the "out" side; we count positively the particles passing in the direction of the normal, and negatively those going the other way (this is the significance of *net*). If the surface is small enough to be effectively plane, the flux per unit area, or flux density is obtained. This density is in fact for a three-dimensional volume in spacetime, like $n$ and $n_{(0)}$, but in the case of the flux density two dimensions are spatial and one is temporal, a number per unit area per unit time. For the case of dust there is a simple formula for the flux density.

Figure 5.7 shows a projection (along $\boldsymbol{E}_0$) into the *space* for frame $K$ of a region of spacetime in which the worldlines are nearly straight. The tracks in time of three particles are shown (many millions of particles would not be unrealistic), together with their locations at $t = 0$. In the course of the time interval $\delta t$, all the particles in the volume will pass through the surface with unit normal $\boldsymbol{S}$. The volume has been chosen for convenience: its top and bottom both have normal $\boldsymbol{S}$ and area $\delta A$; its sides are parallel to $\boldsymbol{V}$, the relative velocity of the particles in the volume. The volume of such a parallelepiped is (orthogonal height between parallel ends)×(area of end), hence $\delta A \delta t \boldsymbol{V} \cdot \boldsymbol{S}$, provided $\boldsymbol{V} \cdot \boldsymbol{S} > 0$. If the $K$-density for the number of particles is $n$ then the $K$-flux density is $n\boldsymbol{V} \cdot \boldsymbol{S}$ (the net number of particles crossing per unit area per unit time).

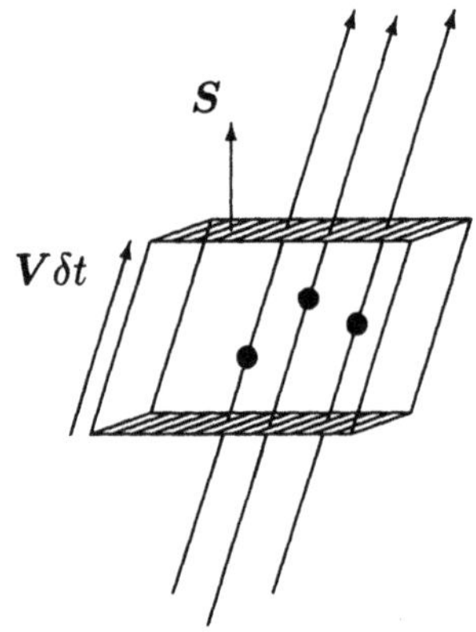


**Figure 5.7**

Electromagnetism, in its frame-dependent version, uses a spatial three-current vector $\boldsymbol{J}$ such that $\boldsymbol{J} \cdot (\delta t \delta A \boldsymbol{S})$ gives the (net) flow of charge through an area $\delta A$ (in the direction of $\boldsymbol{S}$) in time $\delta t$. Using the $K$-density of charge $\rho = qn$, the flow is $\rho \delta A \delta t \boldsymbol{V} \cdot \boldsymbol{S}$, and so $\boldsymbol{J} = \rho \boldsymbol{V} = qn\boldsymbol{V}$. (Note that if $\boldsymbol{V} \cdot \boldsymbol{S} < 0$, the physical interpretation applies as before but with a positive flow in the direction $-\boldsymbol{S}$. As a flow in the $+\boldsymbol{S}$−direction, the flux and the flow are regarded as negative.) The final relation for $\boldsymbol{J}$ is just what is given by (5.34), $\boldsymbol{J} = qn\boldsymbol{V} = qn_{(0)}\gamma\boldsymbol{V}$. This verifies for the case of dust, and so it is presumably true for any source, that the four-current density is *geometrical,* so the charge density and three-current for an arbitrary frame $K$ may be obtained by decomposing $\boldsymbol{j}$ (for dust, (5.33)) with respect to the $K$-basis.

For a physical situation involving several particle types, the currents for the individual components are added. The appropriate model for a physical mixture of different particle types is simply a sum of components of the form (5.33) for each different particle type with different charges and different spacetime velocity vectors for the superposed flows. From the physical point of view the superposition might need to be viewed critically: are the densities of the different components large enough that significant interference will occur?

### 5.5.4 Conservation of Charge

Conservation of electric charge is expressed in a frame-independent differential form by (5.32). If the current density is decomposed with respect to $K$ as in (5.34), then the corresponding equation expressing conservation in terms of components is

$$\frac{\partial \rho}{\partial t} + \frac{\partial J^m}{\partial x^m} = 0. \tag{5.37}$$

The simplest integral consequence is (4.111)

$$\int_V [\rho](t_1)\, d^3x = \int_V [\rho](t_2)\, d^3x + \int_{t_1}^{t_2} \int_S J^m dS_m dt, \tag{5.38}$$

which states that any change in the charge content of a spatial volume $V$ must be accounted for by net flow over its boundary surface $S$. The interpretation of the integral formula follows from the general interpretation of the components of the current: for $\boldsymbol{v}$, a future-pointing timelike unit vector, $-\boldsymbol{v}\cdot\boldsymbol{j}$ is the charge density with respect to any frame whose time basis vector is $\boldsymbol{v}$, and if $\boldsymbol{S}$ is a spacelike unit vector orthogonal to $\boldsymbol{v}$, then $\boldsymbol{S}\cdot\boldsymbol{j}$ is the net outward flux of electric charge, per unit area, per unit time, with respect to the same frame, through a surface with outward normal $\boldsymbol{S}$. Like any other vector, the current density $\boldsymbol{j}$ is determined by its scalar products with all timelike unit vectors $\boldsymbol{v}$. Hence, at each point, the current density vector $\boldsymbol{j}$ is determined by the (inertial frame) charge densities $\rho = -\boldsymbol{v}\cdot\boldsymbol{j}$ at the point.

If the current satisfies the differential conservation formula (5.32), and if it actually vanishes in sufficiently distant spacelike directions, then the last term in (5.38) drops out for volumes $V$ so large that no charge crosses its bounding surface. Then the total charge, as measured by $K$, is a constant in time,

$$\int_V [\rho](t_1)\, d^3x = \int_V [\rho](t_2)\, d^3x. \tag{5.39}$$

The constant total charge is in fact the same for all inertial observers. When the current vanishes in distant spatial directions, the integral

$$\int \chi \boldsymbol{\partial}\cdot\boldsymbol{j}\, d^4x = 0$$

may be integrated simply by parts. Choosing $\chi = \theta(-x^0) - \theta(-x'^0)$, where $x^0$ and $x'^0$ are time coordinates for two *different* inertial frames, one obtains

$$\int_{(x^0=0)} d^3x(-\boldsymbol{E}_0\cdot\boldsymbol{j}) = \int_{(x'^0=0)} d^3x'(-\boldsymbol{E}'_0\cdot\boldsymbol{j}),$$

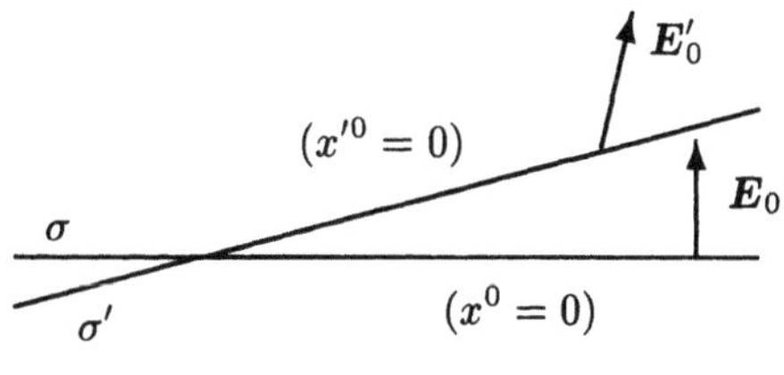


**Figure 5.8**

or, using $\sigma$, $\sigma'$ to denote any spatial hyperplanes for the two inertial systems (Figure 5.8),

$$\int_{\sigma} dV \rho = \int_{\sigma'} dV' \rho'. \tag{5.40}$$

The total charge as measured by $K$, at any $K$-time, is the same as the total charge as measured by $K'$, at any $K'$-time. The final formula expresses the existence of a total charge, the same for all inertial frames and at all times.

### 5.5.5 Conservation of Charge Along a Worldline

The way that current conservation works for the one-component dust model (5.33) can be understood geometrically quite simply. One focuses on an individual worldline (or integral curve) for the velocity field $\boldsymbol{v}(P)$. The worldline $z(\tau)$, parameterised by proper time, satisfies

$$\dot{\boldsymbol{z}}(\tau) = \dot{z}^{\mu}(\tau)\boldsymbol{E}_{\mu} = \boldsymbol{v}(z(\tau)) = v^{\mu}(z(\tau))\boldsymbol{E}_{\mu}. \tag{5.41}$$

Hence, when the equation expressing the conservation of (5.33), given by

$$\boldsymbol{\partial} \cdot \boldsymbol{j} = 0 = \boldsymbol{v} \cdot \boldsymbol{\partial} n_{(0)} + n_{(0)} \boldsymbol{\partial} \cdot \boldsymbol{v} = v^{\mu} \partial_{\mu} n_{(0)} + n_{(0)} \boldsymbol{\partial} \cdot \boldsymbol{v}, \tag{5.42}$$

is restricted to the single worldline and so may be regarded as an equation in the single variable $\tau$, it reads

$$0 = \frac{\partial n_{(0)}}{\partial \tau} + n_{(0)} \boldsymbol{\partial} \cdot \boldsymbol{v}. \tag{5.43}$$

If $\boldsymbol{\partial} \cdot \boldsymbol{v} > 0$ (or $\boldsymbol{\partial} \cdot \boldsymbol{v} < 0$), the density $n_{(0)}$ drops (or increases) along the worldline. The solution of (5.43) may be written simply

$$n_{(0)}(\tau) = n_{(0)}(0) \exp[-\int_0^{\tau} \boldsymbol{\partial} \cdot \boldsymbol{v} \, d\tau], \tag{5.44}$$

showing how the local rest frame density of particles is determined by the cumulative effect of the divergence of the spacetime velocity along the worldline. (There is no restriction on the variation of $n_{(0)}$ from one worldline to another.)

## 5.6 Point Particle: A Singular Source

We consider here the current density and electromagnetic field generated by a point charge $q$ with a timelike worldline $P(\tau)$. The worldline in spacetime is regarded here as given; it is the solution of the physical equations of electrodynamics (not yet written down). The theory is in large part intrinsic, built up from variables constructed from the worldline itself.

The worldline is parameterised with proper time as in Section 2.6 (Figure 5.9), so

$$\overrightarrow{P(\tau)P(\tau+\delta\tau)} \cdot \overrightarrow{P(\tau)P(\tau+\delta\tau)} = -(\delta\tau)^2.$$

When an origin (zero point) for proper time has been chosen, $P(\tau)$ is a well defined function $\mathcal{R} \to \mathcal{M}$. The tangent vector

$$\boldsymbol{v}(\tau) = \lim_{\delta\tau\to 0} \frac{\overrightarrow{P(\tau)P(\tau+\delta\tau)}}{\delta\tau}$$

is then also a definite function $\mathcal{R} \to \mathcal{V}$.

It turns out (we will presently understand this fact in several different ways) that the appropriate four-current density for the point charge $q$ is

$$\boldsymbol{j}(x) = q \int \boldsymbol{v}(\tau)\delta^{(4)}(x - z(\tau))\, d\tau. \tag{5.45}$$

One can understand that this is the correct current both from a spacetime point of view and from the relative point of view of an inertial frame. In eq. (5.45) $x$ and $z$ stand for the inertial coordinates of points $P$ and $P(\tau)$ in $\mathcal{M}$, and $\delta^{(4)}(x-z)$ is the four-dimensional Dirac delta-function. The latter (formally) satisfies

$$\phi(z) = \int \delta^{(4)}(x-z)\phi(x)\, d^4x \tag{5.46}$$

with any inertial coordinates and any smooth test function $\phi$. If no misunderstanding is likely, the superscript $^{(4)}$ is often omitted. Note that if the (inertial) coordinates in (5.46) are changed to a different set of inertial coordinates, the Jacobian is one, and so the delta-function may be treated as if it were a scalar function. The expression (5.45) is particularly attractive for the current because it is elegant and because it displays the current for the whole of spacetime.

If a particular inertial frame $K$ is selected (5.45) may be converted to a $K$-relative form. Using $K$-coordinates, at some given $K$-time $x^0$,

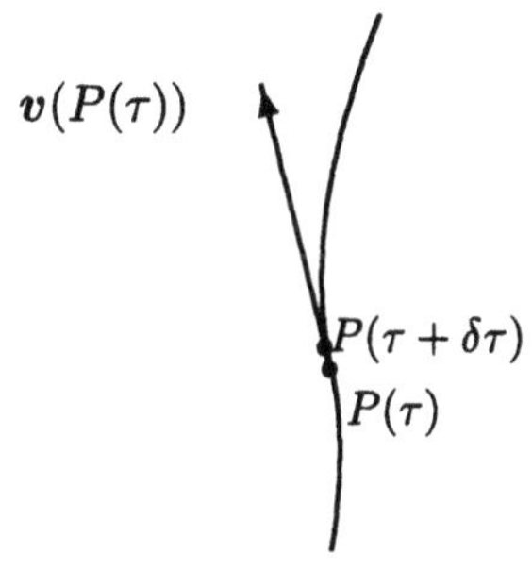


**Figure 5.9**

$$\begin{aligned} \boldsymbol{j}(x^0,x^k) &= q\int d\tau \boldsymbol{v}(\tau)\delta(x^0 - z^0(\tau))\delta^{(3)}(x^k - z^k(\tau)) \\ &= q\int dz^0 \frac{1}{dz^0/d\tau}\boldsymbol{v}(\tau)\delta(x^0 - z^0(\tau))\delta^{(3)}(x^k - z^k(\tau)) \\ &= q\boldsymbol{v}\left[\frac{\delta^{(3)}(x^k - z^k)}{\gamma}\right], \end{aligned} \tag{5.47}$$

where, in the final form, $\boldsymbol{v}$, $\gamma$ and $z^k$ are evaluated at the value of $\tau$ which satisfies $x^0 - z^0(\tau) = 0$ for the given $x^0$. The three-dimensional delta-function, $\delta^{(3)}(x^k - z^k)$, is the number density $n$ for the $K$-frame. The $\gamma$-factor is determined by the speed of the particle with respect to the $K$-frame. So the quantity in the square bracket is $n_{(0)} = n/\gamma$, the *rest* frame number density, expressed as it is in (5.36). Thus, eq. (5.47) can be understood as the point particle version of (5.33), $\boldsymbol{j} = qn_{(0)}\boldsymbol{v}$. On the other hand, substituting in (5.47) its $K$-decomposition (2.52) for $\boldsymbol{v}$ gives

$$\boldsymbol{j}(x^0,x^k) = q\delta^{(3)}(x^k - z^k)\boldsymbol{E}_0 + q\delta^{(3)}(x^k - z^k)\boldsymbol{V}, \tag{5.48}$$

which may be written in terms of the $K$-density $\rho = q\delta^{(3)}(x^k - z^k)$,

$$\boldsymbol{j}(x^0,x^k) = \rho\boldsymbol{E}_0 + \rho\boldsymbol{V}, \tag{5.49}$$

exactly the same expression as (5.34) for a flow of charged dust. The delta-function density allows us to write a single particle as a special case of dust.

If $\boldsymbol{v}(x)$ is a vector field which reduces to $\boldsymbol{v}(\tau)$ when $x = z(\tau)$, then the point current (5.45) may be written

$$\boldsymbol{j}(x) = q\boldsymbol{v}(x)\int \delta^{(4)}(x - z(\tau))\, d\tau = q\boldsymbol{v}n_{(0)}, \tag{5.50}$$

the same form as (5.33). This gives a useful form for the rest frame density $n_{(0)}$ for the single particle.

### 5.6.1 The Intrinsic Variables

We now describe the intrinsic geometry which is used to write the electromagnetic field for the point charge. There is an astonishingly simple formula for the Liénard-Wiechert potential in terms of the intrinsic variables. The geometrical construction was included in Minkowski's famous address in 1908, which historically was the birth of spacetime as a unified concept. The address is included in the collection of early (translated) reprints, *The Principle of Relativity.* Rohrlich's book on point charges includes a long account of the variables and their uses.

For each point $P$ in spacetime the backward lightcone from $P$ cuts the worldline of the charge at a unique point $z(\tau(P))$. This defines a real valued

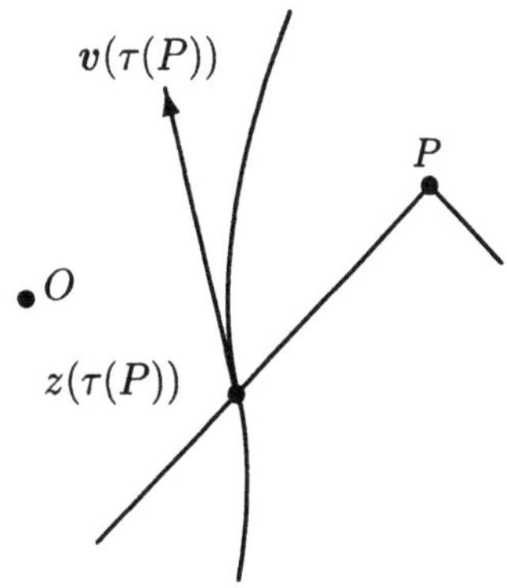


**Figure 5.10**

scalar field $\tau(P)$ in spacetime (Figure 5.10). Then the vector field $\boldsymbol{R}(P)$ is defined by

$$\boldsymbol{R}(P) \equiv \overrightarrow{z(\tau(P))P} = \boldsymbol{x}(P) - \boldsymbol{z}(\tau(P)), \tag{5.51}$$

where in the second form, $\boldsymbol{x}(P) = \overrightarrow{OP}$ and $\boldsymbol{z}(\tau(P)) = \overrightarrow{Oz(\tau(P))}$, in terms of a fixed reference point $O$.

The important new variables are $\tau(P)$ and $\boldsymbol{R}(P)$, defined at each point in spacetime. We have constructed the values of these *fields*, scalar and vector respectively, at the point $P$, in terms of the particle's worldline at the point $z(\tau(P))$, the so-called retarded point. Light signals from $z(\tau(P))$ can reach $P$, and this fact eventually becomes the reason why the variables are ideal for the expression of retarded (causal) electromagnetic fields. The fixed point $O$ is arbitrary and is not intrinsic; it is included only to make subsequent processes of differentiation simple.

Differential relations in the electromagnetism of a point particle all finally depend on knowing the gradients $\boldsymbol{\partial}\tau$ and $\boldsymbol{\partial} \otimes \boldsymbol{R}$. First, using any inertial coordinates with origin $O$, so $\boldsymbol{x} = \overrightarrow{OP} = x^\mu \boldsymbol{E}_\mu$, and the definitions (4.63) and (4.34),

$$\boldsymbol{\partial} \otimes \boldsymbol{x} = \eta^{\mu\nu} \boldsymbol{E}_\mu \otimes \boldsymbol{E}_\lambda \partial_\nu x^\lambda = \boldsymbol{E}_\mu \otimes \boldsymbol{E}_\lambda \eta^{\mu\lambda} = \overset{\Rightarrow}{\eta}. \tag{5.52}$$

Also, using the easily proved identity,

$$\boldsymbol{\partial} \otimes \boldsymbol{A}(\psi(P)) = \boldsymbol{\partial}\psi \otimes \frac{\partial \boldsymbol{A}}{\partial \psi}, \tag{5.53}$$

together with $\boldsymbol{v}(\tau) = \frac{d\boldsymbol{z}(\tau)}{d\tau}$, we have

$$\boldsymbol{\partial} \otimes \boldsymbol{z}(\tau) = \boldsymbol{\partial}\tau \otimes \boldsymbol{v}. \tag{5.54}$$

The two unknown gradients are now related by

$$\boldsymbol{\partial} \otimes \boldsymbol{R} = \boldsymbol{\partial} \otimes \boldsymbol{x} - \boldsymbol{\partial} \otimes \boldsymbol{z} = \overset{\Rightarrow}{\eta} - \boldsymbol{\partial}\tau \otimes \boldsymbol{v}. \tag{5.55}$$

A simple argument permits us to find both.

Recall that, by construction, $\boldsymbol{R}$ is future-pointing lightlike, so

$$\boldsymbol{R} \cdot \boldsymbol{R} = 0. \tag{5.56}$$

When (5.56) is differentiated each factor makes the same contribution and

$$\frac{1}{2}\boldsymbol{\partial}(\boldsymbol{R} \cdot \boldsymbol{R}) = \boldsymbol{0} = (\boldsymbol{\partial} \otimes \boldsymbol{R}) \cdot \boldsymbol{R}. \tag{5.57}$$

Hence, the dot product with $\boldsymbol{R}$ on the right of (5.55) vanishes,

$$\boldsymbol{0} = \overset{\Rightarrow}{\eta} \cdot \boldsymbol{R} - \boldsymbol{\partial}\tau(\boldsymbol{v} \cdot \boldsymbol{R}) = \boldsymbol{R} - \boldsymbol{\partial}\tau(\boldsymbol{v} \cdot \boldsymbol{R}), \tag{5.58}$$

and this allows us to solve explicitly for $\boldsymbol{\partial}\tau$. It is useful to define a new scalar field $\rho(P)$ (another of the ubiquitous intrinsic variables, this one is the simplest to understand physically) by

$$\rho \equiv -\boldsymbol{v} \cdot \boldsymbol{R}. \tag{5.59}$$

This is the time interval (or equally, the spatial interval), in the retarded rest frame, between the field point $P$ and the associated retarded point $z(\tau(P))$. In terms of it,

$$\boldsymbol{\partial}\tau = -\frac{\boldsymbol{R}}{\rho}. \tag{5.60}$$

Substituting this expression in (5.55) gives the explicit formula for $\boldsymbol{\partial} \otimes \boldsymbol{R}$,

$$\boldsymbol{\partial} \otimes \boldsymbol{R} = \overset{\Rightarrow}{\eta} + \frac{\boldsymbol{R} \otimes \boldsymbol{v}}{\rho}. \tag{5.61}$$

The retarded rest frame (the frame with time basis vector $\boldsymbol{v}(\tau(P))$), the frame in which $\rho$ has such a simple geometrical meaning, is the (changing) frame in which point-particle electrodynamics has almost the simplicity of electrostatics. For discussing the electromagnetic fields at $P$ the retarded rest frame is picked out by a purely geometrical construction in terms of the particle's worldline. The fact that the points $P$ and $z(\tau(P))$ may be widely separated is of no consequence since the rest frame basis vectors are specified by *directions,* not positions. The variable $\rho > 0$ is the time component of $\boldsymbol{R}$ in the retarded basis and the single space component if an appropriate unit spatial vector $\boldsymbol{u}(\tau)$ is chosen. Then, exactly as in (2.66),

$$\boldsymbol{R}(P) = \rho(P)(\boldsymbol{v}(\tau) + \boldsymbol{u}(\tau)), \tag{5.62}$$

where $\boldsymbol{u} \cdot \boldsymbol{u} = 1$ and $\boldsymbol{u} \cdot \boldsymbol{v} = 0$. The interval in time is the same as the interval in space since the displacement $\boldsymbol{R}$ is lightlike.

Two further derivatives follow immediately from $\boldsymbol{\partial}\tau$ and $\boldsymbol{\partial}\otimes\boldsymbol{R}$. In terms of the retarded spacetime acceleration $\boldsymbol{a}(\tau)=\frac{d\boldsymbol{v}(\tau)}{d\tau}$,

$$\boldsymbol{\partial}\otimes\boldsymbol{v}=\boldsymbol{\partial}\otimes\boldsymbol{v}(\tau(\tau))=\boldsymbol{\partial}\tau\otimes\boldsymbol{a}=-\frac{\boldsymbol{R}\otimes\boldsymbol{a}}{\rho} \tag{5.63}$$

and then

$$\begin{aligned}\boldsymbol{\partial}\rho &= -\boldsymbol{\partial}(\boldsymbol{v}\cdot\boldsymbol{R})=-(\boldsymbol{\partial}\otimes\boldsymbol{v})\cdot\boldsymbol{R}-(\boldsymbol{\partial}\otimes\boldsymbol{R})\cdot\boldsymbol{v}\\ &= (\boldsymbol{a}\cdot\boldsymbol{R})\frac{\boldsymbol{R}}{\rho}-\boldsymbol{v}+\frac{\boldsymbol{R}}{\rho}.\end{aligned} \tag{5.64}$$

All the derivatives that are necessary for electrodynamics have now been done, and in fact any further calculations are of such a routine nature that they can be computer-programmed in terms of those now available.

### 5.6.2 Electromagnetic Field for a Point Charge

The important relation to establish here is that a vector potential for the retarded field of a point particle with charge $q$ is, at the event $P$,

$$\boldsymbol{A}(P)=\frac{q\boldsymbol{v}(P)}{\rho(P)}, \tag{5.65}$$

in terms of the variables defined above. It will be necessary to check that Maxwell's equations, in the form (5.30), are satisfied with the current $\boldsymbol{j}$ for a point charge whose history is an arbitrary timelike worldline.

For the moment we assume that (5.65) is the correct potential; we will return to the proof after calculating the retarded electromagnetic field. In fact, if (5.65) is decomposed in an inertial basis, the scalar and three-vector parts are the famous Liénard-Wiechert potentials.

Using the two relations (5.63) and (5.64) we get

$$\boldsymbol{\partial}\otimes\left(\frac{\boldsymbol{v}}{\rho}\right)=-\frac{\boldsymbol{R}\otimes\boldsymbol{a}}{\rho^2}-\frac{1}{\rho^2}\left(\frac{\boldsymbol{a}\cdot\boldsymbol{R}\boldsymbol{R}\otimes\boldsymbol{v}}{\rho}-\boldsymbol{v}\otimes\boldsymbol{v}+\frac{\boldsymbol{R}\otimes\boldsymbol{v}}{\rho}\right). \tag{5.66}$$

Taking the trace, (4.66), and using $\boldsymbol{R}\cdot\boldsymbol{v}=-\rho$ and $\boldsymbol{v}\cdot\boldsymbol{v}=-1$, we see that the Lorentz condition for the potential, a part of (5.30), is satisfied,

$$\boldsymbol{\partial}\cdot\left(\frac{\boldsymbol{v}}{\rho}\right)=0. \tag{5.67}$$

The electromagnetic field tensor is given in terms of the vector potential by (5.29). After supplying the factor $q$, this is obtained from (5.66) by subtracting its transpose in order to antisymmetrise; one gets the simple form

$$\vec{\vec{F}} = \partial \wedge \boldsymbol{A} = \frac{q}{\rho^3}\boldsymbol{v} \wedge \boldsymbol{R} + \frac{q}{\rho^2}\boldsymbol{a} \wedge \boldsymbol{R} + \frac{q\boldsymbol{a} \cdot \boldsymbol{R}}{\rho^3}\boldsymbol{v} \wedge \boldsymbol{R}. \tag{5.68}$$

Only the first term in (5.68) appears for a nonaccelerating charge. Decomposing $\vec{\vec{F}}$ in the rest frame ($\tilde{\boldsymbol{E}}_0 = \boldsymbol{v}$) for such a charge gives a Coulomb electric field (and no magnetic field)

$$\vec{\vec{F}} = \tilde{\boldsymbol{E}}_0 \wedge \left(\frac{q\boldsymbol{u}}{\rho^3}\right).$$

Decomposing $\vec{\vec{F}}$ in any other inertial basis gives the electric and magnetic fields for a uniformly moving charge.

### 5.6.3 Electric and Magnetic Fields in the Retarded Rest Frame

It may be of interest to derive from (5.68) expressions for the electric and magnetic fields in the retarded rest frame. We are dealing with the fields at a general point $P$ but the inertial basis used for the decomposition (5.11) is that for the rest frame at $z(\tau(P))$, for which the time basis vector is $\boldsymbol{v}(\tau)$. The electric and magnetic fields are obtained from (5.11),

$$\vec{\vec{F}} = \boldsymbol{v} \wedge \boldsymbol{E} - (\boldsymbol{v} \wedge \boldsymbol{B})^* = \frac{q}{\rho^3}\boldsymbol{v} \wedge \boldsymbol{R} + \frac{q}{\rho^2}\boldsymbol{a} \wedge \boldsymbol{R} + \frac{q\boldsymbol{a} \cdot \boldsymbol{R}}{\rho^3}\boldsymbol{v} \wedge \boldsymbol{R},$$

where it must be remembered that the vectors $\boldsymbol{E}$ and $\boldsymbol{B}$ are spatial in the retarded rest frame, $\boldsymbol{E} \cdot \boldsymbol{v} = \boldsymbol{B} \cdot \boldsymbol{v} = 0$. Both $\boldsymbol{a}$ and $\boldsymbol{u}$ in $\boldsymbol{R} = \rho(\boldsymbol{v} + \boldsymbol{u})$ are spatial, so we can identify

$$\boldsymbol{E} = \frac{q}{\rho^2}\boldsymbol{u} - \frac{q}{\rho}(\boldsymbol{a} - \boldsymbol{a} \cdot \boldsymbol{u}\boldsymbol{u}) \tag{5.69}$$

$$\boldsymbol{B} = \frac{q}{\rho}\boldsymbol{a} \times \boldsymbol{u}, \tag{5.70}$$

where the three-dimensional vector product notation can be used for the algebra of spatial vectors with respect to a specific frame. The expression comes from $(\boldsymbol{v} \wedge \epsilon^{mnl}\boldsymbol{E}_l)^* = -(\boldsymbol{E}_m \wedge \boldsymbol{E}_n)$, which is equivalent to (4.90). Notice the electric Coulomb field (the radial component of $\boldsymbol{E}$) and the relation $\boldsymbol{B} = \boldsymbol{u} \times \boldsymbol{E}$. Notice also the extreme elegance and simplicity of the expressions for the fields in the retarded rest frame as compared with those in a general frame (the laboratory) below.

### 5.6.4 Electric and Magnetic Fields in the Laboratory

To get the individual electric and magnetic fields $\boldsymbol{E}_L$ and $\boldsymbol{B}_L$ in the laboratory, the field tensor (5.68) must be decomposed as in (5.11) using the time basis vector $\boldsymbol{E}_0$ for the lab:

$$\vec{F} = \boldsymbol{E}_0 \wedge \boldsymbol{E}_L - (\boldsymbol{E}_0 \wedge \boldsymbol{B}_L)^* = \frac{q}{\rho^3}\boldsymbol{v} \wedge \boldsymbol{R} + \frac{q}{\rho^2}\boldsymbol{a} \wedge \boldsymbol{R} + \frac{q\boldsymbol{a} \cdot \boldsymbol{R}}{\rho^3}\boldsymbol{v} \wedge \boldsymbol{R}. \quad (5.71)$$

The vectors $\boldsymbol{v}$, $\boldsymbol{R}$ and $\boldsymbol{a}$ must be decomposed parallel and orthogonal to $\boldsymbol{E}_0$ in order to identify $\boldsymbol{E}_L$ and $\boldsymbol{B}_L$. To this end

$$\begin{aligned} \boldsymbol{R} &= \rho(\boldsymbol{u} + \boldsymbol{v}) \equiv r(\boldsymbol{N} + \boldsymbol{E}_0) \\ \boldsymbol{v} &= \gamma(\boldsymbol{E}_0 + \boldsymbol{V}) \\ \boldsymbol{a} &= \gamma^2 \boldsymbol{A} + \gamma^4 \boldsymbol{V} \cdot \boldsymbol{A}(\boldsymbol{E}_0 + \boldsymbol{V}), \end{aligned}$$

where $\boldsymbol{N}$ is a lab-spatial unit vector in the lab-direction of the field point from the retarded point, $\boldsymbol{V}$ is the (lab-spatial) relative velocity of the particle at the retarded point and $\boldsymbol{A}$ is the relative acceleration at the same point (see eq. (2.62)). The relation between the distance from the retarded point to the field point in the retarded rest frame ($\rho$) and in the laboratory ($r$) is given by

$$\rho = -\boldsymbol{v} \cdot \boldsymbol{R} = \gamma r(1 - \boldsymbol{V} \cdot \boldsymbol{N}) \equiv \gamma r \kappa.$$

Substituting the new forms for the variables in (5.71) and simplifying gives

$$\begin{aligned} \vec{F} &= \frac{q}{\gamma^2 \kappa^3 r^2}\left[\boldsymbol{E}_0 \wedge (\boldsymbol{N} - \boldsymbol{V}) + \boldsymbol{V} \wedge \boldsymbol{N}\right] \\ &\quad + \frac{q}{\kappa^3 r}\left[\boldsymbol{E}_0 \wedge (-\kappa \boldsymbol{A} - \boldsymbol{N} \cdot \boldsymbol{A}\boldsymbol{V} + \boldsymbol{N} \cdot \boldsymbol{A}\boldsymbol{N}) \right. \\ &\qquad \left. + (\kappa \boldsymbol{A} + \boldsymbol{N} \cdot \boldsymbol{A}\boldsymbol{V}) \wedge \boldsymbol{N}\right] \end{aligned}$$

from which one extracts

$$\boldsymbol{E}_L = \frac{q}{\gamma^2 \kappa^3 r^2}(\boldsymbol{N} - \boldsymbol{V}) + \frac{q}{\kappa^3 r}[-\kappa \boldsymbol{A} + \boldsymbol{N} \cdot \boldsymbol{A}(\boldsymbol{N} - \boldsymbol{V})] \quad (5.72)$$

$$\boldsymbol{B}_L = \frac{q}{\gamma^2 \kappa^3 r^2}(\boldsymbol{V} \times \boldsymbol{N}) + \frac{q}{\kappa^3 r}[\kappa(\boldsymbol{A} \times \boldsymbol{N}) + (\boldsymbol{N} \cdot \boldsymbol{A})(\boldsymbol{V} \times \boldsymbol{N})]. \quad (5.73)$$

These are to be compared with the fields in the retarded rest frame above. The extra complication in the present case is because the description is in an arbitrary frame with no *physical* relevance.

### 5.6.5 Maxwell's Equations for a Point Charge

In order to prove that Maxwell's equations (5.30) are satisfied, it remains to check $\boldsymbol{\partial} \cdot (\boldsymbol{\partial} \otimes \boldsymbol{A}) = \partial^2 \boldsymbol{A} = -4\pi \boldsymbol{j}$. We will show

$$\partial^2 \boldsymbol{A} = \partial^2 \left(\frac{q\boldsymbol{v}}{\rho}\right) = -4\pi q \int \boldsymbol{v}(\tau) \delta^{(4)}(x - z(\tau))\, d\tau, \quad (5.74)$$

where the point-particle current on the right-hand side applies for an arbitrary timelike worldline. The current is zero except along the worldline $\rho = 0$.

Using the decomposition $\boldsymbol{x} = \boldsymbol{z}+\boldsymbol{R} = \boldsymbol{z}+\rho(\boldsymbol{v}+\boldsymbol{u})$, the argument of the delta-function is zero only when $\boldsymbol{R} = \boldsymbol{0}$ or $\rho = 0$. To check (5.74), a straightforward calculation will verify

$$\partial^2 \left(\frac{\boldsymbol{v}}{\rho}\right) = \boldsymbol{0}, \qquad \rho \neq 0. \tag{5.75}$$

A more subtle argument is needed to confirm that the source term on the right-hand side is correct.

### 5.6.6 The Region Off the Worldline: Empty Space

To confirm (5.75), a streamlined notation is helpful. We drop the arrows and the tensor product symbols. There will be no ambiguity. The meaning of every symbol is known, but care must be taken to preserve the correct order of factors. In the new notation (5.66) takes the form

$$\partial\left(\frac{v}{\rho}\right) = -\frac{Ra}{\rho^2} + \frac{vv}{\rho^2} - \frac{Rv}{\rho^3}(1 + a \cdot R).$$

From this we must calculate the *triadic* $\partial\partial(\frac{v}{\rho})$, and then take the appropriate trace (which in this case is a vector), $\mathrm{tr}\partial\partial(\frac{v}{\rho}) = \partial\cdot\partial(\frac{v}{\rho}) = \partial^2(\frac{v}{\rho})$. The general identities for differentiating (the vector gradient of a dyadic) are listed in the Problems to Chapter 4. The specific derivatives that are used are (5.60), (5.61) and (5.64). To make the calculation clear, in the equation below the contributions from differentiating vector $R$ are on the first line, from vectors $a$ and $v$ on the second, from $\rho$ on the third and from $a \cdot R$ on the last:

$$\begin{aligned}
\partial\partial\left(\frac{v}{\rho}\right) = & - \left(\eta + \frac{Rv}{\rho}\right)\frac{a}{\rho^2} - \left(\eta + \frac{Rv}{\rho}\right)\frac{v}{\rho^3}(1 + a \cdot R) \\
& + \frac{RR\dot{a}}{\rho^3} - \frac{Rav + Rva}{\rho^3} + \frac{RRa}{\rho^4}(1 + a \cdot R) \\
& + [(1 + a \cdot R)\frac{R}{\rho} - v][\frac{2Ra - 2vv}{\rho^3} + \frac{3Rv}{\rho^4}(1 + a \cdot R)] \\
& + \frac{RRv}{\rho}\frac{\dot{a} \cdot R}{\rho^3} - \left(\eta + \frac{Rv}{\rho}\right) \cdot a\frac{Rv}{\rho^3}.
\end{aligned}$$

The next step is to calculate the trace to get $\partial \cdot \partial(\frac{v}{\rho})$. Using $\mathrm{tr}\eta = 4$, $\eta \cdot X = X$, $v \cdot a = 0$, $v \cdot v = -1$, $R \cdot R = 0$, $R \cdot v = -\rho$, everything is found to cancel and one gets $\partial \cdot \partial(\frac{v}{\rho}) = 0$ in the region off the worldline.

### 5.6.7 The Electromagnetic Field as a Distribution

The current density (5.45) for a point charge involves a delta-function, which mathematically is a singular distribution (see, for example, the book by

Gel'fand and Shilov for the mathematical subtleties). To calculate $\partial^2 \left(\frac{q\boldsymbol{v}}{\rho}\right)$ accurately in a region including the worldline one must take account of the fact that the potential is singular for $\rho = 0$. The way to cope with this circumstance is to regard the potential also as a distribution, defined by its integrals over smooth test functions which vanish in distant regions of spacetime

$$(\boldsymbol{A}, \phi) = \int \phi(x) \boldsymbol{A}(x)\, d^4x, \tag{5.76}$$

or their generalisations. Significant generalisations (called regularisations) are needed for functions with nonintegrable singularities, but the vector potential $\boldsymbol{A} = \left(\frac{q\boldsymbol{v}}{\rho}\right)$ is only slightly singular, and the definition (5.76) may be used just by omitting the one-dimensional worldline from the four-dimensional integral. Derivatives of distributions can be given a new definition

$$(\boldsymbol{\partial} \otimes \boldsymbol{A}, \phi) = -\int \boldsymbol{\partial}\phi \otimes \boldsymbol{A}\, d^4x, \tag{5.77}$$

which, by an integration by parts, agrees with the old one when the old one exists. Applying the definition twice, and taking the trace, gives

$$(\partial^2 \boldsymbol{A}, \phi) = \int \partial^2 \phi \boldsymbol{A}\, d^4x. \tag{5.78}$$

This new definition gives meaning to the Laplacian $\partial^2$ applied to the singular vector potential.

### 5.6.8 Change of Variables in Spacetime Integrals

It will be useful to express integrals over the whole of spacetime using variables $\tau$, $\rho$ and $\boldsymbol{u}$ adapted to the worldline. The transformation is most easily done in stages. Begin with the spacetime integral expressed using any one of the inertial coordinate systems (it does not matter which):

$$\int \psi(x)\, d^4x = \int \psi(x)\, dx^0 dx^1 dx^2 dx^3. \tag{5.79}$$

Now, with the help of the decomposition $\boldsymbol{x} = \boldsymbol{z}(\tau) + \boldsymbol{R}$, change variables to $\tau$ and the three *spatial* displacements along the lightcone $R^1, R^2, R^3$ in the given inertial frame:

$$\begin{aligned} x^m &= z^m(\tau) + R^m \\ x^0 &= z^0(\tau) + R^0. \end{aligned}$$

Since $R^0 = (R^m R^m)^{1/2}$, the Jacobian for the transformation is

$$J = \frac{\partial(x^0, x^1, x^2, x^3)}{\partial(\tau, R^1, R^2, R^3)} = \begin{vmatrix} v^0 & v^1 & v^2 & v^3 \\ R^1/R^0 & 1 & 0 & 0 \\ R^2/R^0 & 0 & 1 & 0 \\ R^3/R^0 & 0 & 0 & 1 \end{vmatrix} = \frac{-v^\mu R^\nu \eta_{\mu\nu}}{R^0}$$
$$= \frac{\rho}{R^0}, \qquad (5.80)$$

and the integral takes the form

$$\int \psi(x)\, d^4x = \int \psi(z + R) \frac{\rho}{R^0}\, d\tau dR^1 dR^2 dR^3. \qquad (5.81)$$

On the right-hand side is an integral over all proper time $\tau$ of the integrals over the future lightcone with vertex $z(\tau)$. The lightcone integral is expressed in terms of the spatial components of the lightlike vector $\boldsymbol{R}$, with respect to the previously chosen inertial frame. The next step in the transformation of the integral is to change, for each $\tau$, from variables $R^m$ to similar variables $R'^m$(say) with respect to a rest frame at $z(\tau)$, that is, with time basis vector $\boldsymbol{E}'_0 = \boldsymbol{v}(\tau)$. This transformation can be regarded (in several different ways) as a product of a simple boost in the $x^1$-direction and two rotations. Only the boost has a Jacobian that differs from one, and an easy calculation gives

$$\frac{\partial(R^1, R^2, R^3)}{\partial(R'^1, R'^2, R'^3)} = \frac{R^0}{R'^0}. \qquad (5.82)$$

In the rest frame, $R'^0 = \rho$, so the Jacobian of the previous stage is cancelled, and the integral, *now ordered,* takes a very simple form

$$\int \psi\, d^4x = \int d\tau \int \psi\, dR'^1 dR'^2 dR'^3$$
$$= \int d\tau \int \psi(z + \rho(v + u))\rho^2\, d\rho d\Omega; \qquad (5.83)$$

in the last version, polar coordinates in the rest frame at $z(\tau)$ have been used.

### 5.6.9 Maxwell's Equations Along the Worldline

To calculate $\partial^2 \left(\frac{qv}{\rho}\right)$, apply (5.78). The immediate region around the worldline in the integral is omitted by inserting a factor $\theta(\rho - \epsilon)$, which vanishes for $\rho < \epsilon$.

$$(\partial^2 \left(\frac{qv}{\rho}\right), \phi) = \int \partial^2 \phi \left(\frac{qv}{\rho}\right) \theta(\rho - \epsilon)\, d^4x \qquad (5.84)$$

The plan is to take the limit $\epsilon \to 0$ after integrating by parts. Because the singularity is integrable, this procedure gives the correct integral. Integrating once by parts gives

$$-\int \partial_\mu\phi[(\partial^\mu\rho)\delta(\rho-\epsilon)(\frac{q\boldsymbol{v}}{\rho})]\,d^4x - \int \partial_\mu\phi[\theta(\rho-\epsilon)\partial^\mu(\frac{q\boldsymbol{v}}{\rho})]\,d^4x.$$

The $\delta(\rho-\epsilon)$ integral vanishes in the $\epsilon \to 0$ limit because the integrand is of order $\frac{1}{\rho}$ but the Jacobian supplies a factor $\rho^2$ (see (5.83)). Integrating the remaining term by parts again, and using (5.75) in the region $\rho \neq 0$, leaves

$$(\partial^2\left(\frac{q\boldsymbol{v}}{\rho}\right), \phi) = +\int \phi[(\partial_\mu\rho)\delta(\rho-\epsilon)\partial^\mu(\frac{q\boldsymbol{v}}{\rho})]\,d^4x. \tag{5.85}$$

Calculating $\boldsymbol{\partial}\rho\cdot\boldsymbol{\partial}\otimes(\frac{q\boldsymbol{v}}{\rho})$ from (5.64) and (5.66), and omitting the terms which are less singular than $\frac{1}{\rho^2}$ (they will vanish in the limit), gives finally, with the help of the change of variable in (5.83),

$$\begin{aligned}(\partial^2\left(\frac{q\boldsymbol{v}}{\rho}\right), \phi) &= -\int \phi[\delta(\rho-\epsilon)(\frac{q\boldsymbol{v}}{\rho^2})]\,d^4x \\ &= -\int d\tau \int \phi(z+\rho(v+u))[\delta(\rho-\epsilon)(\frac{q\boldsymbol{v}}{\rho^2})]\rho^2\,d\rho d\Omega \\ &= -\int d\tau\phi(z)q\boldsymbol{v}(\tau)\,d\Omega = -4\pi q\int \phi(z)\boldsymbol{v}(\tau)\,d\tau.\end{aligned}$$

The last expression may be written as a (formal) integral in terms of the particle's four-current

$$\int \phi(x)\left[-4\pi q\int \boldsymbol{v}(\tau)\delta^{(4)}(x-z(\tau))\,d\tau\right]\,d^4x = (-4\pi\boldsymbol{j}, \phi),$$

and this at last establishes Maxwell's equations (5.74) along the worldline.

## 5.7 Plane Waves

Plane electromagnetic waves arise in free space as the simple solutions to (5.30), with $\boldsymbol{j} = \boldsymbol{0}$, of the form

$$\boldsymbol{A}(x) = \boldsymbol{a}\,\psi(\xi), \tag{5.86}$$

where $\xi = \boldsymbol{k}\cdot\boldsymbol{x}$, for some constant vector $\boldsymbol{k}$. (Points $P$ of spacetime are here specified by $x$, which stands for an element $x^\mu \in \mathcal{R}^4$, the coordinates with respect to an arbitary inertial frame with spacetime origin $O$. The vector $\boldsymbol{x}$ locates $P$ with reference to $O$ by $\boldsymbol{x} = \overrightarrow{OP} = x^\mu\boldsymbol{E}_\mu$.) With $\boldsymbol{a}$ constant,

$$\boldsymbol{\partial}\otimes\boldsymbol{A} = \boldsymbol{\partial}\psi\otimes\boldsymbol{a} = \frac{d\psi}{d\xi}\boldsymbol{\partial}\xi\otimes\boldsymbol{a} = \psi'\,\boldsymbol{k}\otimes\boldsymbol{a},$$

and then

$$\boldsymbol{\partial} \otimes \boldsymbol{\partial} \otimes \boldsymbol{A} = \psi'' \boldsymbol{k} \otimes \boldsymbol{k} \otimes \boldsymbol{a}.$$

Consequently, $\partial^2 \boldsymbol{A} = 0$ and $\boldsymbol{\partial} \cdot \boldsymbol{A} = 0$ if the constant vectors $\boldsymbol{a}$ and $\boldsymbol{k}$ satisfy $\boldsymbol{k} \cdot \boldsymbol{k} = 0$ and $\boldsymbol{a} \cdot \boldsymbol{k} = 0$. Hence the vector $\boldsymbol{k}$ is lightlike, and $\boldsymbol{a}$ lies in the lightlike hyperplane that $\boldsymbol{k}$ determines. The electromagnetic field dyadic is given by (5.29)

$$\vec{\vec{F}} = \boldsymbol{\partial} \wedge \boldsymbol{A} = \boldsymbol{k} \wedge \boldsymbol{a} \frac{d\psi}{d\xi}(\xi). \tag{5.87}$$

Maxwell's equations are satisfied in the form above with any differentiable function $\psi$, but such solutions are most used with sines and cosines. The importance of the sines and cosines, which give harmonic plane waves, arises partly from the elementary processes of quantum electrodynamics and partly from their use as orthogonal basis functions for the expansion of general solutions. When sines and cosines are used for $\psi$, and $\xi$ is decomposed with respect to a frame, $\boldsymbol{k} \cdot \boldsymbol{x} = k^m x^m - k^0 x^0$, we see that $k^0$ is the *angular* frequency of the wave, that is $k^0 = 2\pi f = 2\pi/T$. In the present notation, the momentum vector for the corresponding photons (denoted $\boldsymbol{k}$ in Chapter 3) is $hf(\boldsymbol{E}_0 + \boldsymbol{S}) = hf(\boldsymbol{k}/k^0)$ and so $hf = hk^0/(2\pi) \equiv \hbar k^0$. The two traditional meanings for the symbol $\boldsymbol{k}$ merge when units such that $\hbar = 1$ are used.

We obtain the relative electric and magnetic fields when (5.87) is decomposed with respect to an inertial basis. (Actually, according to (5.11) only the time basis vector $\boldsymbol{E}_0$ is really necessary to determine the geometical, but frame-dependent fields $\boldsymbol{E}$ and $\boldsymbol{B}$.) Given a lightlike, future-pointing $\boldsymbol{k}$ and a time basis vector $\boldsymbol{E}_0$, we can choose the spatial basis vectors so that $\boldsymbol{k} = k^0(\boldsymbol{E}_0 + \boldsymbol{E}_1)$. Then the most general possible $\boldsymbol{a}$ satisfying $\boldsymbol{a} \cdot \boldsymbol{k} = 0$ has the form

$$\boldsymbol{a} = a^0(\boldsymbol{E}_0 + \boldsymbol{E}_1) + a^2 \boldsymbol{E}_2 + a^3 \boldsymbol{E}_3. \tag{5.88}$$

Without changing the earlier choices of basis vectors $\boldsymbol{E}_0$ and $\boldsymbol{E}_1$, one can rotate the spatial basis around $\boldsymbol{E}_1$ so that in the new configuration we have

$$\boldsymbol{a} = (\frac{a^0}{k^0})\boldsymbol{k} + a\boldsymbol{E}_2, \qquad \boldsymbol{k} = k^0(\boldsymbol{E}_0 + \boldsymbol{E}_1). \tag{5.89}$$

with the component $a > 0$.

### 5.7.1 Plane Polarised Waves

Choosing a harmonic wave, and domesticating[2] the notation somewhat ($x^1 \mapsto x$, $x^0 \mapsto t$, $k^0 \mapsto \omega$), $\psi(\xi) = \sin \xi = \sin k^0(x^1 - x^0) = \sin \omega(x - t)$, so the electromagnetic field dyadic, for the case (5.89), becomes

[2] For ease of reading, and to help visualisation and the intuition, it is always convenient, when the general arguments involving the summation convention are finished, to revert to a more familiar notation.

$$\vec{F} = \boldsymbol{k} \wedge \boldsymbol{a} \cos\xi = \omega(\boldsymbol{E}_0 + \boldsymbol{E}_1) \wedge a\boldsymbol{E}_2 \cos\xi = \boldsymbol{E}_0 \wedge \boldsymbol{E} - (\boldsymbol{E}_0 \wedge \boldsymbol{B})^*. \quad (5.90)$$

The relative fields in this case, a plane polarised wave, are

$$\boldsymbol{E} = \omega a \cos\omega(x-t)\boldsymbol{E}_2, \qquad \boldsymbol{B} = \omega a \cos\omega(x-t)\boldsymbol{E}_3. \quad (5.91)$$

The relative orientations of the direction of propagation ($\boldsymbol{E}_1$), the direction of the electric field ($\boldsymbol{E}_2$) and the direction of the magnetic field ($\boldsymbol{E}_3$) are significant: they form a right-handed system in the order given.

### 5.7.2 Circularly Polarised Waves

By taking suitable superpositions of the harmonic, plane polarised waves, circularly polarised waves can be formed. For *left* circularly polarised waves (so-called "positive helicity"),

$$\vec{F} = \boldsymbol{k} \wedge a(\boldsymbol{E}_2 \cos\xi - \boldsymbol{E}_3 \sin\xi). \quad (5.92)$$

Here, the relative fields, on the spatial plane $x = 0$, are

$$\boldsymbol{E} = a\omega(\cos\omega t\boldsymbol{E}_2 + \sin\omega t\boldsymbol{E}_3), \quad \boldsymbol{B} = a\omega(-\sin\omega t\boldsymbol{E}_2 + \cos\omega t\boldsymbol{E}_3). \quad (5.93)$$

These vectors rotate anticlockwise about $\boldsymbol{E}_1$, the direction of propagation. The relative fields rotate with the opposite sense in the case of *right* circularly polarised waves (negative helicity)

$$\vec{F} = \boldsymbol{k} \wedge a(\boldsymbol{E}_2 \cos\xi + \boldsymbol{E}_3 \sin\xi). \quad (5.94)$$

### 5.7.3 Change of Basis: A Boost in the Direction of Propagation

It is of interest to compare the relative electric and magnetic fields with respect to other inertial frames. If we consider a frame $K'$ reached by a boost in precisely the $K$-direction of motion of the plane wave, then

$$\boldsymbol{E}_0' = \gamma(\boldsymbol{E}_0 + V\boldsymbol{E}_1), \quad \boldsymbol{E}_1' = \gamma(\boldsymbol{E}_1 + V\boldsymbol{E}_0), \quad \boldsymbol{E}_2' = \boldsymbol{E}_2, \quad \boldsymbol{E}_3' = \boldsymbol{E}_3,$$

and so

$$\boldsymbol{E}_0' + \boldsymbol{E}_1' = \sqrt{\frac{1+V}{1-V}}(\boldsymbol{E}_0 + \boldsymbol{E}_1). \quad (5.95)$$

A plane polarised wave of the form

$$\vec{F} = \boldsymbol{k} \wedge \boldsymbol{a} \cos\xi = k^0(\boldsymbol{E}_0 + \boldsymbol{E}_1) \wedge (a^2\boldsymbol{E}_2 + a^3\boldsymbol{E}_3)\cos\xi, \quad (5.96)$$

when expressed in terms of the $K'$-basis, takes the form

$$\vec{F} = k^0\sqrt{\frac{1-V}{1+V}}(\boldsymbol{E}_0' + \boldsymbol{E}_1') \wedge (a^2\boldsymbol{E}_2' + a^3\boldsymbol{E}_3')\cos\xi. \quad (5.97)$$

By comparing (5.12) and (5.15) one sees that the electric and magnetic fields with respect to $K'$ are just

$$\boldsymbol{E}' = \sqrt{\frac{1-V}{1+V}}\boldsymbol{E}, \qquad \boldsymbol{B}' = \sqrt{\frac{1-V}{1+V}}\boldsymbol{B}. \tag{5.98}$$

These equations relate the relative fields at a single point in spacetime. As the point changes we become interested in the different expressions for $\xi$ :

$$\begin{aligned} \xi = \boldsymbol{k}\cdot\boldsymbol{x} = \eta_{\mu\nu}k^\mu x^\nu = k^0(x^1 - x^0) &= \eta_{\mu\nu}k'^\mu x'^\nu \\ &= k^0\sqrt{\frac{1-V}{1+V}}(x'^1 - x'^0). \end{aligned} \tag{5.99}$$

The $K'$-frequency is reduced by a factor $\sqrt{(1-V)/(1+V)}$ compared to the $K$-frequency.

### 5.7.4 Change to a General Moving Frame

We now consider the relation with a more general frame $K'$, travelling at relative velocity $\boldsymbol{V}$ with respect to $K$. The spatial directions of $K'$ are not general, however. They will be chosen for convenience in describing the plane wave. The $\boldsymbol{E}_1$ spatial basis vector for $K$ is again chosen so $\boldsymbol{k} = k^0(\boldsymbol{E}_0 + \boldsymbol{E}_1)$, and by spatially rotating around it we choose $\boldsymbol{E}_2$ so $\boldsymbol{V} = V\cos\vartheta\boldsymbol{E}_1 + V\sin\vartheta\boldsymbol{E}_2$ for some angle $\vartheta$. Then the $K'$-time basis vector is

$$\boldsymbol{E}'_0 = \gamma(\boldsymbol{E}_0 + V\cos\vartheta\boldsymbol{E}_1 + V\sin\vartheta\boldsymbol{E}_2). \tag{5.100}$$

The $K'$-basis vector $\boldsymbol{E}'_1$ is conveniently chosen (so far only $\boldsymbol{E}'_0$ has been fixed, the two frames are not connected by a standard boost and no restrictions on the spatial basis vectors have been made) so that $\boldsymbol{k} = k'^0(\boldsymbol{E}'_0 + \boldsymbol{E}'_1)$. Hence, solving for $\boldsymbol{E}'_1$,

$$\boldsymbol{E}'_1 = -\gamma(\boldsymbol{E}_0 + V\cos\vartheta\boldsymbol{E}_1 + V\sin\vartheta\boldsymbol{E}_2) + \frac{\boldsymbol{E}_0 + \boldsymbol{E}_1}{\gamma(1 - V\cos\vartheta)}, \tag{5.101}$$

or, more usefully, decomposing this vector into $K'$-spatial unit vectors,

$$\begin{aligned} \boldsymbol{E}'_1 &= \frac{\cos\vartheta - V}{1 - V\cos\vartheta}\gamma(V\boldsymbol{E}_0 + \cos\vartheta\boldsymbol{E}_1 + \sin\vartheta\boldsymbol{E}_2) \\ &\quad + \frac{\sin\vartheta}{\gamma(1 - V\cos\vartheta)}(\sin\vartheta\boldsymbol{E}_1 - \cos\vartheta\boldsymbol{E}_2). \end{aligned}$$

The $K$-basis vector $\boldsymbol{E}_3$ forms, with the two orthogonal unit vectors on the right, an orthogonal basis for $K'$-space. By a rotation around $\boldsymbol{E}'_1$ we can arrange that $\boldsymbol{E}'_3 = \boldsymbol{E}_3$. The last basis vector $\boldsymbol{E}'_2$ is now fixed. To get an expression for it expanded in the basis $\boldsymbol{E}_\mu$, recall that $\boldsymbol{E}'_2 \cdot \boldsymbol{k} = \boldsymbol{E}'_2 \cdot \boldsymbol{E}_3 = 0$; hence, $\boldsymbol{E}'_2 = \alpha(\boldsymbol{E}_0 + \boldsymbol{E}_1) + \beta \boldsymbol{E}_2$. Using $\boldsymbol{E}'_2 \cdot \boldsymbol{E}'_0 = 0$, and $\boldsymbol{E}'_2 \cdot \boldsymbol{E}'_2 = 1$, we find

$$\boldsymbol{E}'_2 = \frac{V \sin\vartheta}{1 - V\cos\vartheta}(\boldsymbol{E}_0 + \boldsymbol{E}_1) + \boldsymbol{E}_2 \tag{5.102}$$

(the orientation $\beta > 0$ comes from the $V = 0$ case). If the plane polarised wave (5.96) is expressed in the $K'$-basis it becomes

$$\vec{\vec{F}} = k^0\gamma(1 - V\cos\vartheta)(\boldsymbol{E}'_0 + \boldsymbol{E}'_1) \wedge (a^2\boldsymbol{E}'_2 + a^3\boldsymbol{E}'_3)\cos\xi. \tag{5.103}$$

## 5.8 Problems

1. Consider uniform current flowing in an infinitely long straight wire (along the $x$-axis of the laboratory, frame $K$). Model this situation as a continuous, linear, positive charge density of ions at rest in the laboratory, together with a continuous, linear, negative charge density of electrons at constant drift velocity $V$ down the wire, which neutralises (in the lab) the charge density of the ions. Write appropriate four-current densities for the two components and for the sum. Decompose the components (and the sum) in the rest frame of the drifting electrons (frame $K'$) and offer your reflections on the nonzero charge density.
Now suppose that the return wire, carrying electrons at drift speed $V$ in the negative $x$-direction, also lies along the $x$-axis. Write the four-current densities for the two components of charge in the return wire. Decompose these components in the frame $K'$, and combine them finally with those of the original wire.

2. $K$ and $K'$ are the standard boost-related frames, with common spacetime origin $O$. A ball of charge is at rest with respect to $K'$, and centred on its spatial origin; the $K'$-radius is $R$ and the uniform charge density (with respect to $K'$) is $\rho_0$. Find the electric and magnetic fields with respect to both $K$ and $K'$ at the events $P$ and $Q$ defined by their $K'$-coordinates: $P(t' = 0, x' = R, y' = 0, z' = 0)$, $Q(0, 0, R, 0)$.

3. If $E^n = E\delta^{n1}$, $B^n = B\delta^{n3}$ are the components of the electric and magnetic fields with respect to a frame $K$, where $E > B > 0$, find basis vectors for another inertial frame in which the magnetic field vanishes. What is the strength of the electric field in the second frame?

4. Calculate $\epsilon^{\mu\nu\lambda\rho}F_{\mu\nu}F_{\lambda\rho}$ in terms of the electric and magnetic fields in the implied frame. Show that if the electric field is (Euclidean-)orthogonal to the magnetic field in one frame, the same is true in all frames.

5. The standard boost-related frames are $K$ and $K'$. Obtain the relations between the components $E^m$, $B^n$ and $E'^m$, $B'^n$ of the relative electric and magnetic fields for the two frames in two ways: first, by replacing $\boldsymbol{E}_\mu$ by $a^\nu{}_\mu \boldsymbol{E}'_\nu$ in

$$\begin{aligned}\vec{\vec{F}} &= E^1(\boldsymbol{E}_0 \wedge \boldsymbol{E}_1) + \dots + B^1(\boldsymbol{E}_2 \wedge \boldsymbol{E}_3) + \dots \\ &= E'^1(\boldsymbol{E}'_0 \wedge \boldsymbol{E}'_1) + \dots + B'^1(\boldsymbol{E}'_2 \wedge \boldsymbol{E}'_3) + \dots\end{aligned}$$

and reading off $E'^m$, $B'^n$; second, by using the matrix transformation law (5.19) for the components of $\vec{\vec{F}}$, $F' = aFa^T$.

6. When the electromagnetic field tensor is decomposed in two ways with two different unit timelike vectors $\boldsymbol{v}$ and $\boldsymbol{w}$ there are two sets of relative electric and magnetic fields satisfying

$$\vec{\vec{F}} = \boldsymbol{v} \wedge \boldsymbol{E} - (\boldsymbol{v} \wedge \boldsymbol{B})^* = \boldsymbol{w} \wedge \boldsymbol{E}_w - (\boldsymbol{w} \wedge \boldsymbol{B}_w)^*.$$

Introduce relative velocities by $\boldsymbol{v} = \gamma(\boldsymbol{w} + \boldsymbol{V}) = \gamma(\boldsymbol{w} + V\boldsymbol{N})$ and $\boldsymbol{w} = \gamma(\boldsymbol{v} + \boldsymbol{U}) = \gamma(\boldsymbol{v} + U\boldsymbol{M})$, where $\boldsymbol{V} \cdot \boldsymbol{w} = 0 = \boldsymbol{U} \cdot \boldsymbol{v}$ and $\boldsymbol{N}$, $\boldsymbol{M}$ are unit vectors. Use the latter to decompose

$$\boldsymbol{B} = \boldsymbol{M}\boldsymbol{M} \cdot \boldsymbol{B} + \boldsymbol{B}_\perp,$$

$$\boldsymbol{E} = \boldsymbol{M}\boldsymbol{M} \cdot \boldsymbol{E} + \boldsymbol{E}_\perp.$$

Verify the transformation formulae

$$\boldsymbol{B}_w = \gamma \boldsymbol{B}_\perp - \boldsymbol{N}(\boldsymbol{M} \cdot \boldsymbol{B}) + \gamma \boldsymbol{V} \times \boldsymbol{E}_\perp,$$

$$\boldsymbol{E}_w = \gamma \boldsymbol{E}_\perp - \boldsymbol{N}(\boldsymbol{M} \cdot \boldsymbol{E}) - \gamma \boldsymbol{V} \times \boldsymbol{B}_\perp.$$

7. Find the components of $\boldsymbol{\partial} \wedge \boldsymbol{A}$ and $(\boldsymbol{\partial} \wedge \boldsymbol{A})^*$ with respect to an inertial basis as derivatives of the components of $\boldsymbol{A}$. In the explicit form, verify $\boldsymbol{\partial} \cdot (\boldsymbol{\partial} \wedge \boldsymbol{A})^* = \boldsymbol{0}$. Verify too that when components with respect to a frame $K$ are used, the $K$-electric and magnetic fields are expressed in terms of the spatial components $A^m$ and $A^0$ as by the three-dimensional vector potential and scalar potential $\phi$ in $K$.

8. Use the relation

$$\partial_\mu F_{\lambda\rho} + \partial_\rho F_{\mu\lambda} + \partial_\lambda F_{\rho\mu} = 0$$

to show that the derivatives of

$$\psi_\nu(x^0, x^1, x^2, x^3) = x^\mu \int_0^1 t F_{\mu\nu}(tx^0, tx^1, tx^2, tx^3)\, dt$$

satisfy

$$F_{\lambda\rho} = \partial_\lambda \psi_\rho - \partial_\rho \psi_\lambda.$$

9. Two frames $K$ and $K'$ are related by a standard boost, as is a third frame $K''$, "parallel" to the previous and with relative velocity $U\boldsymbol{E}_1$ with respect to $K$, $W\boldsymbol{E}'_1$ with respect to $K'$. In $K''$ there is a constant and uniform charge density $\rho_{(0)}$ at rest in the frame. Verify that the charge density in $K$ is $\rho = \rho_{(0)}\gamma_U$ and that the $\boldsymbol{E}_1$-component of the current density is $j^1 = \rho_{(0)}U\gamma_U$. Write down the similar expressions for $\rho'$, $j'^1$ in $K'$ and confirm that the $K$- and $K'$-components are related by the appropriate instance of $j'^\mu = a^\mu{}_\nu j^\nu$.

10. In a region of spacetime two streams of *freely moving* charged particles are superposed (imagine that the two streams do not interfere with each other). The (constant) unit spacetime velocity vectors for the two streams are $\boldsymbol{v}_1$ and $\boldsymbol{v}_2$, respectively, with $\gamma_{12} = -\boldsymbol{v}_1 \cdot \boldsymbol{v}_2$. The number density for the particles in each stream in its own rest frame is a constant $n_{(0)}$.
    If the particles in both streams each have charge $+q$, find the charge density in an inertial frame in which the flux of charge vanishes.
    Alternatively, if the particles of one stream have charge $+q$ and those of the other $-q$, find the flux in an inertial frame in which the charge density vanishes.

11. In the relative version of electromagnetism the charge density $\rho$ and three-current $\mathbf{J}$ satisfy the conservation equation $\frac{\partial \rho}{\partial t} + \nabla \cdot \mathbf{J} = 0$. This equation is identically satisfied by the static sources for polarisable media in the form $\rho = -\nabla \cdot \mathbf{P}$, $\mathbf{J} = \frac{\partial \mathbf{P}}{\partial \mathbf{t}}$, and for the sources of permeable media in the form $\rho = 0$, $\mathbf{J} = \nabla \times \mathbf{M}$. Suppose the implied inertial frame here is $K$. Show that the charge densities and three-currents for static media arise as a special case of an (identically conserved) spacetime current of the form
$$\boldsymbol{j} = \boldsymbol{\partial} \cdot \overset{\Rightarrow}{S}, \quad \overset{\Rightarrow}{S} = -\overset{\Rightarrow}{S}{}^T.$$
    Finally, find the charge density and three-current in the (usual) boosted frame $K'$.

12. Let $\boldsymbol{v}(P)$ be an arbitrary, timelike future-pointing unit vector *field*. Then $\boldsymbol{j} = \rho \boldsymbol{v}$ is a conserved current provided $\rho$ satisfies $\boldsymbol{v} \cdot \boldsymbol{\partial}\rho + \rho(\boldsymbol{\partial} \cdot \boldsymbol{v}) = 0(*)$. Let $\sigma(\tau)$ be an integral curve of the field $\boldsymbol{v}$, that is, a curve $\sigma(\tau) = P(\tau)$ located by the vector $\boldsymbol{\sigma} = \overrightarrow{OP}$ which satisfies $\frac{d\boldsymbol{\sigma}}{d\tau}(\tau) = \boldsymbol{v}(\sigma(\tau))$. Equation $(*)$ can be solved along the curve: show that the solution is
$$\rho(\sigma(\tau)) = \rho(\sigma(\tau_0)) \exp[-\int_{\tau_0}^{\tau} (\boldsymbol{\partial} \cdot \boldsymbol{v})\, d\tau].$$

13. Given a uniform distribution of freely moving particles, write the number current density in terms of the spacetime velocity and the rest frame number density. Decompose the current with respect to frame $K$ to get

the number density for $K$ and the three-current for $K$. Specialise to the case in which all particles move at speed $U$ in the direction of $\cos\theta \boldsymbol{E}_1 + \sin\theta \boldsymbol{E}_2$. If $K'$ is the usual boosted frame, show that the relation between the number densities in $K$ and $K'$ is $n' = n\gamma(1 - UV\cos\theta)$.

14. Let $K$ and $K'$ be the usual boost-related frames with coincident spacetime origins and $V\boldsymbol{E}_1$ be the relative velocity of $K'$ with respect to $K$. A particle of charge $q$ is fixed at the spatial origin of $K'$. Write down the $K'$-components of the $K'$-electric and magnetic fields at an event $P$ with $K'$-coordinates $(t', x', y', z')$. Then use the absolute character of the electromagnetic field dyadic $\overset{\Rightarrow}{F}$ to obtain the $K$-components of the $K$ electric and magnetic fields at "observation events" $Q(t)$ with $K$-coordinates $(t, 0, a, 0)$.
    Verify your results using the general formula (5.68) for the field dyadic of a point charge.

15. $K$ and $K'$ are the usual boost-related frames, for which the relative velocity of the spatial origin of $K'$ with respect to $K$ is $V > 0$ in the $x$-direction, spacetime origins coinciding. In $K'$ a total current $I$ circulates anticlockwise (to a viewer on the positive $z'$-axis) uniformly on the circle $x'^2 + y'^2 = R^2$. Find the electric and magnetic fields in $K'$ at events with $K'$-coordinates $(t', 0, 0, d)$. Hence find an expression for the electromagnetic field dyadic at these events, and from it the electric and magnetic fields in $K$ at event $P$ with $K$-coordinates $(0, 0, 0, d)$. Find an approximation, to first order in $V$, for the charge density in $K$ at $t = 0$ and show that the instantaneous Coulomb electric field it creates at $P$ is half the total electric field there.

16. Let point $O$ be the spacetime origin of an inertial frame $K$ with orthonormal basis $\boldsymbol{E}_\mu$. A straight timelike worldline through $O$ and parallel to the constant, future-pointing timelike unit vector $\boldsymbol{v} = v^\mu \boldsymbol{E}_\mu$ is the centre of a charge distribution in its rest frame.
    Define simple vector fields $\boldsymbol{x}(P) \equiv \overrightarrow{OP} = x^\mu \boldsymbol{E}_\mu$, and $\boldsymbol{\xi}(P) \equiv \boldsymbol{x} + \boldsymbol{v}(\boldsymbol{v}\cdot\boldsymbol{x}) = \xi^\mu \boldsymbol{E}_\mu$. Call $\xi^2 = \boldsymbol{\xi}\cdot\boldsymbol{\xi}$. Verify the following relations, which will be of help in further calculations: $\partial^\mu x^\nu = \eta^{\mu\nu}$, $\xi^\mu v_\mu = 0$, $\partial_\mu \xi^\nu = \delta^\nu_\mu + v^\nu v_\mu$, $\partial_\mu \xi^2 = 2\xi_\mu$, $\boldsymbol{\partial}\cdot\boldsymbol{\xi} = 3 = \partial_\mu \xi^\mu$, $\boldsymbol{v}\cdot\boldsymbol{\partial}\boldsymbol{\xi} = \boldsymbol{0} = (v^\mu \partial_\mu \xi^\nu)\boldsymbol{E}_\nu$.
    If the electromagnetic vector potential is given by $\boldsymbol{A} = q\boldsymbol{v} f(\xi^2)$, where $f \sim (\xi^2)^{-\frac{1}{2}}$ as $\xi^2 \to \infty$, and the field tensor is given by $F_{\mu\nu} = \partial_\mu A_\nu - \partial_\nu A_\mu$, show that Maxwell's equations are satisfied with the current $\boldsymbol{j} = -(q\boldsymbol{v}/2\pi)[2\xi^2 f'' + 3f']$.
    [This question is continued in Problem 3 of Chapter 6.]

17. Use the frames $K$ and $K'$ related by the standard boost. Suppose that the electric and magnetic fields relative to $K$ are those of a plane wave:

$$\boldsymbol{E} = \boldsymbol{E}_2 A \cos\omega(t - x), \qquad \boldsymbol{B} = \boldsymbol{E}_3 A \cos\omega(t - x),$$

where $A$ and $\omega$ are constants.

Find the electromagnetic field tensor $\overset{\Rightarrow}{F}$, and from it find expressions for the electric and magnetic fields relative to $K'$.

Show that

$$\overset{\Rightarrow}{F} = \frac{A}{\omega}\boldsymbol{E}_2 \wedge \boldsymbol{\partial} \sin\omega(t - x),$$

and hence check that the source-free Maxwell's equations are satisfied.

18. Consider a uniform current in a plane, circular loop of radius $r$. Suppose the loop is in the $xy$-plane of the laboratory, with centre at the origin, and one is interested in the electromagnetic field at an event $P$ on the $z$-axis with $z = b$ at some time $t$. Model the current as follows: a large number $N$ of positive ions uniformly distributed round the loop, and a compensatingly large number of negative electrons uniformly distributed and drifting with constant speed $V$ round the loop. Calculate $\overset{\Rightarrow}{F}$ by using

$$\overset{\Rightarrow}{F}(P) = \frac{q}{\rho^3}\boldsymbol{v} \wedge \boldsymbol{R} + \frac{q}{\rho^2}(\boldsymbol{a} + \frac{\boldsymbol{a}\cdot\boldsymbol{R}}{\rho}\boldsymbol{v}) \wedge \boldsymbol{R}$$

for each particle separately and integrating the result in a continuum approximation. It is interesting to see how the total electric field vanishes in the laboratory, leaving only the magnetic field of a current loop.

19. Two charges $q_1$ and $q_2$ are made to rotate at opposite ends of a diameter of a circle of radius $b$ in the $xy$-plane of the laboratory. The trajectory $P_1(t)$ of $q_1$, parameterised by laboratory time $t$, is given by

$$\overrightarrow{OP_1}(t) = \boldsymbol{z}(t) = t\boldsymbol{E}_0 + b(\boldsymbol{E}_1 \cos\omega t + \boldsymbol{E}_2 \sin\omega t),$$

with constant lab angular velocity $\omega > 0$. The trajectory $P_2(t)$ of $q_2$ is given by the same formula with the substitution $b \mapsto -b$.

If we are interested in the force on $q_1$ at $P_1(t)$ due to $q_2$, then to calculate the relevant fields we need the appropriate retarded position $P_2'(t)$ of $q_2$. Show that $P_2'(t)$ lags behind $P_2(t)$ by an angle $2\phi$ (subtended at the centre of the circle) determined by $\phi = \omega b \cos\phi$. Find the electric and magnetic fields in the laboratory, $\boldsymbol{E}_L$ and $\boldsymbol{B}_L$, at the event $P_1(t)$. The electric field has no component perpendicular to the circle. Its component in the outward radial direction may be written

$$\frac{q_2}{4b^2}\left\{\frac{(\phi\sin 2\phi + \cos^2\phi)(1 + \phi^2\cos 2\phi\sec^2\phi) - 2\phi^2(1 + \phi\tan\phi)\cos 2\phi}{(\cos\phi + \phi\sin\phi)^3}\right\},$$

and the component in the tangential direction of motion is

$$\frac{q_2}{4b^2}\left\{\frac{(\phi\cos 2\phi - \sin\phi\cos\phi)(1 + \phi^2\cos 2\phi\sec^2\phi) + 2\phi^2(1 + \phi\tan\phi)\sin 2\phi}{(\cos\phi + \phi\sin\phi)^3}\right\}.$$

The result is quoted this way in a paper of Rivera and Villarroel.

## References

I.M. Gel'fand and G.E. Shilov, Generalized Functions, Volume 1 (Academic, 1964)

J.D. Jackson, Classical Electrodynamics, Second Edition (Wiley, 1975)

H.A. Lorentz, A. Einstein, H. Minkowski, H. Weyl, The Principle of Relativity (Dover Publications reprint, 1952, of Methuen and Co., 1923)

R. Rivera and D. Villarroel, An exactly solvable two-body problem with retarded interactions and radiation reaction in classical electrodynamics, J. Math. Phys. **38** (1997) 5690-5706

F. Rohrlich, Classical Charged Particles (Addison-Wesley, 1965)

# 6. The Energy Tensor

*Flow of Momentum and Local Conservation*

The energy tensor, or, called by its longer name, the stress-energy-momentum tensor, is one of the key concepts in theoretical physics.

- The energy tensor $\vec{\vec{T}}$ is a geometrical dyadic in a physical theory which describes the densities and flows of four-momentum (hence of energy and three-momentum) for a continuum and, by using delta-functions, for assemblies of point particles as well.
- The vanishing of its divergence is the local expression of four-momentum conservation. Using the symmetry of the energy tensor one has also conservation of angular momentum.
- It generates equations of motion (to satisfy momentum conservation).
- In general relativity the energy tensor is the source of the gravitational curvature of spacetime.

The energy tensor functions, with respect to momentum, just like the current vector with respect to electric charge. Hence, for $\boldsymbol{E}_0$ a future-pointing timelike unit vector, $-\boldsymbol{E}_0 \cdot \vec{\vec{T}}$ is the density, with respect to any frame whose time basis vector is $\boldsymbol{E}_0$, of absolute four-momentum; and if $\boldsymbol{S}$ is a spacelike unit vector orthogonal to $\boldsymbol{E}_0$, $\boldsymbol{S} \cdot \vec{\vec{T}}$ is the net flux of four-momentum, per unit area, per unit time, with respect to the same frame, through a surface with normal $\boldsymbol{S}$.

Conservation of momentum is very like conservation of charge. Conservation of charge means much more than just the existence of a total charge in the universe, the same for all observers and for each observer at all times, though that is true. It means also that in any given spatial three-volume, a discrepancy in the charge content at two different times must be accountable by a flux of charge over the surface of the volume during the time interval between the times. This means there can be no creation or destruction of charge in the arbitrary volume. In special relativity, energy-momentum behaves in an analogous way, but the conserved quantity is now a vector. The differential expression of charge conservation is $\boldsymbol{\partial} \cdot \boldsymbol{j} = 0$. Correspondingly, the differential expression of conservation of four-momentum is $\boldsymbol{\partial} \cdot \vec{\vec{T}} = \boldsymbol{0}$.

The conservation of charge may be regarded as a consistency requirement for the Maxwell equation, $\boldsymbol{\partial} \cdot \overleftrightarrow{F} = -4\pi \boldsymbol{j}$, with the antisymmetric field dyadic $\overleftrightarrow{F}$. One may regard the conservation of energy-momentum in a similar way as a consistency requirement for Einstein's equation $\overleftrightarrow{G} = 8\pi\overleftrightarrow{T}$ in general relativity (although this analogy involves stepping outside the domain of special relativity).

## 6.1 The Energy Tensor for Dust

We consider again a dense cloud of dust particles each of mass $m$, but treated as a continuum for the sake of mathematical simplicity. We suppose that the particles move like points in a continuum, yet without dynamical contact, because the conditions for motion are continuous. There are no collisions or heat vibrations. There are no direct inter-particle interactions, but if the particles were charged they could interact ***indirectly*** through the electromagnetic field that they create. The state of the dust cloud is described by a smoothly averaged scalar number density $n_{(0)}(P)$ in the local rest frame, and by a spacetime velocity field $\boldsymbol{v}(P)$ which smoothly interpolates between the velocity vectors of the particles.

For the dust everything is simply visualisable. We set down what proves to be the energy tensor,

$$\overleftrightarrow{K}(P) = mn_{(0)}\boldsymbol{v} \otimes \boldsymbol{v}, \tag{6.1}$$

and examine its properties. (The symbol $\overleftrightarrow{K}$ is a conventional notation for the energy tensor of ***particles***.)

- $-\overleftrightarrow{K} \cdot \boldsymbol{v} = (m\boldsymbol{v})n_{(0)}$ is the density of four-momentum in a local rest frame since $\boldsymbol{v} = \tilde{\boldsymbol{E}}_0$ is also the time basis vector for any local rest frame. (In the electromagnetic analogy, with current $\boldsymbol{j} = qn_{(0)}\boldsymbol{v}$,this density corresponds to the charge density $qn_{(0)} = \rho_{(0)} = -\boldsymbol{v} \cdot \boldsymbol{j}$ in the local rest frame.)
- $-\overleftrightarrow{K} \cdot \boldsymbol{E}_0 = (m\boldsymbol{v})\gamma n_{(0)}$ (where $\gamma = -\boldsymbol{v} \cdot \boldsymbol{E}_0$ appears in the decompositon $\boldsymbol{v} = \gamma(\boldsymbol{E}_0 + \boldsymbol{V})$ for the frame $K$ implied by $\boldsymbol{E}_0$) is the $K$-density of four-momentum, since $n = \gamma n_{(0)}$ is the relation between the number density in $K$ and the number density in the local rest frame. (This density corresponds to the charge density $\rho = qn = q\gamma n_{(0)}$ for the frame $K$.)
- $+\overleftrightarrow{K} \cdot \boldsymbol{S} = (m\boldsymbol{v})\gamma n_{(0)}(\boldsymbol{V} \cdot \boldsymbol{S})$ is the flux density in $K$, for a surface with normal in the direction of the unit spatial vector $\boldsymbol{S}$, of four-momentum. Multiplied by an area and a time interval it gives the net momentum of the particles crossing the area in the time. (This density corresponds to the component $\boldsymbol{J} \cdot \boldsymbol{S}$ of the electric three-current in $K$.)

Each of the densities above is a vector, of the form $\overleftrightarrow{K} \cdot \boldsymbol{E}_\mu$. They arise when one wants to analyse the flows of ***absolute momentum*** with respect to a particular frame. If *relative* quantities with respect to the frame are being considered, more detailed numerical components will arise:

- $+\boldsymbol{E}_0 \cdot \overleftrightarrow{K} \cdot \boldsymbol{E}_0 = (m\gamma)\gamma n_{(0)} = (m\gamma)n$ is the $K$-density of $K$-energy (total energy $m\gamma$ per particle).
- $-\boldsymbol{E}_j \cdot \overleftrightarrow{K} \cdot \boldsymbol{E}_0 = (m\gamma V^j)n$ is the $K$-density of the $j$-component of three-momentum in $K$.
- $-\boldsymbol{E}_0 \cdot \overleftrightarrow{K} \cdot \boldsymbol{E}_j = (m\gamma)nV^j$ is the flux density in $K$, for the direction of $\boldsymbol{S} = \boldsymbol{E}_j$, of the $K$-energy. Note that this is exactly the same as the previous component, by the symmetry of the energy dyadic. With all quantities referred to $K$, the density of the $j$-component of momentum equals the $j$-flux of energy.
- $+\boldsymbol{E}_j \cdot \overleftrightarrow{K} \cdot \boldsymbol{E}_k = (m\gamma V^j)nV^k$ is the flux density in the $k$-direction of the $j$-component of momentum, and again there is a symmetry under interchange of $j$ and $k$.

## 6.2 The Energy Tensor in General

We make the fundamental assumption that for every physical system, not just dust, there is a symmetric dyadic with properties analogous to those of $\overleftrightarrow{K}$ for dust. The energy tensor for any physical system is a symmetric[1] dyadic $\overleftrightarrow{T}$, whose components with respect to the basis vectors of an inertial frame $K$ have the following interpretations.

$$-\boldsymbol{E}_0 \cdot \overleftrightarrow{T} = -\overleftrightarrow{T} \cdot \boldsymbol{E}_0 = K\text{-density of absolute four-momentum}$$
$$\text{(absolute four-momentum per unit } K\text{-spatial volume)} \tag{6.2}$$

$$\begin{aligned} +\boldsymbol{E}_i \cdot \overleftrightarrow{T} &= +\overleftrightarrow{T} \cdot \boldsymbol{E}_i \\ &= K\text{-flux of absolute four-momentum in } \boldsymbol{E}_i\text{-direction} \end{aligned}$$
$$\text{(net four-momentum crossing unit } K\text{-area in normal}$$
$$\boldsymbol{E}_i\text{-direction per unit } K\text{-time)} \tag{6.3}$$

We do not expect to be able to make general statements about densities in a local rest frame, because not all physical systems have such rest frames (a plane electromagnetic wave, moving at the speed of light, being an important example).

[1] There is, however, a small but longstanding and continuing research area devoted to nonsymmetric energy tensors.

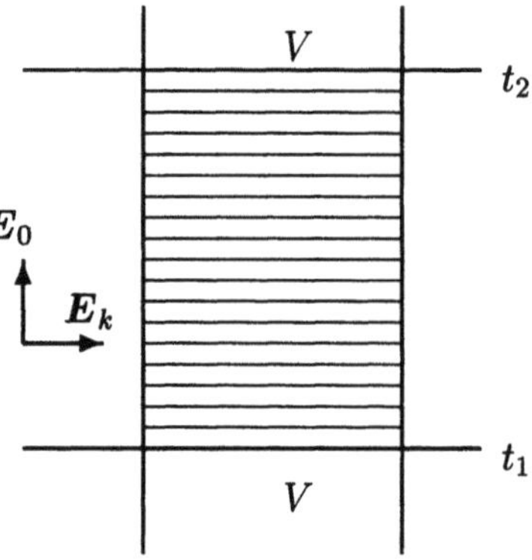


**Figure 6.1**

### 6.2.1 Conservation of Four-momentum

The physical interpretations of eqs (6.2) and (6.3) are exactly what is required to give meaning to the integral of $\boldsymbol{\partial} \cdot \vec{\vec{T}}$ over a three-dimensional volume $V$ for a time interval $[t_1, t_2]$ (with respect to the particular $K$) (Figure 6.1)

$$\begin{aligned}\int \boldsymbol{\partial} \cdot \vec{\vec{T}}\, d^4x &= \int_{t_1}^{t_2} \int_V [-\partial_0(\boldsymbol{E}_0 \cdot \vec{\vec{T}}) + \partial_m(\boldsymbol{E}_m \cdot \vec{\vec{T}})] d^3x\, dt \\ &= \int_V [-\boldsymbol{E}_0 \cdot \vec{\vec{T}}]_{t_1}^{t_2}\, d^3x + \int_{t_1}^{t_2} \int_S \boldsymbol{S} \cdot \vec{\vec{T}} dS dt,\end{aligned}$$

where $\boldsymbol{S}$ is the unit spatial outward normal to the two-dimensional boundary of $V$. Conservation of four-momentum in the volume $V$ over the time interval $[t_1, t_2]$ may be expressed as

$$\int_V [-\boldsymbol{E}_0 \cdot \vec{\vec{T}}](t_1)\, d^3x = \int_V [-\boldsymbol{E}_0 \cdot \vec{\vec{T}}](t_2)\, d^3x + \int_{t_1}^{t_2} \int_S \boldsymbol{S} \cdot \vec{\vec{T}} dS dt. \qquad (6.4)$$

Figure 6.2 shows the total four-momentum in the volume $V$ at time $t_1$. It equals the four-momentum in $V$ at time $t_2$ plus the net momentum which outflows over the boundary $S$ during the time interval $[t_1, t_2]$. By the previous integral formula this relation will be true for all $V$, and all time intervals, and all frames $K$, if and only if

$$\boldsymbol{\partial} \cdot \vec{\vec{T}} = \boldsymbol{0}. \qquad (6.5)$$

Equation (6.5) is the general differential expression of total four-momentum conservation. It has the same form as charge conservation, except that the

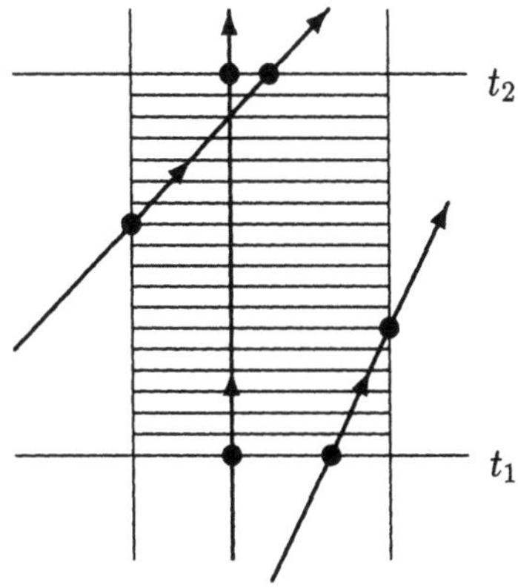


**Figure 6.2**

conserved quantity is a vector rather than a scalar. According to special relativity, only physical theories which possess an energy tensor satisfying (6.5) have any claim to credibility.

If (6.4) is always true, and $V$ is taken to be so large that no momentum crosses its boundary (assuming that $\vec{\vec{T}}$ vanishes sufficiently fast in spacelike directions to make this possible), then the total four-momentum, as measured by $K$,

$$\boldsymbol{p} = -\int \boldsymbol{E}_0 \cdot \vec{\vec{T}} dV \tag{6.6}$$

(an integration over the whole of $K$-space is implied), is independent of the time at which it is measured. It is a constant vector. If the total momentum is decomposed with respect to the $K$-basis,

$$\boldsymbol{p} \equiv p^0 \boldsymbol{E}_0 + p^m \boldsymbol{E}_m \equiv E\boldsymbol{E}_0 + \boldsymbol{P}, \qquad \boldsymbol{E}_0 \cdot \boldsymbol{P} = 0, \tag{6.7}$$

then the total $K$-energy is $E = p^0$ and the total three-momentum in $K$ is $\boldsymbol{P} = p^m \boldsymbol{E}_m$; both of these relative quantities are constant.

In fact the constant vector four-momentum is the same for all inertial observers. Assuming that $\vec{\vec{T}}$ vanishes in distant spatial directions, the integral

$$\int \chi \boldsymbol{\partial} \cdot \vec{\vec{T}} \, d^4x = 0$$

may be integrated simply by parts. Choosing $\chi = \theta(-x^0 + t_0) - \theta(-x'^0 + t'_0)$, where $x^0$ and $x'^0$ are time coordinates for two *different* inertial frames (Figure 6.3), one obtains, by using in the integral different inertial coordinates for the two terms,

$$\int_\sigma d^3x(-\boldsymbol{E}_0 \cdot \vec{\vec{T}}) = \int_{\sigma'} d^3x'(-\boldsymbol{E}'_0 \cdot \vec{\vec{T}}),$$

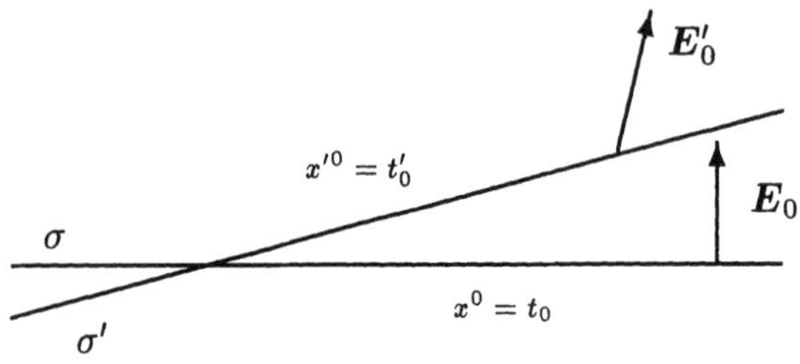


**Figure 6.3**

hence for *any* spacelike hyperplane $\sigma$ with its unit, future-pointing timelike normal $\boldsymbol{v}$,

$$\boldsymbol{p}(\sigma) = -\int_\sigma \boldsymbol{v} \cdot \vec{\vec{T}}\, dV \tag{6.8}$$

is constant and independent of the hyperplane, $\boldsymbol{p}(\sigma_1) = \boldsymbol{p}(\sigma_2) = \boldsymbol{p}$. The total, absolute momentum as measured by $K$, at any $K$-time, is the same as the total, absolute momentum as measured by $K'$, at any $K'$-time. (Of course, the *components* with respect to the two frames will be different. The different components correspond to the different splitting of the total momentum into energy and three-momentum for the different frames.) The final formula expresses the existence of a total momentum, the same for all inertial frames and at all times.

## 6.3 The Variational Principle

It appears that the most important physical theories can be expressed in terms of variational principles,

$$\delta I = 0, \tag{6.9}$$

where, in the relativistic case, the action $I$ is the integral of a scalar (invariant) Lagrangian density $\mathcal{L}$,

$$I = \int \mathcal{L}\, d^4x. \tag{6.10}$$

This action is a generalisation to field theories of the time integral of the nonrelativistic particle Lagrangian,

$$\int (T - U)\, dt = \int (\frac{1}{2} mV^2 - U)\, dt, \tag{6.11}$$

from which Newton's equations follow by Hamilton's Principle. For many physical theories, for example, electromagnetism, the equations of motion were known before a search was made for a Lagrangian. The Lagrangian does indeed generate the equations of motion, but its most important aspect, in the present context, is that for an invariant action it is possible to extract an expression for the energy tensor.

The integral (6.10) is coordinate-independent. It is true that the expression employs an unspecified system of inertial coordinates, but that is only to make it more homely. It would take exactly the same form in any other system of inertial coordinates, and it could be written in entirely geometrical form by using (4.105).

On the other hand, (6.11) is not frame-independent; it is written with respect to a particular inertial frame. The integral is over the time in the particular frame, and the integrand uses the kinetic and potential energies for the particular frame. In the case $U = 0$, it is easy to find an invariant action which reduces to (6.11) in the nonrelativistic limit. If the path of the particle, parameterised by proper time, is given by $z^\mu(\tau)$, and $\mathcal{L}$ is chosen to be

$$\mathcal{L} = -\int m\delta^{(4)}(x - z(\tau))\, d\tau, \tag{6.12}$$

then the action is

$$\begin{aligned} I_{\text{particle}} &= \int \mathcal{L}\, d^4x = -\int m d\tau = -\int m\frac{dt}{dt/d\tau} = -\int m\frac{dt}{\gamma} \quad (6.13) \\ &= -\int m\sqrt{1 - V^2}\, dt = \int (-m + \frac{1}{2}mV^2 + \ldots)\, dt, \quad (6.14) \end{aligned}$$

which, apart from the mass term, agrees with (6.11) at the nonrelativistic level of accuracy. The Lagrangian $\mathcal{L}$, a density for the action in spacetime, is recognised in this case as (minus) the energy density in the local rest frame. This interpretation is clear either from $I_{\text{particle}} = -\int m d\tau$, or by reducing (6.12) by integration to the form $-m\delta^{(3)}(\underset{\sim}{x} - \underset{\sim}{z}(0))$. (The notation $\underset{\sim}{x}$ implies the triple of coordinates; it is used also with components and functions, as appropriate.)

The action density (6.12) will be used in a moment to generate equations of motion once an interaction is available. A simpler case to begin with is that of the free scalar field $\phi$, whose action, regarded as a functional of the field $\phi$, is

$$I(\phi) = -\frac{1}{2}\int \left(\eta^{\mu\nu}\partial_\mu\phi\partial_\nu\phi + m^2\phi^2\right) d^4x. \tag{6.15}$$

If $\phi$ is varied slightly, $\phi \mapsto \phi + \delta\phi$, where the variation $\delta\phi$ vanishes on the boundary of the region of integration, then the variation in the action may be written to lowest order

$$\begin{aligned}\delta I &= I(\phi + \delta\phi) - I(\phi)\\ &= -\int \left(\eta^{\mu\nu}\partial_\mu\delta\phi\partial_\nu\phi + m^2\phi\delta\phi\right)\, d^4x.\end{aligned}$$

Integrating by parts, and using the fact that the variation vanishes on the boundary, one gets

$$\delta I = \int \left(\eta^{\mu\nu}\partial_\mu\partial_\nu\phi - m^2\phi\right)\delta\phi\, d^4x. \tag{6.16}$$

The variation of the action vanishes for all $\delta\phi$ if the equations of motion, the Euler-Lagrange equations

$$\eta^{\mu\nu}\partial_\mu\partial_\nu\phi - m^2\phi = (\partial^2 - m^2)\phi = 0, \tag{6.17}$$

are satisfied.

### 6.3.1 Action for the Electromagnetic Field

The simplest way to include electromagnetism in the set of theories derived from a variational principle is to regard the covector $\underset{\sim}{A}$ as the fundamental field and the field tensor $\underset{\approx}{F} = \underset{\sim}{\partial}\wedge\underset{\sim}{A}$ as derived from it. As a consequence, the homogeneous Maxwell equations, in the form (5.26), are automatically satisfied. The variational principle then need generate only the inhomogeneous Maxwell equations. The action for the electromagnetic field is a functional of $\underset{\sim}{A}$, the Lagrangian density being a function of the antisymmetrised partial derivatives $\partial_\mu A_\nu$,

$$I_{EM} = \frac{1}{16\pi}\int \mathrm{tr}(\underset{\approx}{F}\cdot\underset{\approx}{F})d^4x = \frac{1}{16\pi}\int F_{\mu\nu}F^{\nu\mu}d^4x. \tag{6.18}$$

(The sign and the numerical factor are chosen to produce the appropriate energy tensor by a method to be explained in the next but one section; ultimately, all such factors are chosen to be consistent with (6.11).) For a variation of $A_\mu$, in view of the antisymmetry of $F_{\mu\nu}$,

$$\delta\int F_{\mu\nu}F^{\nu\mu}d^4x = 4\int \partial_\mu\delta A_\nu F^{\nu\mu}\, d^4x,$$

hence, after an integation by parts in which $\delta A_\mu$ vanishes on the boundary of integration,

$$\delta I_{EM} = -\frac{1}{4\pi}\int \delta A_\nu\partial_\mu F^{\nu\mu}\, d^4x. \tag{6.19}$$

The variation of the action vanishes for all $\delta A_\nu$ if $\partial_\mu F^{\mu\nu} = 0$, which is the inhomogeneous Maxwell equation for the case of zero electric current.

If an external electric current is present ("external" implies that we regard it as *given*, and that we do not plan to analyse its dynamics), an extra term is added to the $I_{EM}$ in (6.18):

$$\begin{aligned} I &= \frac{1}{16\pi}\int \mathrm{tr}(\underset{\Rightarrow}{F}\cdot\underset{\Rightarrow}{F})d^4x + \int \underset{\rightarrow}{A}(\underset{\rightarrow}{j})\,d^4x & (6.20) \\ &= \frac{1}{16\pi}\int F_{\mu\nu}F^{\nu\mu}d^4x + \int A_\nu j^\nu\,d^4x. & (6.21) \end{aligned}$$

For this new action, with the interaction term, the variation is zero if

$$\partial_\mu F^{\mu\nu} = -4\pi j^\nu,$$

the inhomogeneous Maxwell equation.

### 6.3.2 Action for a Charged Particle in an External Field

If an appropriate form of the interaction term that was added to $I_{EM}$ in (6.21) is instead added to $I_{\text{particle}}$ we get the action for a charged particle in an external electromagnetic field (here, it is $\underset{\rightarrow}{A}$ instead of the current that is regarded as given and nondynamical):

$$I = -m\int d\tau + q\int \underset{\rightarrow}{A}(\underset{\rightarrow}{v})d\tau. \tag{6.22}$$

The current density (5.45) has been used. In order to calculate the variation of the action it is convenient to introduce a general parameterisation for the worldline (it does not matter which one),

$$I = \int_{\lambda_1}^{\lambda_2}\left(-m\sqrt{-\eta_{\mu\nu}\frac{dz^\mu}{d\lambda}\frac{dz^\nu}{d\lambda}} + qA_\mu(z)\frac{dz^\mu}{d\lambda}\right)d\lambda. \tag{6.23}$$

Definite limits have been put on this integral, which mark the boundary outside of which the variation of the worldline, $\delta z(\lambda)$, is to vanish. The variation of the action due to a variation of the worldline may now be written, to lowest order,

$$\delta I = \int_{\lambda_1}^{\lambda_2}\left(\frac{m\eta_{\mu\nu}\frac{d\delta z^\mu}{d\lambda}\frac{dz^\nu}{d\lambda}}{\sqrt{-\eta_{\alpha\beta}\frac{dz^\alpha}{d\lambda}\frac{dz^\beta}{d\lambda}}} + qA_\mu(z)\frac{d\delta z^\mu}{d\lambda} + q\delta z^\nu\partial_\nu A_\mu\frac{dz^\mu}{d\lambda}\right)d\lambda.$$

Integrating by parts (the variations vanish at the boundary) and relabelling the scripts in the last term gives

$$\delta I = \int_{\lambda_1}^{\lambda_2}\left(-m\frac{d}{d\lambda}\frac{\eta_{\mu\nu}\frac{dz^\nu}{d\lambda}}{\sqrt{-\eta_{\alpha\beta}\frac{dz^\alpha}{d\lambda}\frac{dz^\beta}{d\lambda}}} - q\frac{dz^\nu}{d\lambda}\partial_\nu A_\mu + q\partial_\mu A_\nu\frac{dz^\nu}{d\lambda}\right)\delta z^\mu\,d\lambda.$$

The variation of the action vanishes for all choices of $\delta z^\mu$ if and only if

$$m\frac{d}{d\lambda}\frac{\eta_{\mu\nu}\frac{dz^\nu}{d\lambda}}{\sqrt{-\eta_{\alpha\beta}\frac{dz^\alpha}{d\lambda}\frac{dz^\beta}{d\lambda}}} = q(\partial_\mu A_\nu - \partial_\nu A_\mu)\frac{dz^\nu}{d\lambda} = qF_{\mu\nu}\frac{dz^\nu}{d\lambda}. \tag{6.24}$$

(Note that the left-hand side is orthogonal to $dz^\mu/d\lambda$.) Dividing through by the square root, which is just $d\tau/d\lambda$, allows us to express the equation in terms of the parameterisation with *proper time* and the unit vector $v^\mu = dz^\mu/d\tau$,

$$m\frac{d}{d\tau}v^\mu = qF^\mu{}_\nu v^\nu, \tag{6.25}$$

the equation of motion for a charged point particle subject to the Lorentz force of an external electromagnetic field. The geometrical form is even more compact,

$$m\boldsymbol{a} = q\overset{\Rightarrow}{F}\cdot\boldsymbol{v}. \tag{6.26}$$

The equation of motion for a free point particle is obtained by setting $q = 0$.

### 6.3.3 Action for Charged Particles Interacting Electromagnetically

The complete action for a charged point particle interacting with an electromagnetic field to which it contributes would appear to be

$$I = \frac{1}{16\pi}\int F_{\mu\nu}F^{\nu\mu}d^4x - m\int d\tau + q\int A_\mu(z)v^\mu\, d\tau, \tag{6.27}$$

but this expression is to some extent only formal because the integrals are not well defined. The problem is the singularity in the field due to the point charge. If $\underset{\rightarrow}{A}$ and $\underset{\Rightarrow}{F}$ satisfy Maxwell's equations with the point charge source, then $\underset{\rightarrow}{A}$ will have a $\frac{1}{\rho}$ singularity and $\underset{\Rightarrow}{F}$ a $\frac{1}{\rho^2}$ singularity. (Work devoted to resolving the problems created by these singularities, and the worse ones when the particles have a magnetic moment, is reported in a paper by Rowe and Rowe.) If the source is instead a continuum model for an incoherent dust of charged particles, with number density $n_{(0)}$ in the local rest frame, then both the mass current $mn_{(0)}\boldsymbol{v}$ and the electric current $qn_{(0)}\boldsymbol{v}$ are smooth. The action for the whole system is well defined for this case:

$$\begin{aligned} I =& \frac{1}{16\pi}\int F_{\mu\nu}F^{\nu\mu}d^4x - m\int n_{(0)}(x)\, d^4x \\ &+ q\int A_\mu(x)n_{(0)}(x)v^\mu(x)\, d^4x. \end{aligned} \tag{6.28}$$

The derivation of the equations of motion for charged dust from this action is somewhat subtle, and requires the techniques used for a fluid in Section 6.8. It is left as a problem.

## 6.4 Noninertial Coordinates

Although inertial coordinates are at the very heart of special relativity and help to form its structure, other coordinate systems arise in special problems and for special procedures. The derivation of the energy tensor from the Lagrangian, by the method of the next section, requires that the action be expressed in general coordinates.

Noninertial coordinates are coordinates for spacetime that are not one of the privileged group of inertial coordinates. The coordinates are invertible mappings $\mathcal{M} \to \mathcal{R}^4, \quad P \mapsto \xi^\mu(P)$. (Often such mappings may not be defined for the whole of spacetime, but only for a patch of it. This subtlety need not concern us if we work in a local region within one patch, or if we simply understand the existence of a set of overlapping patches if required.) The relations between the coordinates $x^\mu$ of some inertial frame and the new, noninertial $\xi^\mu$ are

$$\xi^\mu = f^\mu(x^\nu), \qquad x^\mu = g^\mu(\xi^\nu). \tag{6.29}$$

The symbol $\xi$ is being used to signal that the coordinates may be noninertial, curvilinear coordinates; and more importantly, the symbol $x$ is saved for use *only* with inertial coordinates. To make the new coordinates as simple as possible, we assume that the orientation does not change, so

$$\det \frac{\partial x}{\partial \xi} > 0. \tag{6.30}$$

### 6.4.1 New Basis Vectors

New basis vectors, associated with the new, noninertial coordinates, are defined *at each point* in spacetime. If $P$ and $Q$ are infinitesimally close (Figure 6.4), and their noninertial coordinates differ by $\delta\xi^\mu$,

$$\overrightarrow{PQ} \equiv \delta\xi^\mu \, \boldsymbol{\partial}_\mu(P), \tag{6.31}$$

or, in terms of the change in inertial coordinates,

$$\overrightarrow{PQ} = \delta(x^\nu \boldsymbol{E}_\nu) = \delta x^\nu \boldsymbol{E}_\nu = \frac{\partial x^\nu}{\partial \xi^\mu} \delta\xi^\mu \boldsymbol{E}_\nu. \tag{6.32}$$

Therefore, the noninertial basis vectors, decomposed in terms of the inertial basis vectors, are

$$\boldsymbol{\partial}_\mu(P) = \frac{\partial g^\nu}{\partial \xi^\mu}(\xi) \boldsymbol{E}_\nu = \frac{\partial x^\nu}{\partial \xi^\mu} \boldsymbol{E}_\nu. \tag{6.33}$$

This equation reduces to the relation between basis vectors for different inertial frames in the appropriate circumstances, when $\xi^\mu$ is actually inertial. A comparison can be made with (2.9), which reads $\boldsymbol{E}_\nu = a^\mu{}_\nu \boldsymbol{E}'_\mu$. The structure of the two equations is the same since $a^\mu{}_\nu = \partial x'^\mu / \partial x^\nu$. From (6.29), the

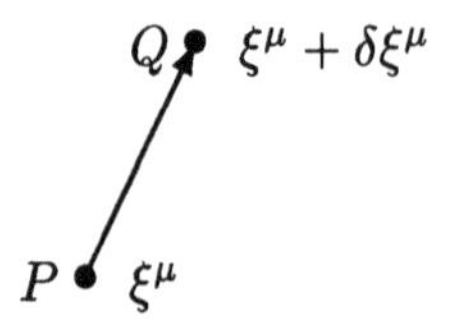


**Figure 6.4**

equation $x^\mu = g^\mu(f^\nu(x^\lambda))$ is identically satisfied, hence

$$\frac{\partial x^\mu}{\partial x^\lambda} = \delta^\mu_\lambda = \frac{\partial g^\mu}{\partial \xi^\nu}(\xi)\frac{\partial f^\nu}{\partial x^\lambda}(x) = \frac{\partial x^\mu}{\partial \xi^\nu}\frac{\partial \xi^\nu}{\partial x^\lambda}. \tag{6.34}$$

Similarly, the equation with $x^\mu$ and $\xi^\mu$ interchanged. Therefore the inverse of (6.33) may be written

$$\frac{\partial f^\mu}{\partial x^\lambda}\boldsymbol{\partial}_\mu = \frac{\partial \xi^\mu}{\partial x^\lambda}\frac{\partial x^\nu}{\partial \xi^\mu}\boldsymbol{E}_\nu = \boldsymbol{E}_\lambda. \tag{6.35}$$

The new basis vectors are not in general unit vectors, nor are they orthogonal. The scalar product of two of them is

$$\boldsymbol{\partial}_\mu \cdot \boldsymbol{\partial}_\nu = (\frac{\partial x^\rho}{\partial \xi^\mu}\boldsymbol{E}_\rho)\cdot(\frac{\partial x^\tau}{\partial \xi^\nu}\boldsymbol{E}_\tau) = \frac{\partial x^\rho}{\partial \xi^\mu}\frac{\partial x^\tau}{\partial \xi^\nu}\eta_{\rho\tau} \equiv g_{\mu\nu}. \tag{6.36}$$

These vary from point to point. Note also that in general $\boldsymbol{\partial}_0$ will not be timelike.

So far as the dual space $\mathcal{V}^*$ is concerned, the covectors

$$\underset{\sim}{d}\xi^\mu = \frac{\partial \xi^\mu}{\partial x^\nu}\underset{\sim}{d}x^\nu \tag{6.37}$$

constitute a cobasis associated with the basis $\boldsymbol{\partial}_\mu$ for $\mathcal{V}$. Using (6.34) for the inverse, and (4.26),

$$\underset{\sim}{d}\xi^\mu(\boldsymbol{\partial}_\nu) = \frac{\partial \xi^\mu}{\partial x^\rho}\underset{\sim}{d}x^\rho(\frac{\partial x^\tau}{\partial \xi^\nu}\boldsymbol{E}_\tau) = \delta^\mu_\nu.$$

### 6.4.2 New Components

When an arbitrary vector field $\boldsymbol{V}$ is expanded in an *inertial* basis, $\boldsymbol{V}(P) = V^\mu(x)\boldsymbol{E}_\mu$, the components depend on the coordinates of $P$ but the basis vectors do not depend on the point. New components arise when the inertial basis is replaced using (6.35),

$$\boldsymbol{V}(P) = V^\mu(x)\boldsymbol{E}_\mu = V^\mu\frac{\partial \xi^\nu}{\partial x^\mu}\boldsymbol{\partial}_\nu(P) \equiv V^\nu_\xi(P)\boldsymbol{\partial}_\nu(P). \tag{6.38}$$

Now, *both* components and basis vectors depend on the point. The relation between the inertial and noninertial components is

$$V_\xi^\nu(\xi) = V^\mu(x)\frac{\partial \xi^\nu}{\partial x^\mu}, \qquad V^\mu(x) = V_\xi^\nu(\xi)\frac{\partial x^\mu}{\partial \xi^\nu}. \tag{6.39}$$

The subscript $\xi$ on the components $V_\xi^\nu$ is just a reminder that they are associated with the basis $\boldsymbol{\partial}_\nu$. The same reminder will be used on other components with respect to a noninertial basis.

Similarly, new components for covectors may be introduced by expanding the inertial cobasis $\underset{\sim}{d}x^\mu$ in terms of the new cobasis. For example,

$$\underset{\sim}{A} = A_\mu \underset{\sim}{d}x^\mu = A_\mu \frac{\partial x^\mu}{\partial \xi^\nu}\underset{\sim}{\partial}\xi^\nu \equiv A_\nu^\xi \underset{\sim}{\partial}\xi^\nu. \tag{6.40}$$

Therefore,

$$A_\nu^\xi(\xi) = A_\mu(x)\frac{\partial x^\mu}{\partial \xi^\nu}, \qquad A_\mu(x) = A_\nu^\xi(\xi)\frac{\partial \xi^\nu}{\partial x^\mu}. \tag{6.41}$$

The new components for other tensors, expanded using tensor products, are straightforward to develop, for example,

$$\underset{\approx}{F} = F_{\mu\nu}\underset{\sim}{d}x^\mu \otimes \underset{\sim}{d}x^\nu = F_{\mu\nu}\frac{\partial x^\mu}{\partial \xi^\rho}\frac{\partial x^\nu}{\partial \xi^\tau}\underset{\sim}{\partial}\xi^\rho \otimes \underset{\sim}{\partial}\xi^\tau \equiv F_{\rho\tau}^\xi \underset{\sim}{\partial}\xi^\rho \otimes \underset{\sim}{\partial}\xi^\tau. \tag{6.42}$$

Similarly, the new components of the metric tensor $\underset{\approx}{\eta} = \eta_{\mu\nu}\underset{\sim}{d}x^\mu \otimes \underset{\sim}{d}x^\nu$ are

$$\eta_{\rho\tau}^\xi = \eta_{\mu\nu}\frac{\partial x^\mu}{\partial \xi^\rho}\frac{\partial x^\nu}{\partial \xi^\tau} \quad (\equiv g_{\rho\tau}). \tag{6.43}$$

The notation $g_{\rho\tau}$, first used in (6.36), is the alternative which we use to write formulae which are more common in their general relativistic form. The determinant of the matrix $\eta_{\mu\nu}$ is $-1$. Writing $g$ for the determinant of $g_{\mu\nu}$, one gets from the expression in (6.43), regarded as a matrix product,

$$g = -\left(\det \frac{\partial x^\mu}{\partial \xi^\nu}\right)^2. \tag{6.44}$$

## 6.5 Construction of the Energy Tensor

Conservation of four-momentum is expressed locally by the vanishing of the divergence of the energy tensor. In the present section it is shown how the energy tensor may be found from the Lagrangian density, and how its divergence may be proved to vanish as a consequence of the equations of motion in variational form. The argument is adapted from the Lagrangian form of the theory of gravity, in which the energy tensor is coupled to the spacetime metric.

The procedure which produces the energy tensor is analogous to one which identifies the electromagnetic current by varying the electromagnetic potential in the interaction term of the action (6.21),

$$\delta \int A_\nu j^\nu \, d^4x = \int j^\nu \delta A_\nu \, d^4x. \tag{6.45}$$

The current $j^\nu$ is identified as the coefficient of the variation $\delta A_\nu$ in the integrand. To show that the current is conserved one uses the postulated gauge invariance of the action, that is, its invariance under $A_\nu \mapsto A_\nu + \partial_\nu \chi$. The first term in (6.21), a functional of $F_{\mu\nu}$, is gauge invariant because $F_{\mu\nu}$ alone is gauge invariant. The second term is gauge invariant if for all $\chi$,

$$\int j^\nu \partial_\nu \chi \, d^4x = -\int \chi \partial_\nu j^\nu \, d^4x = 0, \tag{6.46}$$

that is, if $\partial_\nu j^\nu = 0$. The argument which identifies the energy tensor $T^{\mu\nu}$ and proves that it is conserved has just this structure. In place of the gauge invariance of the action, however, the argument uses the *form* invariance of the Lagrangian when the variables are changed (a change of variables induced by a change of coordinates, a change to arbitrary noninertial coordinates, not just a different choice of inertial coordinates) in the action integral.

The argument is most easily understood by taking the example of a Lagrangian depending on a few fields in spacetime, say, $\phi$, $\underset{\sim}{A}, \ldots$. As we have seen in several examples, $\mathcal{L}$ is a simple sum of terms which are scalar invariants formed from the fields and their derivatives, terms such as

$$I_1 = \phi^2, \quad I_2 = \underset{\sim}{\partial}\phi \cdot \underset{\sim}{\partial}\phi, \quad I_3 = \mathrm{tr}(\underset{\approx}{F} \cdot \underset{\approx}{F}), \quad \ldots. \tag{6.47}$$

If the argument is to be fully understood one needs to be especially clear about the functions that are involved.

The $I_k$ are scalar functions $\mathcal{M} \to \mathcal{R}$, with $P \mapsto I_k(P)$, formed from the geometical fields $\phi$, $\underset{\sim}{\partial}\phi$, $\underset{\sim}{A}$, $\underset{\approx}{F} = \underset{\sim}{\partial} \wedge \underset{\sim}{A}, \ldots$. The action, expressed geometrically as an integral over a region $\mathcal{D} \subset \mathcal{M}$ of spacetime, is

$$I = \int_{\mathcal{D}} \mathcal{L}(I_1, I_2, \ldots) \, \underset{\sim}{d}x^0 \wedge \underset{\sim}{d}x^1 \wedge \underset{\sim}{d}x^2 \wedge \underset{\sim}{d}x^3. \tag{6.48}$$

When the action is written in terms of some inertial coordinates, those of $K$, say, it takes the form of a numerical integral over the coordinate range $K(\mathcal{D}) \subset \mathcal{R}^4$,

$$I = I^K \equiv \int_{K(\mathcal{D})} \mathcal{L}(I_1^K, I_2^K, \ldots) \, dx^0 dx^1 dx^2 dx^3, \tag{6.49}$$

where now the scalars, $I_i^K$, are functions $\mathcal{R}^4 \mapsto \mathcal{R}$ of the $K$-coordinates of points in $\mathcal{M}$. Each scalar is now expressed in terms of the $K$-components of the fields of which it is composed. With $x \in \mathcal{R}^4$ as the $K$-coordinates of $P$,

$$\begin{aligned} I_1(P) &= I_1^K(x) = (\phi^K(x))^2, \\ I_2(P) &= I_2^K(x) = \partial_\mu \phi^K(x) \partial_\nu \phi^K(x) \eta^{\mu\nu}, \ \ldots \end{aligned} \tag{6.50}$$

The action with respect to any other frame $K'$ has just the form (6.49) (with (6.50)) in which, everywhere, $K \to K'$. This identity of functional form is the *form invariance* of the action for the set of *inertial* frames.

If *noninertial* coordinates, $\xi = f(x)$, as in (6.29), are used to express the action, there is a striking difference in the appearance of the integral. Since the Jacobian for the transformation from the inertial form is no longer unity, it must be present explicitly. By (6.44), it is $J = \sqrt{-g}$ (assumed to be positive) and so (6.48), expressed in the $\Xi$-coordinates, is

$$I = I^\Xi \equiv \int_{\Xi(\mathcal{D})} \mathcal{L}(I_1^\Xi, I_2^\Xi, \ldots) \sqrt{-g}\, d\xi^0 d\xi^1 d\xi^2 d\xi^3. \tag{6.51}$$

The invariants (6.47), when expressed in terms of the $\Xi$-components,

$$\phi^\Xi(\xi) = \phi^K(x), \quad \partial_\mu^\xi \phi^\Xi = \left(\frac{\partial x^\nu}{\partial \xi^\mu}\right) \partial_\nu \phi^K, \quad \text{and so on,} \tag{6.52}$$

are exactly the same functions[2] of the $\Xi$-components as the $I_i^K$ in (6.50) are of the $K$-components,

$$I_1(P) = (\phi^\Xi(\xi))^2, \quad I_2(P) = \partial_\mu^\xi \phi^\Xi(\xi) \partial_\nu^\xi \phi^\Xi(\xi) g^{\mu\nu}, \ldots, \tag{6.53}$$

*provided* that the metric tensor, with $K$-components $\eta^{\mu\nu}$ and $\Xi$-components $g^{\mu\nu}$, is included, as if it were a physical field (of course, in general relativity, it *is* a physical field). Then, if a factor $\sqrt{-\eta}$, with $\eta \equiv \det(\eta_{\mu\nu})$, is inserted into the integral (6.49) for $I^K$, numerically nothing is changed because $\eta = -1$, but it allows us to view the integrands in $I^K$ and $I^\Xi$ (equation (6.51)) as *exactly* the same function of components with respect to the two systems.

One now takes the difference between $I^\Xi$ and $I^K$, which are numerically the same, for the case when the function $\xi = f(x)$ is the identity outside a bounded region. The integrand will then vanish outside the bounded region. Using the symbol $x$ instead of the symbol $\xi$ for a general point in the $I^\Xi$ integral, the difference may be written

[2] The tensor components always appear in the scalar functions $I_k$ in the same way, but if the $K$-components involve *derivatives*, it is only in some simple cases that the $\Xi$-components have the same form of derivatives. These simple cases, $\partial_\mu \phi$, $\partial_\mu A_\nu - \partial_\nu A_\mu$ are just the ones in which we are presently interested. In cases with more complicated derivatives there are extra terms which require an extension of the argument below for the energy tensor.

$$0 = \int \left(\mathcal{L}(I_1^{\Xi}, I_2^{\Xi}, \ldots)\sqrt{-g} - \mathcal{L}(I_1^K, I_2^K, \ldots)\sqrt{-\eta}\right) dx^0 dx^1 dx^2 dx^3. \quad (6.54)$$

Because exactly the same function of components occurs in each member of the difference, we can view the $\Xi$-components as a variation of the $K$-components, just as was used in the variational principle, *but now extended to the components of the metric as well.*

By combining the two forms of the action for two different coordinate systems as in (6.54) we have obtained the difference between the values of two functions at each point $x \in \mathcal{R}^4$. The same coordinate $x \in \mathcal{R}^4$ refers to different points in spacetime $\mathcal{M}$ in the two terms because the terms arise from different coordinate systems. For the rest of the argument it is sufficient that the transformation $\xi^\mu = f^\mu(x^\nu)$ be close to the identity everywhere. It is convenient to write

$$\xi^\mu = x^\mu + \epsilon^\mu(x^\nu), \quad (6.55)$$

with $\epsilon^\mu(x^\nu)$ small everywhere (and actually zero outside a bounded region). The definition of the function $\phi^\Xi$ is

$$\phi^\Xi(\xi) = \phi^\Xi(x + \epsilon(x)) = \phi^K(x),$$

so that *to first order,* the difference in the two functions at a given numerical coordinate $x$ is

$$\delta\phi^K(x) \equiv \phi^\Xi(x) - \phi^K(x) = \phi^K(x - \epsilon(x)) - \phi^K(x) = -\epsilon^\nu(x)\partial_\nu\phi^K(x).$$

Also, $\delta\partial_\mu\phi^K = \partial_\mu\delta\phi^K$, with analogous formulae for components of $\underset{\sim}{A}$, $\underset{\sim}{E}$, ....
All these variations vanish outside the bounded region in which $\epsilon^\mu$ is nonzero.

The specific form for the difference of the metric components will prove to be crucial for the conservation argument below. From (6.43), together with (6.55), to first order,

$$\begin{aligned} g_{\mu\nu}(x+\epsilon) &= (\delta_\mu^\rho - \partial_\mu\epsilon^\rho)(\delta_\nu^\tau - \partial_\nu\epsilon^\tau)\eta_{\rho\tau} = \eta_{\mu\nu} - \partial_\mu\epsilon_\nu - \partial_\nu\epsilon_\mu \\ &= g_{\mu\nu}(x) + \epsilon^\lambda\partial_\lambda g_{\mu\nu}(x). \end{aligned}$$

Note that the "lowering" $\epsilon_\mu = \eta_{\mu\rho}\epsilon^\rho$ is done with the inertial frame metric (which in fact is the same as $g_{\mu\rho}\epsilon^\rho$ to first order). Also, since $g_{\mu\nu} = \eta_{\mu\nu} + O(\epsilon)$, $\partial_\lambda g_{\mu\nu} = O(\epsilon)$ and $\epsilon^\lambda\partial_\lambda g_{\mu\nu}(x) = O(\epsilon^2)$. The difference in the components of the metric at coordinate $x$ is, therefore, *to first order,*

$$\delta g_{\mu\nu}(x) = g_{\mu\nu}(x) - \eta_{\mu\nu} = -\partial_\mu\epsilon_\nu - \partial_\nu\epsilon_\mu. \quad (6.56)$$

This variation is symmetric, and the small vector field $\epsilon^\mu$ is arbitrary.

We think of the integrand in (6.54) as the difference of $\mathcal{L}\sqrt{-g}$ for two sets of components, the inertial components of the physical fields, $\phi^K$, $A_\mu^K$, $\partial_\mu\phi^K$, ..., which satisfy the equations of motion derived from a variational principle as in the earlier section, and the components of fields which deviate slightly from the physical fields $\phi^{\Xi} = \phi^K + \delta\phi^K$, and so on, and finally the metric components. If one writes down the difference of the two forms of the action as a sum of first order terms in the variations of the components, then all the terms associated with the physical fields vanish if the fields satisfy the equations of motion, because the procedure is exactly the same as in the variational principle where the coefficients of the variations vanish. All that then remains of the difference (6.54) are the terms arising from the variation in the components of the metric tensor[3]

$$0 = \delta_g I = \int \frac{\partial}{\partial g_{\mu\nu}} \left(\mathcal{L}\sqrt{-g}\right) \delta g_{\mu\nu}\, dx^0 dx^1 dx^2 dx^3, \tag{6.57}$$

where $\delta g_{\mu\nu}$ is given by (6.56). The great significance of this form is that, as will be proved in a moment, the coefficient of $\delta g_{\mu\nu}$ satisfies a conservation law and so ***it may be identified with a conserved physical quantity.***

Up to a constant, the coefficient of $\delta g_{\mu\nu}$ is, we presume, the component $T^{\mu\nu}$ of the energy tensor, with respect to the inertial system $K$. The interpretation of the conserved symmetric tensor is based on known simple special cases (see below), as well as the crucial connection with the theory of gravity, where this tensor is the source of curvature in spacetime. Despite appearances, the expression depends only on the Minkowski metric $\eta_{\mu\nu}$. When the derivatives with respect to $g_{\mu\nu}$ have been done, the expression is evaluated at $\eta_{\mu\nu}$. The first order deviation is all in $\delta g_{\mu\nu}$. The multiplicative constant is a result of the historically accidental value of the multiplicative constant in (6.11):

$$\begin{aligned} \delta_g I\Big|_{g_{\mu\nu}\to\eta_{\mu\nu}} &= \int \left[\frac{\partial}{\partial g_{\mu\nu}} \left(\mathcal{L}\sqrt{-g}\right)\right]_{g_{\mu\nu}\to\eta_{\mu\nu}} \delta g_{\mu\nu}\, d^4x \\ &\equiv \frac{1}{2}\int T^{\mu\nu} \delta g_{\mu\nu}\, d^4x. \end{aligned} \tag{6.58}$$

It is important for the calculation of $T^{\mu\nu}$ that $\mathcal{L}$ is regarded as a function of the correct components for its geometrical type, for example, $\partial_\mu \phi^K$, for $\underset{\sim}{\partial}\phi$, and not $\partial^\mu \phi^K$, which brings in the metric in a hidden way. When the metric is varied, the appropriate components of the physical fields are kept fixed. The metric itself appears naturally in its covariant manifestation, $g_{\mu\nu}$, or as elements of the inverse matrix, $g^{\mu\nu}$, its contravariant manifestation. Each set is a function of the other; the relation between their corresponding variations follows from $g^{\mu\nu} g_{\nu\lambda} = \delta^\mu_\lambda$. It is

[3] Although the equation has the form of relations in the variational principle, the consequences are quite different because $\delta g_{\mu\nu}$ is not arbitrary, but has the specific (symmetric) form (6.56). Its coefficient, again symmetric, comes from the variation of $g_{\mu\nu}$ and does not vanish. There is no contribution from derivatives of $g_{\mu\nu}$ in the simple cases to which we presently limit ourselves. The two types of situation are referred to in the previous footnote.

$$\delta g^{\mu\nu} = -g^{\mu\rho} g^{\nu\tau} \delta g_{\rho\tau}. \tag{6.59}$$

Hence, the variational integral has an alternative form

$$\delta_g I\Big|_{g_{\mu\nu}\to\eta_{\mu\nu}} = \frac{1}{2}\int T^{\mu\nu}\delta g_{\mu\nu}\, d^4x = -\frac{1}{2}\int T_{\mu\nu}\delta g^{\mu\nu}\, d^4x, \tag{6.60}$$

where $T_{\mu\nu} = \eta_{\mu\alpha}\eta_{\nu\beta}T^{\alpha\beta}$. These relations show how to find the energy tensor from the Lagrangian density.

### 6.5.1 First Example: Energy Tensor for the Free Scalar Field

The action for the scalar field, written as an integral over noninertial coordinates, follows from (6.15):

$$I(\phi) = -\frac{1}{2}\int \left(g^{\mu\nu}\partial^\xi_\mu\phi\partial^\xi_\nu\phi + m^2\phi^2\right)\sqrt{-g}\, d^4\xi. \tag{6.61}$$

The variation of the action requires $\delta g$, the variation in the determinant of the components of the metric. Using the definition of the cofactors and their relation to the inverse matrix, one gets $\delta g = g g^{\mu\nu}\delta g_{\mu\nu}$. Hence, by differentiating (and using (6.59) for the second form),

$$\delta\sqrt{-g} = \frac{1}{2}\sqrt{-g}g^{\mu\nu}\delta g_{\mu\nu} = -\frac{1}{2}\sqrt{-g}g_{\mu\nu}\delta g^{\mu\nu}. \tag{6.62}$$

Varying the action by varying the metric, while holding $\phi$ and its (covariant) derivatives fixed,

$$\delta_g I = -\frac{1}{2}\int\left[\delta g^{\mu\nu}\partial^\xi_\mu\phi\partial^\xi_\nu\phi - \frac{1}{2}(g^{\rho\tau}\partial^\xi_\rho\phi\partial^\xi_\tau\phi + m^2\phi^2)g_{\mu\nu}\delta g^{\mu\nu}\right]\sqrt{-g}\, d^4\xi.$$

After the variation we may replace $g_{\mu\nu} \to \eta_{\mu\nu}$, which means in particular that $g \to -1$ and the coordinates become again inertial. The energy tensor may then be identified from

$$\begin{aligned}\delta_g I\Big|_{g_{\mu\nu}\to\eta_{\mu\nu}} &= -\frac{1}{2}\int\delta g^{\mu\nu}\left[\partial_\mu\phi\partial_\nu\phi - \frac{1}{2}(\eta^{\rho\tau}\partial_\rho\phi\partial_\tau\phi + m^2\phi^2)\eta_{\mu\nu}\right] d^4x\\ &= -\frac{1}{2}\int T_{\mu\nu}\delta g^{\mu\nu}\, d^4x.\end{aligned}$$

Its components with respect to an inertial frame are

$$T_{\mu\nu} = \partial_\mu\phi\partial_\nu\phi - \frac{1}{2}(\eta^{\rho\tau}\partial_\rho\phi\partial_\tau\phi + m^2\phi^2)\eta_{\mu\nu}. \tag{6.63}$$

The geometrical energy tensor is therefore

$$\overset{\Rightarrow}{T} = \boldsymbol{\partial}\phi\otimes\boldsymbol{\partial}\phi - \frac{1}{2}\overset{\Rightarrow}{\eta}(\boldsymbol{\partial}\phi\cdot\boldsymbol{\partial}\phi + m^2\phi^2). \tag{6.64}$$

### 6.5.2 The Total Energy Tensor Has Zero Divergence

To establish the crucial property that the putative energy tensor $\overleftrightarrow{T}$ does indeed have zero divergence, we use the explicit form (6.56) of $\delta g_{\mu\nu}$, and otherwise we replace to zero order $g_{\mu\nu} \to \eta_{\mu\nu}$ in (6.57):

$$0 = \delta_g I = \frac{1}{2}\int T^{\mu\nu}(-\partial_\mu \epsilon_\nu - \partial_\nu \epsilon_\mu)\, d^4x.$$

Because $T^{\mu\nu}$ is symmetric, this gives

$$0 = \int T^{\mu\nu}(-\partial_\mu \epsilon_\nu)\, d^4x = \int \partial_\mu T^{\mu\nu} \epsilon_\nu \, d^4x,$$

the second form after an integration by parts in which $\epsilon_\nu$ vanishes outside a bounded region. Now, because $\epsilon_\nu$ is arbitrary, we can deduce the relation we want:

$$\partial_\mu T^{\mu\nu} = 0. \tag{6.65}$$

The conservation law is a consequence of the field equations, which were assumed to be satisfied in the equation $\delta_g I = 0$ with which we began.

It is the *total* energy tensor which is conserved. If $T^{\mu\nu}$ is a sum of energy tensors for individual physical subsystems, as it often is, the overall conservation will express the balance of four-momentum exchanges between the subsystems.

As a simple example of conservation, we can calculate the divergence of (6.63)

$$\begin{aligned} \partial_\mu T^{\mu\nu} &= \partial_\mu \left[\partial^\mu \phi \partial^\nu \phi - \frac{1}{2}(\eta^{\rho\tau}\partial_\rho \phi \partial_\tau \phi + m^2\phi^2)\eta^{\mu\nu}\right] \\ &= (\partial^2\phi)\partial^\nu \phi + \partial^\mu \phi \partial_\mu \partial^\nu \phi - \partial^\nu \partial_\rho \phi \partial^\rho \phi - m^2 \phi \partial^\nu \phi, \end{aligned}$$

which vanishes if the field equations (6.17) are satisfied.

### 6.5.3 Second Example: Energy Tensor for the Electromagnetic Field

The action for the electromagnetic field, (6.18), rewritten in noninertial coordinates, is

$$I_{EM} = \frac{1}{16\pi}\int F^\xi_{\mu\nu} F^\xi_{\rho\tau} g^{\nu\rho} g^{\mu\tau} \sqrt{-g}\, d^4\xi. \tag{6.66}$$

Varying this action by varying the metric, and then replacing $g_{\mu\nu} \to \eta_{\mu\nu}$ gives

$$\delta_g I_{EM} = \frac{1}{16\pi}\int \delta g^{\nu\rho}\left[2F_{\mu\nu}F_{\rho\tau}\eta^{\mu\tau} - \frac{1}{2}\eta_{\nu\rho}F_{\alpha\beta}F^{\beta\alpha}\right] d^4x.$$

Hence, the inertial frame components of the energy tensor are in this case

$$\Theta_{\nu\rho} = \frac{1}{4\pi}\left[-F_{\mu\nu}F_{\rho\tau}\eta^{\mu\tau} + \frac{1}{4}\eta_{\nu\rho}F_{\alpha\beta}F^{\beta\alpha}\right], \tag{6.67}$$

where $\Theta$ is the symbol conventionally used for this important electromagnetic contribution to the total energy tensor. Its divergence vanishes as a consequence of the source-free Maxwell equations. (The total $T^{\mu\nu}$ is equal to $\Theta^{\mu\nu}$ only when the electromagnetic field is the only physical system; it is then free, and of course there are no sources.) The interpretation of $\overset{\Rightarrow}{\Theta}$ will be given in the next section.

### 6.5.4 Third Example: Energy Tensor for a Point Particle

The action for a point particle is determined by the Lagrangian (6.12). It leads to

$$\begin{aligned} I_{\text{particle}} &= -\int\int m\delta^{(4)}(x - z(\tau))\, d\tau\, d^4x = -\int m d\tau \\ &= -\int\int m\delta^{(4)}(x - z(\lambda))\sqrt{-\eta_{\mu\nu}\frac{dz^\mu}{d\lambda}\frac{dz^\nu}{d\lambda}}\, d\lambda\, d^4x \end{aligned} \tag{6.68}$$

$$= -\int m\sqrt{-\eta_{\mu\nu}\frac{dz^\mu}{d\lambda}\frac{dz^\nu}{d\lambda}}\, d\lambda. \tag{6.69}$$

The present system is not really of the same type as the previous fields in spacetime. To force the theory of a particle into the mould of the theory of a field requires delta-functions to be used. And we must use the unconstrained parameterisation $z(\lambda)$ for the particle, rather than the parameterisation with proper time, as we did for the derivation of the equations of motion by the variational principle.

If the coordinates are changed to the noninertial $x^\mu = g^\mu(\xi)$ with inverse $\xi^\mu = f^\mu(x)$, the worldline of the particle will be given by $\sigma^\mu(\lambda) = f^\mu(z(\lambda))$. Using the formula for the change of variable in the delta-function,

$$\delta^{(4)}(g(\xi) - g(\sigma)) = \frac{\delta^{(4)}(\xi - \sigma)}{\det(\partial g/\partial \xi)},$$

the new form of the action is

$$I_{\text{particle}} = -\int\int m\delta^{(4)}(\xi - \sigma(\lambda))\sqrt{-g_{\mu\nu}\frac{d\sigma^\mu}{d\lambda}\frac{d\sigma^\nu}{d\lambda}}\, d\lambda\, d^4\xi \tag{6.70}$$

$$= -\int m\sqrt{-g_{\mu\nu}(\sigma(\lambda))\frac{d\sigma^\mu}{d\lambda}\frac{d\sigma^\nu}{d\lambda}}\, d\lambda. \tag{6.71}$$

This has just the same form in the new variables as the previous expression in inertial variables. Varying this action by varying the metric, and then replacing $g_{\mu\nu} \to \eta_{\mu\nu}$, gives

$$\delta_g I_{\text{particle}} = \frac{1}{2} \int \left( -\eta_{\rho\tau} \frac{dz^\rho}{d\lambda} \frac{dz^\tau}{d\lambda} \right)^{-\frac{1}{2}} m \frac{dz^\mu}{d\lambda} \frac{dz^\nu}{d\lambda} \delta g_{\mu\nu} \, d\lambda.$$

If we now revert to the proper time parameterisation, we can write

$$\begin{aligned} \delta_g I_{\text{particle}} &= \frac{1}{2} \int m \frac{dz^\mu}{d\tau} \frac{dz^\nu}{d\tau} \delta g_{\mu\nu} \, d\tau \\ &= \frac{1}{2} \int \int m \frac{dz^\mu}{d\tau} \frac{dz^\nu}{d\tau} \delta^{(4)}(x - z(\tau)) d\tau \delta g_{\mu\nu} \, d^4x. \end{aligned}$$

Comparing with (6.58) gives a formula for the energy tensor

$$K^{\mu\nu} = \int m \frac{dz^\mu}{d\tau} \frac{dz^\nu}{d\tau} \delta^{(4)}(x - z(\tau)) \, d\tau \tag{6.72}$$

or, in geometrical form,

$$\overset{\Rightarrow}{K} = \int m\boldsymbol{v} \otimes \boldsymbol{v} \delta^{(4)}(x - z(\tau)) \, d\tau. \tag{6.73}$$

The symbol $\overset{\Rightarrow}{K}$ is a conventional one to denote the kinematic energy tensor for a point particle (as well as for dust).

## 6.6 Energy in the Electromagnetic Field

The energy tensor $\overset{\Rightarrow}{\Theta}$ for the electromagnetic field $\overset{\Rightarrow}{F}$, constructed from its components (6.67), is given by

$$4\pi \overset{\Rightarrow}{\Theta} = -\overset{\Rightarrow}{F} \cdot \overset{\Rightarrow}{F} + \frac{1}{4} \overset{\Rightarrow}{\eta} \operatorname{tr}(\overset{\Rightarrow}{F} \cdot \overset{\Rightarrow}{F}). \tag{6.74}$$

Historically, various components of this four-dimensional dyadic were separately found: the energy density, Poynting's vector and Maxwell's stress dyadic, all with respect to a particular frame. An excellent account of the classical theory is available in Jackson's *Classical Electrodynamics*.

To show how the geometrical and relative points of view are connected, decompose $\overset{\Rightarrow}{F}$ with respect to the $K$-basis vectors $\boldsymbol{E}_\mu$ and the $K$-fields $\boldsymbol{E}$ and $\boldsymbol{B}$. From eqs (5.11) and (5.12) the relations are

$$\overset{\Rightarrow}{F} = \boldsymbol{E}_0 \wedge \boldsymbol{E} - (\boldsymbol{E}_0 \wedge \boldsymbol{B})^* = \boldsymbol{E}_0 \wedge \boldsymbol{E}_k E^k + \epsilon^{kmn} B^k \boldsymbol{E}_m \otimes \boldsymbol{E}_n. \tag{6.75}$$

Then,

$$\begin{aligned}\vec{\vec{F}}\cdot\vec{\vec{F}} &= -(\boldsymbol{E}_0\otimes\boldsymbol{E}_0)(\boldsymbol{E}\cdot\boldsymbol{E}) - (\boldsymbol{E}_m\otimes\boldsymbol{E}_m)(\boldsymbol{B}\cdot\boldsymbol{B})\\ &\qquad +\boldsymbol{E}\otimes\boldsymbol{E}+\boldsymbol{B}\otimes\boldsymbol{B}\\ &\qquad\qquad -\boldsymbol{E}_0\otimes(\boldsymbol{E}\times\boldsymbol{B}) - (\boldsymbol{E}\times\boldsymbol{B})\otimes\boldsymbol{E}_0, \qquad (6.76)\end{aligned}$$

and so

$$\operatorname{tr}\vec{\vec{F}}\cdot\vec{\vec{F}} = +2(\boldsymbol{E}\cdot\boldsymbol{E}) - 2(\boldsymbol{B}\cdot\boldsymbol{B}). \qquad (6.77)$$

Inserting these into (6.74), and using $\vec{\vec{\eta}} = \eta^{\mu\nu}\boldsymbol{E}_\mu\otimes\boldsymbol{E}_\nu$, gives

$$\begin{aligned}\vec{\vec{\Theta}} &= (\boldsymbol{E}_0\otimes\boldsymbol{E}_0)\frac{1}{8\pi}(\boldsymbol{E}\cdot\boldsymbol{E}+\boldsymbol{B}\cdot\boldsymbol{B})\\ &\qquad +\boldsymbol{E}_0\otimes\frac{(\boldsymbol{E}\times\boldsymbol{B})}{4\pi}+\frac{(\boldsymbol{E}\times\boldsymbol{B})}{4\pi}\otimes\boldsymbol{E}_0\\ &\frac{1}{4\pi}\left[-\boldsymbol{E}\otimes\boldsymbol{E}-\boldsymbol{B}\otimes\boldsymbol{B}+\frac{1}{2}\boldsymbol{E}_i\otimes\boldsymbol{E}_i(\boldsymbol{E}\cdot\boldsymbol{E}+\boldsymbol{B}\cdot\boldsymbol{B})\right].\end{aligned} \qquad (6.78)$$

According to (6.2), the density in $K$ of electromagnetic four-momentum is

$$-\boldsymbol{E}_0\cdot\vec{\vec{\Theta}} = \boldsymbol{E}_0\frac{1}{8\pi}(\boldsymbol{E}\cdot\boldsymbol{E}+\boldsymbol{B}\cdot\boldsymbol{B})+\frac{(\boldsymbol{E}\times\boldsymbol{B})}{4\pi}. \qquad (6.79)$$

We recognise the energy density $(\mathbf{E}^2+\mathbf{B}^2)/8\pi$ as the coefficient $\Theta^{00}$ of the time basis vector, and the Poynting vector $(\mathbf{E}\times\mathbf{B})/4\pi$, which is both the three-momentum density and the energy flux, as the spatial part. The last line of (6.78), which describes the flux of electromagnetic three-momentum, is (apart from its sign) Maxwell's stress dyadic.

### 6.6.1 Four-momentum in a Plane Wave

From eq. (5.87), the electromagnetic field for a plane wave has the form

$$\vec{\vec{F}}(x) = \psi'(\boldsymbol{k}\cdot\boldsymbol{x})\boldsymbol{k}\wedge\boldsymbol{a} = \psi'(\boldsymbol{k}\otimes\boldsymbol{a}-\boldsymbol{a}\otimes\boldsymbol{k}), \qquad (6.80)$$

where $\boldsymbol{k}$ is a lightlike momentum vector and $\boldsymbol{a}$ is a constant vector orthogonal to $\boldsymbol{k}$. The function $\psi$ of $\boldsymbol{k}\cdot\boldsymbol{x}$ is arbitrary. Using the scalar products of $\boldsymbol{k}$ and $\boldsymbol{a}$, one obtains

$$\vec{\vec{F}}\cdot\vec{\vec{F}} = (\psi')^2[-\boldsymbol{a}\cdot\boldsymbol{a}\,\boldsymbol{k}\otimes\boldsymbol{k}],$$

whose trace vanishes since $\boldsymbol{k}\cdot\boldsymbol{k}=0$. Therefore,

$$\vec{\vec{\Theta}} = \frac{1}{4\pi}\boldsymbol{a}\cdot\boldsymbol{a}(\psi')^2\boldsymbol{k}\otimes\boldsymbol{k}. \qquad (6.81)$$

It is convenient to use a basis $K$ such that $\boldsymbol{k} = E(\boldsymbol{E}_0+\boldsymbol{E}_1)$. Then, referring to the interpretations in Section 6.1, the $K$-density of four-momentum equals the $K$-flux of four-momentum in the $\boldsymbol{E}_1$-direction,

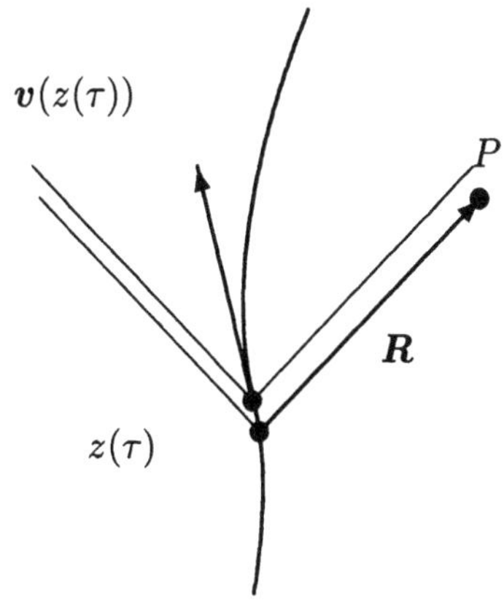


**Figure 6.5**

$$-\boldsymbol{E}_0 \cdot \vec{\vec{\Theta}} = \boldsymbol{E}_1 \cdot \vec{\vec{\Theta}} = \frac{E}{4\pi}\boldsymbol{a} \cdot \boldsymbol{a}(\psi')^2\, \boldsymbol{k}. \tag{6.82}$$

The four-momentum has, of course, the lightlike direction of $\boldsymbol{k}$.

### 6.6.2 Radiation From an Accelerating Point Charge

An accelerating point charge generates an electromagnetic field given by (5.68) in the previous chapter (Figure 6.5). To express $\vec{\vec{F}}$ at a point $P$ one must find the retarded point $z(\tau(P))$ on the worldline of the charge so that $\overrightarrow{z(\tau(P))P} \equiv \boldsymbol{R}$ is lightlike. Then, from (5.68),

$$\begin{aligned} \vec{\vec{F}}(P) &= \boldsymbol{\partial} \wedge \boldsymbol{A} = \boldsymbol{\partial} \wedge \frac{q\boldsymbol{v}}{\rho} \\ &= \frac{q}{\rho^3}\boldsymbol{v} \wedge \boldsymbol{R} + \frac{q}{\rho^2}\boldsymbol{a} \wedge \boldsymbol{R} + \frac{q\boldsymbol{a} \cdot \boldsymbol{R}}{\rho^3}\boldsymbol{v} \wedge \boldsymbol{R}, \end{aligned}$$

where $\rho = -\boldsymbol{v} \cdot \boldsymbol{R}$. To compute $\vec{\vec{\Theta}}$ from (6.74) one must evaluate various scalar products, such as $\boldsymbol{v} \wedge \boldsymbol{R} \cdot \boldsymbol{v} \wedge \boldsymbol{R} = \boldsymbol{R} \otimes \boldsymbol{R} - \rho(\boldsymbol{v} \otimes \boldsymbol{R} + \boldsymbol{R} \otimes \boldsymbol{v})$. One finds

$$\begin{aligned} \frac{4\pi}{q^2}\vec{\vec{\Theta}} &= \frac{1}{2\rho^4}\vec{\vec{\eta}} + (\boldsymbol{v} \otimes \boldsymbol{R} + \boldsymbol{R} \otimes \boldsymbol{v})\left(\frac{1 + \boldsymbol{a} \cdot \boldsymbol{R}}{\rho^5}\right) \\ &\quad + (\boldsymbol{a} \otimes \boldsymbol{R} + \boldsymbol{R} \otimes \boldsymbol{a})\frac{1}{\rho^4} \\ &\quad + \boldsymbol{R} \otimes \boldsymbol{R}\left(\frac{\boldsymbol{a} \cdot \boldsymbol{a}}{\rho^4} - \frac{(\boldsymbol{a} \cdot \boldsymbol{R})^2}{\rho^6} - \frac{1}{\rho^6} - \frac{2\boldsymbol{a} \cdot \boldsymbol{R}}{\rho^6}\right). \end{aligned} \tag{6.83}$$

It is interesting to calculate the total radiated four-momentum from the point charge during an interval $\delta\tau$ of proper time. The calculation is very simple in the retarded rest frame, for which the time basis vector is $\boldsymbol{v}(z(\tau))$. In this frame the radial distance of $P$ from the retarded point is $\rho$, and the unit, spatial vector pointing from it to $P$ is $\boldsymbol{u}$, where $\boldsymbol{R} = \rho(\boldsymbol{v} + \boldsymbol{u})$. The amount of four-momentum radiated into a solid angle $\delta\Omega$ (with vertex at the retarded point), in the proper time interval $\delta\tau$, is the $\rho \to \infty$ limit

$$\delta\tau(\rho^2\delta\Omega)\boldsymbol{u}\cdot\overset{\Rightarrow}{\Theta} = \delta\tau\delta\Omega\left(\frac{q^2}{4\pi}\right)\left(\frac{\boldsymbol{R}}{\rho}\right)(\boldsymbol{a}\cdot\boldsymbol{a} - (\boldsymbol{a}\cdot\boldsymbol{u})^2). \tag{6.84}$$

In the limit of large $\rho$, only two terms, both from the second line of equation (6.83), contribute. Integrating over all solid angles in the retarded rest frame gives

$$\begin{aligned}\delta\tau\int d\Omega(\rho^2\delta\Omega)\boldsymbol{u}\cdot\overset{\Rightarrow}{\Theta} &= \delta\tau\int d\Omega(\boldsymbol{u})\left(\frac{q^2}{4\pi}\right)(\boldsymbol{v}+\boldsymbol{u}) \\ &\qquad\qquad \times(\boldsymbol{a}\cdot\boldsymbol{a} - (\boldsymbol{a}\cdot\boldsymbol{u})^2) \\ &= \frac{2}{3}q^2(\boldsymbol{a}\cdot\boldsymbol{a})\boldsymbol{v}(\delta\tau).\end{aligned} \tag{6.85}$$

With respect to the retarded rest frame, no net three-momentum is radiated, but energy is radiated at the rate $\frac{2}{3}q^2(\boldsymbol{a}\cdot\boldsymbol{a})$ per unit of proper time.

## 6.7 Equations of Motion for Charged Dust

Before looking at the interaction of charged dust with the electromagnetic field, it is helpful to prepare by considering the equations for a free dust. The energy tensor for this medium is conceptually the simplest form of matter and was our point of entry in Section 6.1. There, the energy tensor was discussed from the physical point of view. Below, in Section 6.8, where the perfect fluid is examined, the dust may be regarded as a limiting case of a fluid with zero pressure. In the present section the formula for the energy tensor $\overset{\Rightarrow}{K}(P)$ of incoherent dust will just be set down, (6.86), and used.

### 6.7.1 Uncharged Incoherent Dust

For an incoherent dust of free *uncharged* particles, subject to no forces, the total energy tensor is simply

$$\overset{\Rightarrow}{T} = \overset{\Rightarrow}{K} = mn_{(0)}\boldsymbol{v}\otimes\boldsymbol{v} = K^{\mu\nu}\boldsymbol{E}_\mu\otimes\boldsymbol{E}_\nu = mn_{(0)}v^\mu v^\nu\boldsymbol{E}_\mu\otimes\boldsymbol{E}_\nu. \tag{6.86}$$

The symbol $\overset{\Rightarrow}{K}$ is conventionally reserved for the energy tensor for this simple form of matter, and for that of the limit of a single particle. Conservation of energy-momentum, $\boldsymbol{\partial}\cdot\overset{\Rightarrow}{K} = \boldsymbol{0}$, or in components, $\partial_\mu K^{\mu\nu} = 0$, may now be expressed as

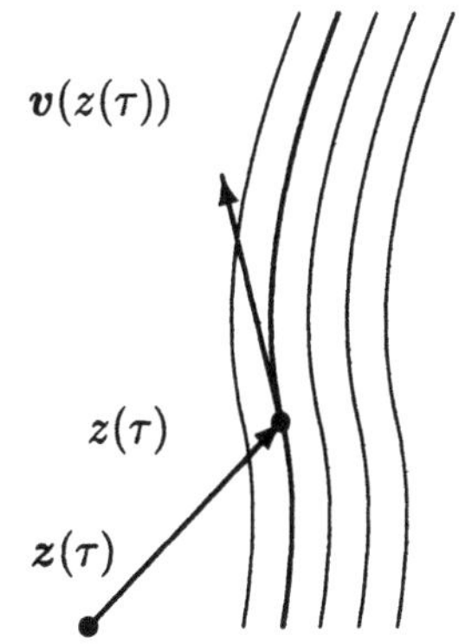


**Figure 6.6**

$$0 = \partial_\mu K^{\mu\nu} = [\partial_\mu (mn_{(0)} v^\mu)] v^\nu + mn_{(0)} v^\mu \partial_\mu v^\nu, \tag{6.87}$$

after judicious grouping of the factors to be differentiated.

The spacetime velocity vector, $\boldsymbol{v} = v^\mu \boldsymbol{E}_\mu$, is a timelike unit vector, so $\boldsymbol{v} \cdot \boldsymbol{v} = -1$, or in components, $v^\rho v^\nu \eta_{\rho\nu} = -1$. Differentiating the latter with respect to $x^\mu$ gives

$$\partial_\mu (v^\rho v^\nu \eta_{\rho\nu}) = 0 = 2v^\rho \eta_{\rho\nu} \partial_\mu v^\nu \equiv 2v_\nu \partial_\mu v^\nu. \tag{6.88}$$

Consequently, if (6.87) is multiplied through by $v_\nu$, the last term vanishes and there remains

$$\partial_\mu (mn_{(0)} v^\mu) = 0, \tag{6.89}$$

an equation which expresses mass conservation for the cloud of dust particles, or alternatively (because the mass $m$ per particle is constant), conservation of the number of particles.

When (6.89) is substituted back into (6.87) then, wherever the number density in the local rest frame is nonzero,

$$v^\mu \partial_\mu v^\nu = 0. \tag{6.90}$$

This equation implies that the spacetime acceleration for each individual particle in the dust cloud vanishes; in other words, the particles are subject to no force, as we assumed at the beginning.

In order to focus on the worldline of a particular particle described by the spacetime velocity field $\boldsymbol{v}(P)$ we consider the timelike curve $z(\tau)$, parameterised with proper time $\tau$, and satisfying

$$\frac{d\boldsymbol{z}(\tau)}{d\tau} = \boldsymbol{v}(z(\tau)) = \frac{dz^\nu(\tau)}{d\tau} \boldsymbol{E}_\nu = v^\nu(z(\tau)) \boldsymbol{E}_\nu. \tag{6.91}$$

Mathematically, $z(\tau)$ is called an integral curve for the vector field $\boldsymbol{v}$ (Figure 6.6).

The spacetime acceleration of the particle whose worldline is the integral curve $z(\tau)$ is obtained by differentiating (6.91) with respect to $\tau$.

$$\frac{d^2\boldsymbol{z}(\tau)}{d\tau^2} = \frac{dz^\mu(\tau)}{d\tau}\partial_\mu v^\nu \boldsymbol{E}_\nu = v^\mu \partial_\mu v^\nu \boldsymbol{E}_\nu \equiv \boldsymbol{a}, \tag{6.92}$$

which, by (6.90), vanishes in the present situation.

### 6.7.2 Charged Dust

This example is extremely important because it allows us to examine in detail the way two physical systems, charged dust and the electromagnetic field, exchange four-momentum (while conserving the total). We are able to derive the traditional equation of motion for the dust particles from the law of conservation of momentum expressed as the vanishing of the divergence of the total energy tensor.

If each particle in the cloud has mass $m$ and charge $q$, the total energy tensor is given by adding those of dust, (6.86), and of the electromagnetic field, (6.74),

$$\begin{aligned} \vec{\vec{T}} &= \vec{\vec{K}} + \vec{\vec{\Theta}} & (6.93)\\ &= mn_{(0)}\boldsymbol{v}\otimes\boldsymbol{v} + \frac{1}{4\pi}\left(-\vec{\vec{F}}\cdot\vec{\vec{F}} + \frac{1}{4}\vec{\vec{\eta}}\,\mathrm{tr}(\vec{\vec{F}}\cdot\vec{\vec{F}})\right) & (6.94)\\ &= \left[mn_{(0)}v^\mu v^\nu + \frac{1}{4\pi}\left(-F^{\mu\alpha}F_\alpha{}^\nu + \frac{1}{4}\eta^{\mu\nu}F^{\alpha\beta}F_{\beta\alpha}\right)\right]\boldsymbol{E}_\mu\otimes\boldsymbol{E}_\nu, & (6.95) \end{aligned}$$

where $\vec{\vec{F}}$ is the appropriate solution to Maxwell's equations with the electric current of the dust $\boldsymbol{j} = qn_{(0)}\boldsymbol{v}$. Both $\boldsymbol{j}$ and $\vec{\vec{F}}$ are nonsingular if $n_{(0)}$ is smooth, as it is for a model of *continuous* dust.

Conservation of four-momentum, $\boldsymbol{\partial}\cdot\vec{\vec{T}} = \boldsymbol{0}$, now gives rise to detailed equations of motion for the dust. After taking into account charge conservation, $\boldsymbol{\partial}\cdot\boldsymbol{j} = \boldsymbol{\partial}\cdot(qn_{(0)}\boldsymbol{v}) = 0$, which is equivalent to mass conservation for the dust since the mass per particle and the charge per particle are both constant, we obtain, using (6.87) and (6.92),

$$\boldsymbol{\partial}\cdot\vec{\vec{T}} = \boldsymbol{0} = \boldsymbol{\partial}\cdot\vec{\vec{K}} + \boldsymbol{\partial}\cdot\vec{\vec{\Theta}} = mn_{(0)}\boldsymbol{a} + \boldsymbol{\partial}\cdot\vec{\vec{\Theta}}. \tag{6.96}$$

We will find by a short calculation (following immediately) that

$$-\boldsymbol{\partial}\cdot\vec{\vec{\Theta}} = \vec{\vec{F}}\cdot\boldsymbol{j} = qn_{(0)}\vec{\vec{F}}\cdot\boldsymbol{v}, \tag{6.97}$$

from which then follows the equation of motion for a single particle of the (continuous) dust

$$m\boldsymbol{a} = q\vec{\vec{F}} \cdot \boldsymbol{v}. \tag{6.98}$$

This is just the equation of motion of a point particle subject to the Lorentz force of an *external* field. Here, it arises as the equation of motion for a "particle" of a *continuum*, for which the mass and charge are vanishingly small (but with constant ratio $q/m$).

In order to verify $\boldsymbol{\partial} \cdot \vec{\vec{\Theta}} = -\vec{\vec{F}} \cdot \boldsymbol{j}$, it is convenient to begin the calculation in the middle, with

$$\boldsymbol{\partial} \cdot \left[\vec{\vec{\eta}}\,\mathrm{tr}(\vec{\vec{F}} \cdot \vec{\vec{F}})\right] = \boldsymbol{\partial}\left[\mathrm{tr}(\vec{\vec{F}} \cdot \vec{\vec{F}})\right],$$

whose components are the vector version of

$$\begin{aligned}
\partial_\lambda[F_{\mu\nu}F^{\nu\mu}] &= (\partial_\lambda F_{\mu\nu})F^{\nu\mu} + F_{\mu\nu}\partial_\lambda(\eta^{\nu\alpha}\eta^{\mu\beta}F_{\alpha\beta}) \\
&= (\partial_\lambda F_{\mu\nu})F^{\nu\mu} + F^{\beta\alpha}(\partial_\lambda F_{\alpha\beta}) = 2(\partial_\lambda F_{\mu\nu})F^{\nu\mu} \\
&= -2(\partial_\nu F_{\lambda\mu} + \partial_\mu F_{\nu\lambda})F^{\nu\mu} && (1) \\
&= -4(\partial_\nu F_{\lambda\mu})F^{\nu\mu} && (2) \\
&= -4\partial_\nu(F_{\lambda\mu}F^{\nu\mu}) + 4F_{\lambda\mu}\partial_\nu F^{\nu\mu} \\
&= +4\partial_\nu(F^{\nu\mu}F_{\mu\lambda}) - 16\pi F_{\lambda\mu}j^\mu. && (3)
\end{aligned}$$

To get (1), one uses the alternative form (5.27) of the homogeneous Maxwell equation; (2) comes from the antisymmetry of $\vec{\vec{F}}$, and a relabelling of dummy scripts. In (3) the right-hand side of the inhomogeneous Maxwell equation (5.8) has been substituted. Raising $\lambda$ by multiplying through by $\eta^{\rho\lambda}$, relabelling and rearranging slightly gives

$$\frac{1}{4\pi}\partial_\lambda\left(-F^{\lambda\mu}F_\mu{}^\rho + \frac{1}{4}\eta^{\lambda\rho}F_{\mu\nu}F^{\nu\mu}\right) = -F^\rho{}_\mu j^\mu, \tag{6.99}$$

the component version of the equation we wished to verify.

### 6.7.3 Frame-dependent Equation of Motion for Charged Dust

If the equation of motion for dust, (6.98), is decomposed with respect to a particular frame $K$, it becomes

$$\begin{aligned}
m\boldsymbol{a} &= m\frac{d\boldsymbol{v}}{d\tau} = m\frac{d}{d\tau}[\gamma(\boldsymbol{E}_0 + \boldsymbol{V})] \\
&= m\frac{d\gamma}{d\tau}\boldsymbol{E}_0 + m\frac{d}{d\tau}(\gamma\boldsymbol{V}) \\
&= q\left[\boldsymbol{E}_0 \wedge \boldsymbol{E} - (\boldsymbol{E}_0 \wedge \boldsymbol{B})^*\right] \cdot \gamma(\boldsymbol{E}_0 + \boldsymbol{V}) \\
&= q\gamma\boldsymbol{E} + q\gamma\boldsymbol{E}_0(\boldsymbol{E}\cdot\boldsymbol{V}) + q\gamma\boldsymbol{V}\times\boldsymbol{B}.
\end{aligned}$$

Equating the coefficients of the time basis vectors, one obtains

$$m\frac{d\gamma}{dt} = q(\boldsymbol{E} \cdot \boldsymbol{V}), \tag{6.100}$$

which expresses the rate of change (with respect to $K$-time) of the $K$-energy of a particle to the rate at which the $K$-electric field works on the particle. The spatial components of the equation of motion give

$$m\frac{d}{dt}(\gamma \boldsymbol{V}) = q\boldsymbol{E} + q\boldsymbol{V} \times \boldsymbol{B}, \tag{6.101}$$

which equates the rate of change of the relativistic three-momentum to the Lorentz force. This is the familiar *relative* relativistic equation for a charged particle in electric and magnetic fields.

## 6.8 Perfect Fluid

The energy tensor for a perfect fluid (a gas, but treated as a continuum) is

$$\overset{\Rightarrow}{T} = (\rho_{(0)} + p)\boldsymbol{v} \otimes \boldsymbol{v} + p\overset{\Rightarrow}{\eta}, \tag{6.102}$$

where $\boldsymbol{v}$ is the velocity field of the "average" rest frame, and $\rho_{(0)}$ and $p$ are the energy density and pressure in the average rest frame. (In the kinetic theory model for a gas, the average rest frame at a point is the frame in which the thermal motion at the point is isotropic.) The model of a perfect fluid is meant to account for heat energy and heat convection but not for heat conduction, nor for viscosity.

It is possible to derive the energy tensor (6.102) by the standard method of Section 6.5, but the argument is both interesting and slightly involved. It will be delayed until after a number of simple consequences have been discussed.

The form (6.102) of the energy tensor is satisfying to the intuition. At each point of the fluid there is an average rest frame, with time axis along $\boldsymbol{v}$, such that the spatial situation, with respect to that frame, is isotropic. In such a frame there is no net flux of particles in any spatial direction. Consequently, the density of three-momentum in the average rest frame is zero. Because of the thermal motion, the energy density is greater than the density of rest mass. There is no net heat conduction and no flux of energy or entropy in the average rest frame. There is no shear stress since there is no viscosity, but there is *pressure,* a normal force density $p$.

The density of four-momentum in the average rest frame is $-\boldsymbol{v} \cdot \overset{\Rightarrow}{T} = \rho_{(0)}\boldsymbol{v}$, which implies an energy density $\rho_{(0)}$ and no three-momentum in the average rest frame. The flux of four-momentum in a direction $\boldsymbol{S}$, spatial with respect to $\boldsymbol{v}$, is $\boldsymbol{S} \cdot \overset{\Rightarrow}{T} = p\boldsymbol{S}$. Thus there is no flux of energy in any such spatial direction, but there is a flux of three-momentum, the same in every direction; this manifests itself as a pressure.

### 6.8.1 Equations of Motion

If the perfect fluid is not interacting with any other physical system, then the tensor $\vec{\vec{T}}$ of (6.102) will be conserved on its own:

$$\boldsymbol{\partial} \cdot \vec{\vec{T}} = \boldsymbol{\partial} \cdot \left[ (\rho_{(0)} + p)\boldsymbol{v} \otimes \boldsymbol{v} + p\vec{\vec{\eta}} \right] = \mathbf{0}. \tag{6.103}$$

Grouping the terms judiciously (using hindsight), the relation becomes

$$\boldsymbol{\partial} \cdot [(\rho_{(0)} + p)\boldsymbol{v}]\boldsymbol{v} + (\rho_{(0)} + p)\boldsymbol{v} \cdot \boldsymbol{\partial v} + \boldsymbol{\partial} p = \mathbf{0}.$$

Since $\boldsymbol{v}$ is a unit vector, $\boldsymbol{v} \cdot \boldsymbol{v} = -1$ again implies $v_\mu \partial_\nu v^\mu = 0$ by (6.88), so that if we take the scalar product of the last equation with $\boldsymbol{v}$, it reduces to

$$-\boldsymbol{\partial} \cdot [(\rho_{(0)} + p)\boldsymbol{v}] + \boldsymbol{v} \cdot \boldsymbol{\partial} p = 0. \tag{6.104}$$

Substituting this back into the previous equation, it simplifies to

$$(\rho_{(0)} + p)\boldsymbol{v} \cdot \boldsymbol{\partial}\, \boldsymbol{v} = -\boldsymbol{\partial} p \cdot (\vec{\vec{\eta}} + \boldsymbol{v} \otimes \boldsymbol{v}), \tag{6.105}$$

which may be called the equation of motion. Both sides are orthogonal to $\boldsymbol{v}$. On the left it is perhaps surprising that the pressure is added to the energy density in the inertial term. On the right appears the spatial part of the gradient of the pressure.

### 6.8.2 First Law of Thermodynamics

It is helpful to consider a slightly more specific case of a perfect fluid: suppose the gas consists of a single type of particle, of rest mass $m$, whose number is conserved (a type of baryon, perhaps, or atomic nucleus). Then the energy density in the average rest frame may be written

$$\rho_{(0)} = n_{(0)}(m + e), \tag{6.106}$$

where $n_{(0)}$ is the number density in the average rest frame, and $e$ is the internal energy (that is, the average rest frame energy excluding rest mass) per particle. Equation (6.104) may be written $\boldsymbol{\partial} \cdot (\rho_{(0)}\boldsymbol{v}) + p\boldsymbol{\partial} \cdot \boldsymbol{v} = 0$, or

$$\boldsymbol{\partial} \cdot [n_{(0)}(m + e)\boldsymbol{v}] + p\boldsymbol{\partial} \cdot \boldsymbol{v} = 0. \tag{6.107}$$

Conservation of the number of particles may be expressed by $\boldsymbol{\partial} \cdot [n_{(0)}\boldsymbol{v}] = 0$, or in expanded form, $n_{(0)}\boldsymbol{\partial} \cdot \boldsymbol{v} + \boldsymbol{v} \cdot \boldsymbol{\partial} n_{(0)} = 0$. Consequently, eq. (6.107) may be rewritten

$$n_{(0)}\boldsymbol{v}\cdot\boldsymbol{\partial}e + p\boldsymbol{\partial}\cdot\boldsymbol{v} = 0 = n_{(0)}\boldsymbol{v}\cdot\boldsymbol{\partial}e - \frac{p}{n_{(0)}}\boldsymbol{v}\cdot\boldsymbol{\partial}n_{(0)},$$

or finally,

$$(\boldsymbol{v}\cdot\boldsymbol{\partial})e + p(\boldsymbol{v}\cdot\boldsymbol{\partial})\left(\frac{1}{n_{(0)}}\right) = 0. \tag{6.108}$$

If we introduce $V_{(0)} \equiv \frac{1}{n_{(0)}}$, the volume per particle in the local average rest frame, the final equation reads

$$(\boldsymbol{v}\cdot\boldsymbol{\partial})e + p(\boldsymbol{v}\cdot\boldsymbol{\partial})V_{(0)} = 0. \tag{6.109}$$

This is the first law of thermodynamics ($\delta e = \delta q - \delta w = \delta q - p\delta V$) for the case when no heat is exchanged ($\delta q = 0$) and it expresses that the flow is isentropic ($\delta s = T^{-1}\delta q = 0$).

### 6.8.3 Label Space

For a material continuum there are two significantly different types of description. If one tags each "particle" and tries to follow it through spacetime, the description is called Lagrangian, or materially based. If one sits at a typical spatial point and describes what happens there, the method is called Eulerian. To get an interesting variational principle, and hence an expression for the energy tensor, it appears that a combination of the two methods must be used. The discussion of the label space that follows owes much to the book by Soper, though the ideas are older.

The history of each "particle" of the fluid flow is a field line, or integral curve, of the timelike velocity field $\boldsymbol{v}$. The timelike curves will intersect any spacelike hypersurface and so may be labelled, once for all, with three labels, the three coordinates (of arbitrary nature), $\underset{\sim}{a} \equiv (a_1, a_2, a_3)$, of the point of intersection with the hypersurface. Then, three *scalar fields* in spacetime may be defined by $\underset{\sim}{R}(P) = \underset{\sim}{a}$, where $\underset{\sim}{a}$ are the coordinates of the curve ("particle") which passes through $P$. At only a very slight risk of confusion, the same symbol, $\underset{\sim}{R}$, is used for the function $\mathcal{R}^4 \to \mathcal{R}^3$ when the points $P$ are specified by their coordinates $x^\mu(P)$ in an inertial frame. The whole worldline of the particle labelled $\underset{\sim}{a}$ is the set $\underset{\sim}{R}^{-1}(\underset{\sim}{a})$.

A model of a set of particles, treated as a continuum, must be able to describe the number of particles in any spatial region of an inertial coordinate frame. This will determine, and be determined by, the density in the *label space*, $n_L(\underset{\sim}{a})\underset{\sim}{d}a_1 \wedge \underset{\sim}{d}a_2 \wedge \underset{\sim}{d}a_3$, so that the number of particles with labels in a region $A \subseteq \mathcal{R}^3$ of label space is

$$n(A) = \int_A n_L(\underset{\sim}{a})\, da_1 da_2 da_3. \tag{6.110}$$

The density $n_L$ in the label space determines the spatial $K$-density at $K$-time $x^0$ by regarding the formulae $R_m(x^0, x^1, x^2, x^3) = a_m$, with fixed $x^0$, as a label change, from $\underset{\sim}{a}$ on the arbitrary hypersurface to $\underset{\sim}{x}$ on the hypersurface with fixed $K$-time $x^0$. In a transformation similar to (4.104), the original label density may be written in terms of the $K$-density

$$\begin{aligned} n_L(\underset{\sim}{a})\mathbf{d}a_1 \wedge \mathbf{d}a_2 \wedge \mathbf{d}a_3 &= n_L(\underset{\sim}{R})\epsilon^{mnl}\left(\frac{\partial R_1}{\partial x^m}\right)\left(\frac{\partial R_2}{\partial x^n}\right)\left(\frac{\partial R_3}{\partial x^l}\right) \\ &\quad \times \mathbf{d}x^1 \wedge \mathbf{d}x^2 \wedge \mathbf{d}x^3, \end{aligned} \tag{6.111}$$

and the determinant that appears here is just the Jacobian $J$ for the transformation of the integral

$$\begin{aligned} n(A) &= \int_A n_L(\underset{\sim}{a})\, da_1 da_2 da_3 \\ &= \int_{\underset{\sim}{R}^{-1}(A)} n_L(\underset{\sim}{R}(x^0, x^1, x^2, x^3))\, J dx^1 dx^2 dx^3, \end{aligned} \tag{6.112}$$

with constant $x^0$. Hence, the $K$-density for the number of particles at $x^0$ is

$$n(x^0, x^1, x^2, x^3) = n_L(\underset{\sim}{R})J = n_L(\underset{\sim}{R})\epsilon^{mnl}\left(\frac{\partial R_1}{\partial x^m}\right)\left(\frac{\partial R_2}{\partial x^n}\right)\left(\frac{\partial R_3}{\partial x^l}\right). \tag{6.113}$$

The $K$-density $n$ of eq. (6.113) may simply be written as the time component $n^0$ of a conserved vector $\boldsymbol{n} = n^\rho \boldsymbol{E}_\rho$. For,

$$\begin{aligned} n &= n_L\epsilon^{0mnl}\left(\frac{\partial R_1}{\partial x^m}\right)\left(\frac{\partial R_2}{\partial x^n}\right)\left(\frac{\partial R_3}{\partial x^l}\right) \\ &= n_L\epsilon^{0\mu\nu\lambda}\left(\frac{\partial R_1}{\partial x^\mu}\right)\left(\frac{\partial R_2}{\partial x^\nu}\right)\left(\frac{\partial R_3}{\partial x^\lambda}\right) = n^0, \end{aligned}$$

where

$$n^\rho(x) \equiv n_L(\underset{\sim}{R}(x))\epsilon^{\rho\mu\nu\lambda}\left(\frac{\partial R_1}{\partial x^\mu}\right)\left(\frac{\partial R_2}{\partial x^\nu}\right)\left(\frac{\partial R_3}{\partial x^\lambda}\right). \tag{6.114}$$

It is easy to see that $\boldsymbol{n}$ is conserved. In calculating $\partial_\rho n^\rho$, there are two sorts of terms, both of which vanish because of the symmetry of partial derivatives and the antisymmetry of $\epsilon^{\rho\mu\nu\lambda}$.

It is evident that $\boldsymbol{n}$ is just the number current $n_{(0)}\boldsymbol{v}$ for the dust since the time components of the two vector fields agree for all inertial frames. Another way to understand this fact is to look at the kinematics.

Apart from its normalisation and orientation, the velocity field $\boldsymbol{v}(x)$ is uniquely determined by the three scalar fields $R_m(x)$. Locating a particle whose worldline is $\underset{\sim}{R}^{-1}(\underset{\sim}{a})$ by its $K$-coordinates at successive events (Figure 6.7),

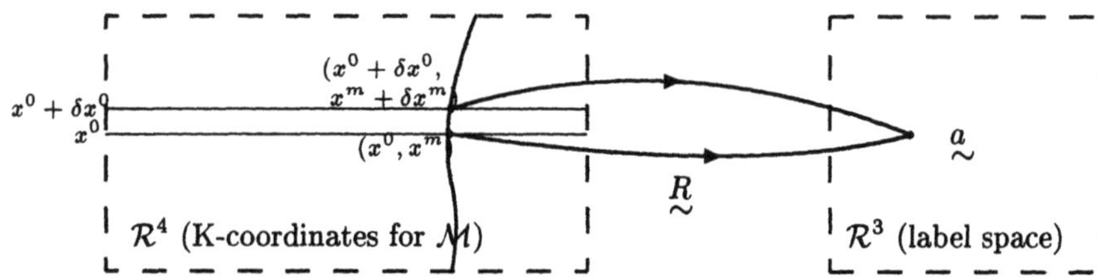


**Figure 6.7**

$$\underset{\sim}{R}(x) = \underset{\sim}{R}(x^0, x^m) = \underset{\sim}{a} = \underset{\sim}{R}(x^0 + \delta x^0, x^m + \delta x^m),$$

so that

$$\delta x^\mu \partial_\mu \underset{\sim}{R}(x) = \underset{\sim}{0} = \delta\tau v^\mu \partial_\mu \underset{\sim}{R}(x). \tag{6.115}$$

Each of these three equations for $v^\mu$ has a solution space with two independent spacelike solutions and one timelike, but the intersection for all three gives a unique timelike direction. Assuming $v^\mu$ is timelike and dividing by $\delta\tau\gamma$, the three equations (6.115) take the form

$$-\partial_0 R_m = V^n \partial_n R_m, \tag{6.116}$$

which have a unique solution for the relative velocity $\boldsymbol{V}$ since the determinant of the matrix on the right is just the nonzero Jacobian that appears in eq. (6.113). Since $\boldsymbol{v} \propto \epsilon^{\mu\nu\lambda\rho}\partial_\nu R_1 \partial_\lambda R_2 \partial_\rho R_3$, which is a multiple of $n^\mu$ of (6.114), satisfies all three equations (6.115) (in each case a determinant is formed with two columns the same), the velocity field $\boldsymbol{v}$ and $\boldsymbol{n} = n^\rho \boldsymbol{E}_\rho$ must be proportional. If $n_{(0)}$ is the number density in the local average rest frame, then

$$\boldsymbol{n} = n_{(0)}\boldsymbol{v} = n^\rho \boldsymbol{E}_\rho, \tag{6.117}$$

and so

$$-(n_{(0)})^2 = \eta_{\mu\nu} n^\mu n^\nu. \tag{6.118}$$

Taking the square root, we have an explicit formula for the rest frame density $n_{(0)}$ in terms of $\underset{\sim}{R}$ (through the label density) and its derivatives.

### 6.8.4 Lagrangian for a Perfect Fluid

A Lagrangian (a density in spacetime) for the perfect fluid may be generalised from the one for a point particle: (minus) the energy density (in spacetime) in the local average rest frame, written

$$-n_{(0)} U\left(\frac{1}{n_{(0)}}, s\right) \underset{\sim}{d}x^0 \wedge \underset{\sim}{d}x^1 \wedge \underset{\sim}{d}x^2 \wedge \underset{\sim}{d}x^3,$$

where $U$ is the energy per particle, a function of volume per particle $V_{(0)} = \frac{1}{n_{(0)}}$ and entropy per particle $s$. The entropy at each particle will be considered as fixed for isentropic flow, and will not enter in the variations. The action, written either as an integral over inertial coordinates or as an integral over curvilinear coordinates, is

$$I = -\int n_{(0)} U\left(\frac{1}{n_{(0)}}, s\right) dx^0 dx^1 dx^2 dx^3 \tag{6.119}$$

$$= -\int n_{(0)} U\left(\frac{1}{n_{(0)}}, s\right) \sqrt{-g}\, d\xi^0 d\xi^1 d\xi^2 d\xi^3, \tag{6.120}$$

where $g$ is the determinant from (6.44). In the first form of the action, the Lagrangian integrand, $-n_{(0)}U$, is regarded as a function of $R_n$ through $s$ and $n_L$, and a function of $\partial_\mu R_n$ through $n^\mu$, where $n_L$ and $n^\mu$ arise from the formula (6.118) for $n_{(0)}$. In the second form, the scalar fields $R_n$ are regarded as functions of the new curvilinear coordinates, with the same values at coordinates appropriate for each spacetime point. The functions $s$ and $n_L$ are unchanged. The scalar $n_{(0)}$ is given by a formula identical to (6.118), namely,

$$-(n_{(0)})^2 = \eta^\xi_{\mu\nu} n^\mu_\xi n^\nu_\xi = g_{\mu\nu} n^\mu_\xi n^\nu_\xi, \tag{6.121}$$

where $\eta^\xi_{\mu\nu} = g_{\mu\nu}$ is given by (6.43) and $n^\mu_\xi$ by (6.39). Therefore

$$n^\mu_\xi = \frac{\partial \xi^\mu}{\partial x^\nu} n^\nu = n_L \epsilon_\xi^{\mu\alpha\beta\gamma} \partial^\xi_\alpha R_1 \partial^\xi_\beta R_2 \partial^\xi_\gamma R_3, \tag{6.122}$$

where

$$\epsilon_\xi^{\mu\alpha\beta\gamma} = \frac{\partial \xi^\mu}{\partial x^\nu}\frac{\partial \xi^\alpha}{\partial x^\rho}\frac{\partial \xi^\beta}{\partial x^\tau}\frac{\partial \xi^\gamma}{\partial x^\lambda} \epsilon^{\nu\rho\tau\lambda}$$

$$= \det\left(\frac{\partial \xi^\mu}{\partial x^\nu}\right) \epsilon^{\mu\alpha\beta\gamma} = \frac{1}{\det\left(\frac{\partial x^\nu}{\partial \xi^\mu}\right)} \epsilon^{\mu\alpha\beta\gamma} = \frac{\epsilon^{\mu\alpha\beta\gamma}}{\sqrt{-g}}.$$

The formula

$$n^\mu_\xi = n_L(\underset{\sim}{R}(\xi)) \frac{\epsilon^{\mu\alpha\beta\gamma}}{\sqrt{-g}} \partial^\xi_\alpha R_1 \partial^\xi_\beta R_2 \partial^\xi_\gamma R_3. \tag{6.123}$$

now enables us to calculate conveniently the variation of $n_{(0)}$ when the components $g_{\mu\nu}$ are varied, with $R_k$ and its derivatives held fixed. From (6.121), and replacing $g_{\mu\nu} \to \eta_{\mu\nu}$ after variation,

$$-2n_{(0)}\delta n_{(0)} = \delta g_{\mu\nu} n^\mu n^\nu + 2\eta_{\mu\nu} n^\mu \delta n^\nu_\xi.$$

One obtains $\delta n^\nu_\xi$ from (6.123) and (6.62), hence

$$\begin{aligned} -2n_{(0)}\delta n_{(0)} &= \delta g_{\mu\nu} n^\mu n^\nu + 2\eta_{\mu\nu} n^\mu (-1) n^\nu \delta\sqrt{-g} \\ &= \delta g_{\mu\nu} n^\mu n^\nu + 2n_{(0)}^2 \left(\frac{1}{2}\right) g^{\mu\nu} \delta g_{\mu\nu}. \end{aligned}$$

Finally,

$$\begin{aligned} \delta n_{(0)} &= -\frac{1}{2}\frac{1}{n_{(0)}} \delta g_{\mu\nu} (n^\mu n^\nu + n_{(0)}^2 \eta^{\mu\nu}) \\ &= -\frac{1}{2} n_{(0)} \delta g_{\mu\nu} (v^\mu v^\nu + \eta^{\mu\nu}), \end{aligned} \tag{6.124}$$

after using the definition (6.117) for the unit vector fluid velocity $\boldsymbol{v}$.

The energy tensor may be found from the action $I$ by varying $g_{\mu\nu} = \eta^\xi_{\mu\nu}$, while keeping $R_n$ and $\partial^\xi_\mu R_n$ fixed:

$$\delta_g I = \delta \int L\sqrt{-g}\, d^4\xi = \int \left(\delta L\sqrt{-g} + L\delta\sqrt{-g}\right) d^4\xi. \tag{6.125}$$

With $L = -n_{(0)} U(V_{(0)}, s)$,

$$\delta_g L\big|_{g_{\mu\nu}\to\eta_{\mu\nu}} = -\left(U - \frac{1}{n_{(0)}}\frac{\partial U}{\partial V_{(0)}}\right) \delta n_{(0)}. \tag{6.126}$$

Substituting from (6.124) gives

$$\delta L = \frac{1}{2}\left(n_{(0)} U - \frac{\partial U}{\partial V_{(0)}}\right)(v^\alpha v^\beta + \eta^{\alpha\beta})\delta g_{\alpha\beta},$$

so that the energy tensor may at last be extracted from (6.60):

$$\begin{aligned} \frac{1}{2}\left(n_{(0)} U - \frac{\partial U}{\partial V_{(0)}}\right)(v^\alpha v^\beta + \eta^{\alpha\beta})\delta g_{\alpha\beta} &- \frac{1}{2} n_{(0)} U \eta^{\alpha\beta})\delta g_{\alpha\beta} \\ &= +\frac{1}{2} T^{\alpha\beta} \delta g_{\alpha\beta}. \end{aligned}$$

Therefore,

$$T^{\alpha\beta} = \left(n_{(0)} U - \frac{\partial U}{\partial V_{(0)}}\right) v^\alpha v^\beta + \left(-\frac{\partial U}{\partial V_{(0)}}\right)\eta^{\alpha\beta}. \tag{6.127}$$

This reproduces exactly eq. (6.102); the energy density in the local average rest frame is $\rho_{(0)} = n_{(0)} U$ and the pressure is

$$p = -\frac{\partial U}{\partial V_{(0)}},$$

as developed in thermodynamics.

## 6.9 Problems

1. For a single particle of mass $m$, with worldline $z(\tau)$ as parameterised by proper time $\tau$ and spacetime velocity $\boldsymbol{v}(\tau)$, the (kinetic-)energy tensor is

$$\vec{\vec{K}}(x) = \int m\boldsymbol{v} \otimes \boldsymbol{v}\delta^{(4)}(x - z(\tau))d\tau.$$

Verify that the components of $\vec{\vec{K}}$ have the correct interpretation, and show that $\boldsymbol{\partial} \cdot \vec{\vec{K}} = \int m\boldsymbol{a}\delta^{(0)}(x - z)d\tau$.

2. Two uniform streams of freely moving particles are superposed in a region of spacetime. The masses of the particles in the two streams are $m_A$ and $m_B$, their rest frame number densities are $n^A_{(0)}$ and $n^B_{(0)}$ and, with respect to $K$, the particles move with relative velocities $U\boldsymbol{E}_1$ and $-U\boldsymbol{E}_1$, respectively. Form the energy tensor $\vec{\vec{K}}$ and find expressions for (i) the $K$-energy density, (ii) the $\boldsymbol{E}_1$-momentum density, (iii) the $K$-energy flux in the $\boldsymbol{E}_1$-direction, (iv) the $\boldsymbol{E}_1$-momentum flux in the $\boldsymbol{E}_1$-direction (all densities and fluxes with respect to $K$).

   Show that with respect to a frame $K'$ which is moving at $V\boldsymbol{E}_1$ with respect to $K$, the $K'$-energy flux in the $\boldsymbol{E}'_1$-direction vanishes if

$$\frac{V}{1+V^2} = \frac{e_A - e_B}{e_A + e_B}\frac{U}{1+U^2},$$

where $e_A$, $e_B$ are the $K$-energy densities (with respect to $K$) of the two streams.

3. This question is a continuation of Problem 16 in Chapter 5, which should be consulted for the setup.

   If the electromagnetic potential is given by $\boldsymbol{A} = q\boldsymbol{v}f(\xi^2)$, where $f \sim (\xi^2)^{-\frac{1}{2}}$ as $\xi^2 \to \infty$, and the field tensor is given by $F_{\mu\nu} = \partial_\mu A_\nu - \partial_\nu A_\mu$, show that Maxwell's equations are satisfied with the current $\boldsymbol{j} = -(q\boldsymbol{v}/2\pi)[2\xi^2 f'' + 3f']$. Verify that the electromagnetic energy tensor is

$$\Theta^{\mu\nu} = \frac{q^2}{\pi}[\xi^2 v^\mu v^\nu - \xi^\mu \xi^\nu + \frac{1}{2}\xi^2\eta^{\mu\nu}][f'(\xi^2)]^2.$$

For the particular case $\boldsymbol{A} = q\boldsymbol{v}/(\xi^2 + a^2)^{\frac{1}{2}}$, we get $\vec{\vec{F}} = q[\boldsymbol{v}, \boldsymbol{\xi}]/(\xi^2 + a^2)^{\frac{3}{2}}$ with $\boldsymbol{j} = (3qa^2/4\pi)\boldsymbol{v}/(\xi^2 + a^2)^{\frac{5}{2}}$. In this case we may notice that

$$\boldsymbol{\partial} \cdot \vec{\vec{\Theta}} = \boldsymbol{\partial} \cdot \left[\vec{\vec{\eta}}\frac{q^2 a^2}{8\pi(\xi^2 + a^2)^3}\right],$$

which reveals the cohesive stresses needed to counteract the Coulomb repulsive forces.

4. The O(3) $\sigma$-model consists of three real scalar fields $\phi_a$ constrained to satisfy $\phi_a\phi_a = 1$ (a summation convention for repeated $a, b..$). The equations of motion are $\partial^2\phi_a = \phi_a\phi_b\partial^2\phi_b$. Show that they may be derived from the variational principle $\delta I = 0$, where the action is given by

$$I = -\int d^4x[\partial_\mu\phi_a\partial^\mu\phi_a + \lambda(\phi_a\phi_a - 1)],$$

in which $\lambda$ is a Lagrange multiplier field. Derive the energy tensor

$$T^{\mu\nu} = 2\partial^\mu\phi_a\partial^\nu\phi_a - \eta^{\mu\nu}\partial_\rho\phi_a\partial^\rho\phi_a,$$

and verify that it is conserved when the field equations are satisfied.

5. If the energy tensor $\overset{\Rightarrow}{T}$ is symmetric and conserved ($\boldsymbol{\partial}\cdot\overset{\Rightarrow}{T} = \mathbf{0}$), then $\partial_\lambda M_O^{\mu\nu\lambda} = 0$, where $M_O^{\mu\nu\lambda} = [x^\mu T^{\nu\lambda} - x^\nu T^{\mu\lambda}]$. This implies that the integral over a spacelike hyperplane

$$\overset{\Rightarrow}{M}_O(\sigma) = -\int_\sigma \boldsymbol{x}\wedge(\overset{\Rightarrow}{T}\cdot\boldsymbol{n})\,dV$$

is independent of the hyperplane $\sigma$. (The unit normal to $\sigma$ is $\boldsymbol{n}$, and the vector $\boldsymbol{x}$ translates from the fixed point $O$ to a variable point on $\sigma$.) If $O'$ is another fixed point, then

$$\overset{\Rightarrow}{M}_O(\sigma) = \overrightarrow{OO'}\wedge\boldsymbol{P}(\sigma) + \overset{\Rightarrow}{M}_{O'}(\sigma).$$

If $O'$ is a point on $\sigma$ such that $\boldsymbol{n}\cdot\overset{\Rightarrow}{M}_{O'}(\sigma) = \mathbf{0}$, show that $O'$ is the centre of energy on $\sigma$. In this case we denote $O'$ by $X(\sigma)$, and the angular momentum about $X(\sigma)$ is determined by an $\boldsymbol{n}$-spatial vector $\boldsymbol{S}(\sigma)$ :

$$\overset{\Rightarrow}{M}_O(\sigma) = \overrightarrow{OX(\sigma)}\wedge\boldsymbol{P}(\sigma) - (\boldsymbol{n}\wedge\boldsymbol{S}(\sigma))^*.$$

6. Two particles travel freely in spacetime, with masses $m_1$ and $m_2$, and four-momenta $\boldsymbol{p}_1$ and $\boldsymbol{p}_2$. Consider the angular momentum $\overset{\Rightarrow}{M}_O$ about a point $O$. It may be written $\overset{\Rightarrow}{M}_O = \boldsymbol{x}_1(\sigma)\wedge\boldsymbol{p}_1 + \boldsymbol{x}_2(\sigma)\wedge\boldsymbol{p}_2$, where $\boldsymbol{x}_1$ and $\boldsymbol{x}_2$ are the locations with respect to $O$ of the two particles on some spacelike hyperplane $\sigma$. Using $\boldsymbol{p}$ to denote the total four-momentum, and $M^2 = -\boldsymbol{p}\cdot\boldsymbol{p}$, choosing the special hyperplane $\Sigma$ containing $O$ and with normal $\boldsymbol{p}/M$, show that

$$\begin{aligned}\overset{\Rightarrow}{M}_O &= \left[\boldsymbol{x}_1(\Sigma)\left(\frac{M^2+m_1^2-m_2^2}{2M^2}\right) + \boldsymbol{x}_2(\Sigma)\left(\frac{M^2-m_1^2+m_2^2}{2M^2}\right)\right]\wedge\boldsymbol{p}\\ &+ \left[\frac{\sqrt{\lambda}}{2M}(\boldsymbol{x}_1(\Sigma)-\boldsymbol{x}_2(\Sigma))\right]\wedge\boldsymbol{q},\end{aligned}$$

where $\boldsymbol{q}$ is the unit vector, spatial in the CM frame, in the direction of the first particle's 3-momentum. In the first term we see the centre of energy with respect to $O$ on the hyperplane $\Sigma$, and the second term is the "internal" spin with respect to the centre of energy.

7. The Proca equation provides a version of electromagnetism in which the "photon" has mass. The theory is derived from the action

$$\begin{aligned} I &= \int d^4x \left\{ -\frac{1}{8\pi} G^{\mu\nu}(\partial_\mu \phi_\nu - \partial_\nu \phi_\mu) + \frac{1}{16\pi} G^{\mu\nu} G_{\mu\nu} \right. \\ &\quad \left. - \frac{1}{8\pi} m^2 \phi^\nu \phi_\nu + j^\nu \phi_\nu \right\}. \end{aligned}$$

The current $j^\nu$ is to be regarded as conserved and externally fixed. Vary independently $\phi_\nu$ and the antisymmetric $G^{\mu\nu}$ to show that $\delta I = 0$ implies

$$G_{\mu\nu} = \partial_\mu \phi_\nu - \partial_\nu \phi_\mu, \qquad \partial_\mu G^{\mu\nu} - m^2 \phi^\nu = -4\pi j^\nu,$$

hence $\partial_\nu \phi^\nu = 0$ and $\partial^2 \phi^\nu - m^2 \phi^\nu = -4\pi j^\nu$. Show that the divergence of the energy tensor

$$\begin{aligned} T^{\mu\nu} =& -\frac{1}{4\pi} G^{\mu\alpha} \eta_{\alpha\beta} G^{\beta\nu} + \frac{1}{16\pi} \eta^{\mu\nu} (G^{\alpha\beta} \eta_{\beta\gamma} \eta_{\alpha\delta} G^{\gamma\delta}) \\ &+ \frac{m^2}{4\pi} (\phi^\mu \phi^\nu - \frac{1}{2} \eta^{\mu\nu} \phi_\rho \phi^\rho) \end{aligned}$$

is $-G^{\nu\beta} j_\beta$.

8. Find the equations of motion for two charged particles interacting according to the Fokker action

$$I = -m_1 \int d\tau_1 - m_2 \int d\tau_2 + q_1 q_2 \int \delta[(\boldsymbol{z}_1 - \boldsymbol{z}_2) \cdot (\boldsymbol{z}_1 - \boldsymbol{z}_2)]\, d\boldsymbol{z}_1 \cdot d\boldsymbol{z}_2.$$

9. At a certain point in spacetime, observers A,B,C,D measure the energy densities in their rest frames. The spacetime velocity vectors and the measurements for each are $\boldsymbol{v}_A = \boldsymbol{E}_0$, $e_A = 1$, $\boldsymbol{v}_B = \frac{5}{3}\boldsymbol{E}_0 + \frac{4}{3}\boldsymbol{E}_1$, $e_B = 25/9$, $\boldsymbol{v}_C = \frac{5}{3}\boldsymbol{E}_0 - \frac{4}{3}\boldsymbol{E}_1$, $e_C = 25/9$, $\boldsymbol{v}_D = \frac{5}{4}\boldsymbol{E}_0 + \frac{3}{4}\boldsymbol{E}_1$, $e_D = 3$. Show that at least one of them must have made a mistake.

10. In the laboratory there is a uniform, static magnetic field $\boldsymbol{B} = B\boldsymbol{E}_3$ and the electric field in that frame is zero. Show that for a particle of mass $m$ and charge $q$ uniform circular motion in the $x^1x^2$-plane is possible (that is, eq. (6.98) is satisfied). Show that the angular velocity is given by $\omega = |q|B/m\gamma$, and that the energy, which is constant, is related to the radius $R$ by

$$m\gamma = \sqrt{m^2 + q^2 B^2 R^2}.$$

11. A $(m, q)$-particle is travelling along the $x$-axis of an inertial frame $K$ at constant speed $V_0$. At $x = 0$ it suddenly encounters an electric field of constant magnitude $E$ in $K$, which points along the $y$-axis. Show that the subsequent path of the particle is given by

$$y = \frac{m\gamma_0}{qE}\left[\cosh\left(\frac{qEx}{mV_0\gamma_0}\right) - 1\right],$$

where $\gamma_0 = (1 - V_0^2)^{-1/2}$.

12. A $(m, q)$-particle is travelling along the $x$-axis of an inertial frame $K$ at constant speed $V_0$. It suddenly encounters a region of $K$-width $L$ in which it is deflected by a transverse $K$-electric field of constant magnitude $E$. Show that the angle of deflection $\theta$ of the particle when it leaves this region is given by

$$\tan\theta = \frac{1}{V_0}\sinh\left(\frac{qEL}{m\gamma_0 V_0}\right),$$

where $\gamma_0 = (1 - V_0^2)^{-1/2}$.

13. Verify that the electromagnetic field tensor $\overset{\Rightarrow}{F}$ and its dual $\overset{\Rightarrow}{F}{}^*$ satisfy

$$\overset{\Rightarrow}{\Theta} = -\frac{1}{8\pi}\left(\overset{\Rightarrow}{F}\cdot\overset{\Rightarrow}{F} + \overset{\Rightarrow}{F}{}^*\cdot\overset{\Rightarrow}{F}{}^*\right)$$

and

$$\frac{1}{2}\eta\,\mathrm{tr}\overset{\Rightarrow}{F}\cdot\overset{\Rightarrow}{F} = \overset{\Rightarrow}{F}\cdot\overset{\Rightarrow}{F} - \overset{\Rightarrow}{F}{}^*\cdot\overset{\Rightarrow}{F}{}^*,$$

where $\overset{\Rightarrow}{\Theta}$ is the electromagnetic energy tensor.
Notice that under the "duality rotation"

$$\overset{\Rightarrow}{F} \mapsto \overset{\Rightarrow}{F}_{new} = \cos\alpha\overset{\Rightarrow}{F} + \sin\alpha\overset{\Rightarrow}{F}{}^*,$$

the electromagnetic energy tensor remains invariant, $\overset{\Rightarrow}{\Theta} \mapsto \overset{\Rightarrow}{\Theta}$.

14. Write the equations of motion and the energy tensor for a system with Lagrangian

$$\mathcal{L} = \frac{1}{16\pi}F_{\mu\nu}F^{\nu\mu} - \frac{1}{2}\overline{\mathcal{D}_\mu\psi}\mathcal{D}^\mu\psi - \frac{1}{2}m^2\overline{\psi}\psi,$$

where $\mathcal{D}_\mu = \partial_\mu - ieA_\mu$. (The overbar denotes the complex conjugate.)

15. If the scalar fields $R_m$ satisfy the Euler-Lagrange equations

$$\frac{\partial\mathcal{L}}{\partial R_m} = \partial_\mu\frac{\partial\mathcal{L}}{\partial\partial_\mu R_m},$$

where $\mathcal{L} = \mathcal{L}(R_m, \partial_\mu R_m)$, show that the energy tensor

$$T_\mu{}^\nu = \delta_\mu{}^\nu \mathcal{L} - \partial_\mu R_m \frac{\partial L}{\partial \partial_\nu R_m}$$

is conserved "on the right", that is, $\partial_\nu T_\mu{}^\nu = 0$.

16. A scalar, special relativistic theory of gravity has a Lagrangian density

$$\mathcal{L} = -\frac{1}{8\pi G}\eta^{\mu\nu}\partial_\mu \Phi \partial_\nu \Phi - m\int e^{\Phi}\delta^{(4)}(x - z(\tau))d\tau$$

for the case of a single gravitating point particle. Find the field equations that follow from the variational principle $\delta I = 0$, and the equations of motion for the particle.

17. Derive the total energy tensor from the action for a dust of charged particles interacting electromagnetically (eq. (6.28)). Derive the equations of motion for the system from the same action.

18. The action for a scalar field, when it is rewritten in noninertial coordinates, takes the form

$$I(\phi) = -\frac{1}{2}\int \left(g^{\mu\nu}\frac{\partial}{\partial \xi^\mu}\phi \frac{\partial}{\partial \xi^\nu}\phi + m^2\phi^2\right)\sqrt{-g}\, d^4\xi.$$

Find the equation of motion in noninertial coordinates by using the variational principle.

19. The two forms of the Minkowski metric, $\eta_{\mu\nu}$ and $\eta^{\mu\nu}$, are (by definition) matrix inverses of one another. The same is true for their component forms in the noninertial bases. Verify that

$$g^{\mu\nu} = \frac{\partial \xi^\mu}{\partial x^\alpha}\frac{\partial \xi^\nu}{\partial x^\beta}\eta^{\alpha\beta}.$$

When the gradient of a vector, or a covector, or any higher rank tensor is expressed in a noninertial basis, the expansion components are more complicated than for the inertial case. For example, show that

$$\underset{\sim}{\partial} \otimes \underset{\sim}{A} = \underset{\sim}{E}^\mu \otimes \underset{\sim}{E}^\nu \partial_\mu A_\nu = \underset{\sim}{\partial\xi}^\mu \otimes \underset{\sim}{\partial\xi}^\nu \left[\frac{\partial A^\xi_\nu}{\partial \xi^\mu} - \frac{\partial^2 x^\gamma}{\partial \xi^\mu \partial \xi^\nu}\frac{\partial \xi^\tau}{\partial x^\gamma}A^\xi_\tau\right].$$

The extra term is commonly written $-\Gamma^\tau_{\mu\nu}A^\xi_\tau$, where

$$\Gamma^\tau_{\mu\nu} = \frac{1}{2}g^{\tau\rho}[\partial^\xi_\mu g_{\rho\nu} + \partial^\xi_\nu g_{\rho\mu} - \partial^\xi_\rho g_{\mu\nu}].$$

## References

J.D. Jackson, Classical Electrodynamics, Second Edition (Wiley, 1975)
E.G.P. Rowe and G.T. Rowe, The Classical Equations of Motion for a

Spinning Point Particle with Charge and Magnetic Moment, Physics Reports **149** (1987) 287-336
D.E. Soper, Classical Field Theory (Wiley, 1976)
S. Weinberg, Gravitation and Cosmology (Wiley, 1972)

# Index

MIX
Papier aus verantwortungsvollen Quellen
Paper from responsible sources
FSC® C105338

If you have any concerns about our products,
you can contact us on
**ProductSafety@springernature.com**

In case Publisher is established outside the EU,
the EU authorized representative is:
**Springer Nature Customer Service Center GmbH**
**Europaplatz 3, 69115 Heidelberg, Germany**

Printed by Libri Plureos GmbH
in Hamburg, Germany